Proceedings of the

Third IEEE
International Symposium on
Requirements Engineering

RE '97

Proceedings of the

Third IEEE
International Symposium on
Requirements Engineering

January 6–10, 1997   Annapolis, Maryland, USA

Sponsored by

IEEE Computer Society
Technical Council on Software Engineering

In cooperation with

ACM SIGSOFT
IFIP Working Group 2.9

IEEE Computer Society Press
Los Alamitos, California

Washington   •   Brussels   •   Tokyo

IEEE Computer Society Press
10662 Los Vaqueros Circle
P.O. Box 3014
Los Alamitos, CA 90720-1264

IEEE Computer Society Press Order Number PR07740
IEEE Order Plan Catalog Number 97TB100086
ISBN 0-8186-7740-6
Microfiche ISBN 0-8186-7742-2
ISSN 1090-705X

Additional copies may be ordered from:

IEEE Computer Society Press
Customer Service Center
10662 Los Vaqueros Circle
P.O. Box 3014
Los Alamitos, CA 90720-1314
Tel: +1-714-821-8380
Fax: +1-714-821-4641
Email: cs.books@computer.org

IEEE Service Center
445 Hoes Lane
P.O. Box 1331
Piscataway, NJ 08855-1331
Tel: +1-908-981-1393
Fax: +1-908-981-9667
misc.custserv@computer.org

IEEE Computer Society
13, Avenue de l'Aquilon
B-1200 Brussels
BELGIUM
Tel: +32-2-770-2198
Fax: +32-2-770-8505
euro.ofc@computr.org

IEEE Computer Society
Ooshima Building
2-19-1 Minami-Aoyama
Minato-ku, Tokyo 107
JAPAN
Tel: +81-3-3408-3118
Fax: +81-3-3408-3553
tokyo.ofc@computer.org

Editorial production by Penny Storms
Cover by Joseph Daigle / Studio Productions
Printed in the United States of America by KNI, Inc.

The Institute of Electrical and Electronics Engineers, Inc.

Table of Contents

vii

Welcome from the General Chair

In a landmark article published in 1987, Fred Brooks states that the "hardest single part" of building a software system is deciding what the system requirements are [1]. In Brooks' view, "No other part of the conceptual work is as difficult as establishing the detailed technical requirements...No other part of the work so cripples the resulting system if done wrong. No other part is as difficult to rectify later." In these few words, Brooks tells us a lot about requirements: not only are they of fundamental importance in software development, they are difficult to produce and hard to fix later on.

Although producing a precise, unambiguous statement of the requirements is difficult, the process of doing so can have tremendous payoff. Focusing on requirements is an effective means of managing the inherent and arbitrary complexity of software systems [2]. Brooks refers to the system requirements as the "essence" of the system to be built [1]. Concentrating on this essence omits layers of complexity. In capturing the essential characteristics of the system to be built, developers can ignore the many details (about algorithms, data representations, etc.) that are needed to produce a running version of the system.

To develop both a solid conceptual foundation and an engineering discipline for constructing requirements, a new series of symposia was initiated in 1993 when the first International Symposium on Requirements Engineering (RE '93) took place in San Diego. Two years later, a second symposium (RE '95) was held in York, England. This year's symposium, RE '97, will feature many of the same events that proved successful during the preceding symposia—the strong technical program, the popular tools exhibit and doctoral consortium, an educational minitutorial, stimulating keynote talks and panel discussions, and two lively workshops.

RE '97 will also feature several new events. One of the most important is an industrial program, the goal of which is to encourage greater industry participation. Stuart Faulk, who has had considerable experience transferring advanced requirements methods to the aerospace industry, is the RE '97 Industrial Chair. He has organized two technical sessions entitled Applications and Tools 1 and 2, which feature presentations (1) by industry representatives and (2) by tool vendors and developers. Industry representatives will describe results and lessons learned applying advanced requirements technology in industry and current problem areas in requirements that are not adequately addressed by available technology. Tool vendors and developers will give short presentations to introduce their tools. Attendees interested in learning more about a given tool are encouraged to visit the RE '97 Tools Exhibit for a tool demonstration. Another significant component of the industrial program is a panel discussing how to tranfer the results of research into practice. To make the overall program attractive to both researchers and attendees from industry, the industrial program has been carefully integrated with the rest of the RE '97 technical program.

Also new this year is a preconference tutorial program. The program features tutorials on measurable requirements, object-oriented requirements methods, the SCR approach to requirements, requirements and traceability, and requirements and safety. Taken together, the tutorial program and the RE '97 technical program should offer many events of interest to both researchers and software practitioners.

In previous years, the research papers presented at the RE symposia were largely focused on requirements acquisition, AI techniques, and business applications. To achieve a broader technical program and to attract a wider audience, we extended a special invitation this year to researchers and developers working on formal methods and safety-critical systems. The result is a technical program which contains research papers in the more traditional areas of requirements as well as new work on formal methods and safety-critical systems.

I extend my sincerest thanks to all members of the organizing committee for their efforts. Moreover, I hope that you enjoy and benefit from the RE '97 program.

Connie Heitmeyer
General Chair

References

[1] Frederick P. Brooks, Jr. No Silver Bullet: Essence and Accidents of Software Engineering. *IEEE Computer*, April 1987, Vol. 20, No. 4, pp. 10-19.

[2] Frederick P. Brooks, Jr. The Computer Scientist as Toolsmith II. *CACM*. March 1996, Vol. 39, No. 3, pp. 61-68.

Foreword by the Program Chair

> Requirements definition is a careful assessment of the needs that a system is to fulfill. It must say why a system is needed, based on current and foreseen conditions, which may be internal operations or an external market. It must say what system features will serve and satisfy this context. And it must say how the system is to be constructed... Doug Ross [3]

The problem of defining requirements for software systems is as old as software. Led by pioneers such as Doug Ross, software engineers had compiled by the mid-70s a wealth of empirical data, confirming that "...the rumored 'requirements problems' are a reality..." [1]. The data suggested that requirements errors were the most numerous and, even more significantly, that they also were the most costly and time-consuming to correct. This recognition of the critical nature of requirements established Requirements Engineering (RE) as an important subfield of Software Engineering. Over the past 4-5 years, this subfield has matured and come of age as an autonomous research area and field of professional practice, with an IEEE conference series (the IEEE Conference on Requirements Engineering), an international journal (the Requirements Engineering Journal, published by Springer-Verlag and edited by Periklis Loucopoulos and Colin Potts), an IFIP Working Group on RE (IFIP WG 2.9, chaired by Stephen Fickas), an international RE network of cooperating research groups (RENOIR, coordinated by Anthony Finkelstein) and – last but not least – this symposium series.

The IEEE International Symposium on Requirements Engineering (ISRE) is establishing itself as an important international forum for the presentation of research results and the exchange of ideas on topics that relate to requirements elicitation, analysis and definition. Now in its third meeting, this forum is attracting participation and attention from major research groups conducting research on requirements world-wide, is producing consensus on research directions and methodology, and is generating considerable enthusiasm and even momentum within the research community that identifies itself with RE.

The technical program of ISRE'97, detailed in these proceedings, was put together with three objectives in mind. Firstly, we wanted to uphold and maintain the high standards in the quality of the technical program set out at the previous ISRE meetings in San Diego, California ('93) and York, UK ('95). Secondly, we wanted to improve on the representation of certain topics on the technical program (formal methods and safety-critical systems are two that come immediately to mind). Finally, we wanted to make sure that industry has a strong and vocal presence at the symposium, providing both direction for future research and the proverbial "reality check" for the results presented and discussed at the symposium.

The first and second objectives were operationalized (...to use a technical term) by putting together a strong program committee that included a good cross-section of both senior, established researchers and young promising ones from around the world. The 50-member committee included key researchers in areas we wanted to encourage, and was asked to complete a grand total of 231 reviews, all of which were actually received in time to offer feedback to authors. About half the committee attended a two-day meeting in Eugene, Oregon, where 21 papers were selected, after a long and laborious discussion. Members of the committee also had many valuable suggestions for invited speakers, panels, workshops and minitutorials, which were integrated into the technical program one finds in these proceedings.

To encourage industrial participation, we included in the symposium program an industrial track of presentations which report on industrial RE practice and experiences. In addition, the technical program features keynote talks, workshops and panels which, we believe, will appeal to RE practitioners as much as to RE academics and researchers. Finally, the technical program has been complemented by a series of tutorials primarily aimed at the RE practitioner. Special thanks are due to Connie Heitmeyer and Stuart Faulk for their excellent work in pulling together these critical components of the ISRE'97 program.

Some statistics on the technical program are next in order. There were 76 submissions to this symposium, down from 99 for ISRE'95. For purposes of reviewing, these were classified into three basic categories according to a scheme proposed by Pamela Zave [4] and used for ISRE'95 as well as this year's symposium:

- Specifying system behavior, covering topics such as: integrating multiple views and representations; evaluating alternative strategies for satisfying requirements; choosing which optional requirements to satisfy; obtaining complete; consistent, and unambiguous specifications; checking that the specified system will satisfy the requirements; obtaining specifications that are well-suited for design and implementation activities.

- Problems of investigating the goals, functions, and constraints of a software system, covering topics such as: overcoming barriers to communication; generating strategies for converting vague goals (e.g., "user-friendliness," "security," "reliability") into specific properties or behavior; generating strategies for allocating requirements among the system and the various agents of its environment; understanding priorities and ranges of satisfaction; estimating costs, risks, and schedules; ensuring completeness.

- Problems of managing evolution of systems and families of systems, including requirements reuse during evolutionary phases, reconstructing requirements, etc.

Of the 76 submissions, 37 (or 49%) dealt primarily with issues of specifying system behavior, 25 (33%) dealt with problems of investigating the goals, functions and constraints of a software system, while only 14 (18%) dealt with problems of evolution. Looking at the issues addressed by this year's submissions – where a paper may address more than one issue – more than one out of every three papers (37%, to be exact) dealt with languages and/or tools for RE, while an almost equal number of papers dealt with methodology and lifecycle issues. Formal method issues were treated in about 20% of the submissions and slightly fewer papers (16%) discussed case studies. Quality requirements (mostly safety), scenarios and reuse are three other topics which had good coverage in the submissions and are well represented on the final program. On the other hand, there were very few papers on social aspects of RE work and fewer still on requirements for user interfaces which would have rounded out the program very nicely (maybe next time).

As in previous meetings, ISRE'97 maintained an international flavor with balanced participation from North America and Europe and lesser participation from the more distant Far East. Unlike ISRE'95 where the UK won gold, the greatest number of submitted papers originated in the US (19 with 10 accepted papers), followed by the UK (15 submitted, 4 accepted), France (8 submitted, 2 accepted), Canada (7 submitted, 1 accepted), Belgium (6 submitted, 3 accepted), Germany (6 submitted), Brazil (4 submitted, 1 accepted) and Italy (4 submitted).

On several occasions we were guided by a steering committee consisting of past organizers of ISRE symposia. Thanks are due for prompt, helpful and always wise advice to Axel van Lamsweerde, Martin Feather, Stephen Fickas, Anthony Finkelstein, Sol Greenspan, Michael Harrison and Pamela Zave.

In summary, we wish to extend a great "THANK YOU!" to all those who generously volunteered their time and expertise to make this symposium a success. The fact that so many busy (and famous!) people have taken the trouble to help us with the organization of this symposium and the formation of its technical program speaks well for the future of ISRE and the field of RE.

In these days of cost-cutting and restructuring, an engineering research field – any engineering field, young or old – can justify itself when three conditions apply. Firstly and most importantly, the field must address a real problem where there is demand for better and more cost-effective solutions. Secondly, the field must have sufficient homogeneity and coherence in research methodology and criteria for judging success. Finally, the field must demonstrate steady tangible progress over the years. There is much evidence that the first condition clearly applies for RE ([1] but also[2] can serve as starting points for readings on the subject). We hope that as the reader familiarizes herself with these proceedings, she will agree that the other two conditions apply for RE as well, and justify its continued existence and growth.

John Mylopoulos
Program Chair

References

[1] T. E. Bell and T. A. Thayer. Software requirements: Are they really a problem? *Proceedings, Second International Conference on Software Engineering*, January 1976, pp. 61-68.

[2] B. Curtis, H. Krasner, and N. Iscoe A Field Study of the Software Design Process for Large Systems *Communications of the ACM 31*, 11, 1268-1287, 1988.

[3] D. Ross. Structured analysis: A language for communicating ideas. *IEEE Transactions on Software Engineering 3*, 1, January 1977.

[4] P. Zave. Classification of research efforts in requirements engineering. *Proceedings, Second International Symposium on Requirements Engineering*. York, UK, March 27-29, 1995. Also available from http://www1126.research.att.com:/people/pamela.

Organizing Committee

General Chair
Constance Heitmeyer
Naval Research Laboratory, USA

Program Chair
John Mylopoulos
University of Toronto, Canada

Industrial Chair
Stuart Faulk
University of Oregon, USA

Finance Chair
James Kirby, Jr.
Naval Research Laboratory, USA

Publicity Chair
Ralph Jeffords
Naval Research Laboratory, USA

Local Arrangements Chair
Ramesh Bharadwaj
Naval Research Laboratory, USA

Doctoral Consortium Chair
Myla Archer
Naval Research Laboratory, USA

Proceedings Chair
Carolyn Gasarch
Naval Research Laboratory, USA

Registration Cochairs
Todd Grimm and Janine Stone
Naval Research Laboratory, USA

Exhibits Cochairs
Charles Payne and Dwight Colby
Secure Computing Corporation, USA

Tutorials Chair
John Marciniak
Kaman Sciences, USA

Social Arrangements Chair
Janine Stone
Naval Research Laboratory, USA

Technical Arrangements Chair
Bruce Labaw
Naval Research Laboratory, USA

Program Committee

William Agresti, *Mitretek Systems, USA*
Mark Ardis, *Bell Laboratories, Lucent Technologies, USA*
Joanne Atlee, *University of Waterloo, Canada*
Daniel Berry, *Technion, Israel*
Alex Borgida, *Rutgers University, USA*
Pere Botella, *Universitat Politecnica de Catalunya, Spain*
Janis Bubenko, *University of Stockholm, Sweden*
Jaelson Castro, *Universidade Federal de Pernambuco, Brazil*
Lawrence Chung, *University of Texas, Dallas, USA*
Alan Davis, *University of Colorado, USA*
Valeria di Antonellis, *Universitá di Ancona, Italy*
Eric Dubois, *University of Namur, Belgium*
Stuart Faulk, *University of Oregon, USA*
Martin Feather, *Computing Services Support Solutions and
Jet Propulsion Laboratory*, USA
Mark Feblowitz, *GTE Laboratories, USA*
Stephen Fickas, *University of Oregon, USA*
Anthony Finkelstein, *City University, London, UK*
Carlo Ghezzi, *Politecnico di Milano, Italy*
Sol Greenspan, *GTE Laboratories, USA*
Michael Harrison, *University of York, UK*
Ian Hayes, *University of Queensland, Australia*
Mats Heimdahl, *University of Minnesota, USA*
Connie Heitmeyer, *Naval Research Laboratory, USA*
Daniel Jackson, *Carnegie Mellon University, USA*
Matthias Jarke, *RWTH Aachen, Germany*
Jeff Kramer, *Imperial College, UK*
Julio Cesar Leite, *PUC-Rio, Brazil*
Peri Loucopoulos, *UMIST, UK*
Robyn Lutz, *Jet Propulsion Laboratory, USA*
Kalle Lyytinen, *University of Jyvaskyla, Finland*
Neil Maiden, *City University, London, UK*
Nazim Madhavji, *McGill University, Canada*
John McDermid, *University of York, UK*
Roland Mittermeir, *Klagenfurt University, Austria*
Bashar Nuseibeh, *Imperial College, UK*
Andreas L. Opdahl, *University of Bergen, Norway*
Barbara Pernici, *Politecnico di Milano, Italy*
Klaus Pohl, *RWTH Aachen, Germany*
Howard Reubenstein, *Concept Five Technologies, USA*

Session 1

Keynote Address

Speaker

Anthony Hall
Praxis Critical Systems

"What's the Use of Requirements Engineering?"

What's the Use of Requirements Engineering?

Anthony Hall
Praxis Critical Systems

Abstract

There are many ideas about how to do requirements engineering and often they conflict with each other. Such conflicts can best be resolved by asking of anything one proposes to do: "What is the use of doing that?". The question demands a thorough understanding of the principles behind different methods, and the answers may surprise those who equate pragmatism with informality. I discuss how applying this rule helps in choosing requirements engineering methods and in dealing with the difficulties that arise in applying these methods.

1 Introduction

In engineering, the ultimate test of a theory is whether it is useful. There are many conflicting theories about what systems developers should do, about what methods they should use, and about how to apply particular methods. The conflicts arise not so much because one method is wrong and another right, but because they make different assumptions about what is useful. These assumptions are rarely explicit and in order to choose between methods we have to understand what their hidden assumptions are. They can often be revealed by asking, of any activity, "what is the use of doing this?" In this talk I will describe some of the methods we find useful for requirements engineering and analyze the characteristics of those methods by looking at their underlying assumptions. In particular I will examine our understanding of the nature of requirements and how to understand and resolve the conflict between the formal and the structured schools of analysis.

2 The World and the Machine

I make no apology for plagiarizing the title of Michael Jackson's invited talk to ICSE '95 [1]. For many years it has been received wisdom that describing the environment is a key aspect of requirements capture, but it was not until Parnas' work on the Four-Variable Model [2] and Michael Jackson and Pamela Zave's work published in 1995 that there was a clear exposition of what exactly this meant. The insights that this series of papers gave us are now fundamental to the way we define requirements.

I cannot overemphasize the importance of this framework. Applying it in practice has helped enormously to understand the problems and to structure our solutions. One of the many insights it has given us is the importance of the domain description. When systems fail, it can often be traced back to a poor or missing domain description, not to the requirements themselves.

It turns out, however, that deciding what is the domain and what is the machine is not always obvious. Frequently it is useful to think of "the machine" as not just the computer system we are building, but also many of the people who will interact with it. For example I recently worked on a safety-critical communications system. We could not guarantee, by the computer alone, that a user was speaking to the person they thought they were. So it was necessary to put in place not just the computer, but also a human protocol for checking that the correct parties were in communication. Conversely, the physical system we are building sometimes has to be regarded also as part of the domain. For example many operational systems have to monitor their own health. Parts of the machine may fail in ways over which we have no control. In general, we find it useful to regard as in the machine those things we can control, whatever form they take, and as in the domain those things we cannot control, even if they are what we are building.

3 The Role of Formality

We want to specify only those properties of the machine that affect the domain. However, we do want to make that specification as useful as possible. This means that the specification must be precise, because an ambiguous specification is bound to be misinterpreted. It must also be expressive, because we want to make the specifications as close to the users' conceptual model as possible, and to say, for example, what the system must not do as well as what it must do. We therefore use mathematical notation – formal methods – as the prime means of system specification. Mathematical notation is both precise and expressive.

There are, of course, some problems with current formal methods. An obvious difficulty is communication: few users understand the formal notation itself. The formal notation must for this reason (as well as many others) be accompanied by natural language or domain-specific notations. This is not, however, a difficult or problematic task. A more fundamental problem is the expressiveness of current formal notations. We have found it much easier to use formal notations for the system specification than we have for the description of the domain or for the statement of requirements. I suspect this is for two reasons. First, writing specifications is what formal notations were invented for. Second, the domain is typically vastly more complex than the machine, so to describe it formally would require a good deal of effort, effort that is not worth while unless the notation is well matched to the domain. This problem is mitigated by the fact that many domains have well understood languages of their own, but these are rarely fully formal. The validation of the specification against the domain properties therefore remains a weak link which can only be strengthened by a dialogue between domain experts and the engineers writing the specification. This is why, regardless of whether one is using a formal notation or not, it is never sufficient to take on trust a customer's written statement of requirements: it is always essential to talk to real system users in great depth. Elicitation and validation with users are therefore an integral part of our requirements engineering process.

Jackson has pointed out that different problems fall into different "problem frames": and a real problem, of course, never falls into a single frame. At the very least one needs to be able to conjoin several descriptions of the same problem. Only a few languages, of which Z is one, allow this freedom. Even that is not enough, however, since some aspects of problems – for example user interfaces – cannot easily be expressed in Z, and so we need to combine different languages in the same specification. This is currently an active area of research. We have often had to combine different notations but we do not yet have a good theory of how to do it.

I do not want to overemphasize the problems: on the contrary, we find formal methods extremely useful in practice. An example is CDIS [3], an ATC information system we built where we found that using formal specification reduced the number of defects in the delivered system. Furthermore the defects that remained were very rarely specification errors. This suggests that formal specifications do indeed help both to build the system right and to build the right system. This improvement was free: CDIS cost no more to build than it would have with conventional methods.

4 Nonformal Approaches

We do not use only formal methods in requirements engineering. On the contrary, we use notations and ideas from structured and object-oriented analysis. However, only some of the ideas of conventional analysis are useful, while others are positively harmful. It is important to understand why that is so.

A typical structured analysis starts with a "context diagram", which shows the system and all the external entities that interact with it. The requirements are then defined by decomposing the system into a number of processes and defining the inputs and outputs of each process. These processes in turn are broken down, until elementary processes are reached and these are then defined using some other notation such as pseudocode. The initial step here is entirely right, and we always draw a context diagram as a step in defining the machine and the domain and in clarifying the boundary between them. From then on, however, structured analysis is not addressing requirements at all – it is sketching a design of the system. The "processes" in the analysis do not correspond to anything real in the domain.

An alternative, more fashionable approach is to use object-oriented analysis. Here instead of using processes as the fundamental units one uses objects. The first step in OOA is to identify the objects that exist in the domain. This, too, I consider a vital part of defining the domain and an object model always forms part of our domain description. Again, however, one can fall into a trap: one can try to define the behaviour of the system in terms of the behaviour of the individual objects, typically treating each object as a state machine. Once again, this is carrying out design, not specification: it can be extremely difficult to understand the behaviour of a system if one is only given the behaviour of its components. There are OOA methods like Fusion which do not fall into this trap: that is because the basic ideas of such methods, if not the notations they use, are the same as those of formal methods.

The common theme to these examples is that describing how a system might be built is not a good way to specify what it does. The hidden assumption in methods where the analysis model is like a design is that such a design is a good model of the system, either because it makes it comprehensible to the users or perhaps because it advances the project towards its goal of an implemented system. We have found, on the contrary, that separation of concerns between the users' conceptual model and the physical design is essential. The characteristics which make a structure comprehensible to a user are completely different from those which make it a good design.

5 Using Requirements

Good requirements are essential if we are to be sure that we are building the system the users want, and that we are not doing more than is needed. They should, of course, continue to play this role not just when the system is initially built, but through its subsequent maintenance and enhancement. In practice there are very few systems where the requirements are good enough to be useful throughout the system's life. All too often the implemented system "supersedes" the requirements: a euphemism for saying that it does not meet its requirements, which were probably not truly requirements in the first place. It is only by a clear understanding of what a requirements definition should be, a clear separation between specification and design and the use of precise specification notations that we can make requirements definitions which are of lasting use.

References

[1] Michael Jackson. The World and the Machine. In *Proceedings, 17th International Conference on Software Engineering*, 1995, pp. 283-292.

[2] D. L. Parnas and J. Madey. Functional Documentation for Computer Systems Engineering (Version 2). Technical Report CRL 237. Telecommunications Research Institute of Ontario, McMaster University, Hamilton Ontario 1991.

[3] Anthony Hall. Using formal methods to develop an ATC Information System. *IEEE Software*, March 1996, pp. 66-76.

3

Session 2A

Reuse

Ten Steps Towards Systematic Requirements Reuse
W. Lam, J.A. McDermid, and A.J. Vickers

Despite several proposals which tackle the problem of requirements reuse from different perspectives, there is little evidence in the literature that reuse can be effectively put in practice. This paper presents interesting results on this subject, which were generated at the Technology Centre of Rolls-Royce University on the domain of aero-engine control systems. Among other things, the paper presents and discusses some criteria towards systematic requirements reuse and supports them with actual examples from the chosen domain. The general approach of this research is refreshingly practical and can be readily adopted for other, similar efforts.

Reusing Operational Requirements: A Process-Oriented Approach
Robert Darimont and Jeanine Souquieres

The key insight offered by this paper is that it pays to keep track and record the trace of decisions that generated a particular requirements specification, because this trace can be used for documentation, and also reused to generate other, similar requirements. This is an idea that has been around as long as software reuse. The paper demonstrates convincingly that the idea can work for operational requirements and explores, among other things, new linguistic features that need to be added to specification languages, so that they can capture not just operational requirements, but also the trace whereby they were generated.
— *John Mylopoulos*

Analogical Reuse of Requirements Frameworks
Philippe Massonet and Axel van Lamsweerde

Reusing similar requirements fragments is a promising approach in software engineering that allows for reducing the system development cost while increasing the quality of requirements specification. The paper presents an interesting approch to reuse, using known techniques from analogical and case-based reasoning. In particular, the authors convincingly demonstrate that a rich requirements meta-model with an expressive formal assertion language can be exploited to improve the effectiveness of analogical reuse.

Ten Steps Towards Systematic Requirements Reuse

W. Lam, J.A. McDermid and A.J. Vickers

Rolls-Royce University Technology Centre
Department of Computer Science
The University of York
Heslington, York Y01 5DD, UK

Tel: +44 01904 433387, Fax: +44 01904 432708
E-mail: {wing, jam, andyv}@minster.york.ac.uk

Abstract

Reusability is widely suggested to be a key to improving software development productivity and quality [1], [2]. It has been further argued that reuse at the requirements level can significantly increase reuse at the later stages of development [3], [4]. However, there is little evidence in the literature to suggest that requirements reuse is widely practised. This paper describes ten practical steps towards systematic requirements reuse based on work at the Rolls-Royce University Technology Centre (UTC) for Rolls-Smiths Engine Controls Ltd (RoSEC) in the domain of aero-engine control systems. We believe these steps have made a significant overall contribution to the 50% reuse figure quoted by the management at RoSEC for current projects within the BR700 family of engine controllers.

1 Introduction

Reusability is widely suggested to be a key to improving software development productivity and quality [1], [2]. It has been further argued that reuse at the requirements level can significantly increase reuse at the later stages of development [3], [4]. However, while there have been successful cases of reuse in companies such as Digital, Motorola and Hewlett Packard [5], [6], [1], there is little evidence in the literature of requirements reuse as part of the normal systems development process.

This paper describes ten practical steps towards systematic requirements reuse, based on work at the Rolls-Royce UTC which has been involved in institutionalising reuse within Rolls-Royce since 1993 [7], [8], [9], [10], [11]. Our experience is largely drawn from the close working relationship we have with RoSEC, a company jointly owned by Rolls-Royce and Smiths Industries Ltd, which was set up to develop and market engine controllers.

2 A Brief History of Requirements Reuse

Why should one attempt to reuse requirements? Although the argument has no documented empirical foundation, it would seem logical that the reuse of requirements in functionally similar systems will bring economic savings. Certainly in the domain of aero-engine control systems, which the UTC has been involved in, where development costs are high, even a small amount of reuse may convert into large financial savings. It can also be argued that by reusing the same set of requirements again and again, one is likely to 'trust' them more than requirements written 'from scratch'.

Requirements reuse has been examined from a number of different perspectives: analogy [12], [13], case-based reasoning [14] and generic modeling [15], [16], [17], [18]. Unfortunately, the above ideas have been restricted to small-scale academic examples, and remain largely untested in a genuine industrial or commercial capacity. More recently, however, the reuse community has reported success in the use of domain-specific approaches to reuse within certain organisations [19], [20]. Central to these domain-specific approaches is the use of domain analysis, which Prieto-Diaz [21] defines as *"a process by which information used in developing software systems is identified, captured, and organised with the purpose of making it reusable when creating new systems"*. Although early domain analysis work appeared more relevant to code reuse [22], there is evidence that domain analysis techniques are now being used during earlier stages of development [23], [24], [25].

In sum, the notion of reuse at the requirements stage is accepted by many within the community as a desirable aim. However, what appears to be missing from the literature is pragmatic advice on achieving requirements reuse as part of regular project practice. This paper addresses this concern.

3 Ten Steps Towards Systematic Requirements Reuse

The theme of the 4th annual workshop on Software Reuse Education and Training was 'Making Reuse Happen –

Factors for Success' [26], emphasising the need for the reuse community to consider the practical implications of their ideas. Such a theme accords with the technology transfer goals of the UTC [27], [28] which has been involved in the reuse of requirements for FADECs (Full Authority Digital Engine Controllers) at Rolls-Royce and RoSEC.

A FADEC is a control system for an aero-engine, taking inputs from sensors located on the engine and aircraft, and producing output in the form of electrical signals to actuators such as fuel valves and igniters. The FADEC is embedded and safety-critical.

In the following, we describe ten practical steps which we feel have brought RoSEC towards systematic requirements reuse, backed up with examples from our experiences. The steps we describe are not meant to be carried out in a fixed sequence – each step represents an idea that can be taken onboard and implemented within an organisation seperately in its own right We have classified the steps into one of two categories: orthodox and nonconformist. Orthodox refers to a step which conforms with generally accepted reuse principles (but which may not have much empirical foundation). Nonconformist refers to a step which suggests revisions or new openings to the current way the community thinks about reuse.

3.1 "Beware of Seductive Generalisations." (nonconformist)

Generic modelling, in one form or another, is often the cornerstone technique of most approaches to reuse. While we do not dispute that generalisations are important in reuse, we do feel that there is a need to examine the usefulness of generalisations more critically (a point also raised in [29]).

To illustrate, in one study of the functional requirements for three different aero-engine starting systems, we calculated (on the basis of a simple count) that only about 30% of requirements could be considered reusable, despite the systems having comparable functionality. This reason why the reuse figure is lower than expected is probably because the high level requirements between systems were similar, giving an overall impression of similarity. In fact, most of the requirements were low-level requirements, often tied in with design, more detailed in nature, and therefore difficult to reuse.

Our point here is that a view of reuse based solely on generalisations can be deceptive, and need to be examined more closely to reach a more realistic estimation of the true amount of reuse that is possible. Within RoSEC, the UTC has encouraged a broader view of reuse beyond

that of generalisations, for example, the use of 'pluggable' requirement parts described in step 3.9 promotes the use of optional and configurable requirements as well as generic ones.

Contribution to RoSEC: A realistic view of requirements reuse is taken at all levels of management within the company. Reuse education via the UTC has helped ensure that manageable but progressive reuse targets have been set for new projects within the BR700 family of engine controllers.

3.2 "Identify System Families to Maximise Reuse."(orthodox)

Code libraries have achieved significant success. Unlike code, however, requirements are context sensitive and are specific to a problem or set of problems. In addition, requirements are often 'knitted' together as part of an overall model, unlike code fragments which can be more modular and 'stand-alone' in nature. A library of individual requirements, therefore, is likely to be difficult to construct and use.

A more promising organisation for requirements from a reuse point of view is that of system 'families' as proposed in [23]. Within a system family, it may be possible to:

- Identify commonalties between the 'parent' system and 'child' system.
- Impose a common or standard requirements engineering process within the organisation.
- Anticipate certain kinds of change and specialisations.
- Reuse domain knowledge.
- Recognise working patterns which aid project planning.

A simple tree diagram can be used to depict a family of systems, and help identify sub-families where the requirements between family members may be even more closely aligned. Figure 1 shows the family structures for FADECs which is composed of marks and variants.

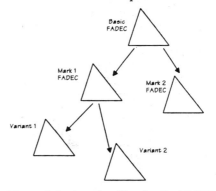

Figure 1 System families for the FADEC

A mark is a FADEC for a specific engine within a series, such as the medium thrust BR710 within the BR700 series. Clearly there is reuse potential between the 'basic' (or parent) FADEC and its marks (or children). A variant is a mark that is produced to the specific requirements of an airframer (such as Boeing or Airbus). Again, there is further reuse potential between the a mark and its variants.

Contribution to RoSEC: The development of a reuse programme which maximises reuse based upon a 'parent' and 'child' view of engine controllers. This is best shown in the document structure used by RoSEC for the BR700 family of engine controllers. Here, a set of generic requirements documents currently exists for the BR710 engine controller which are referenced by variants of the BR710 engine controller.

3.3 "Evaluate Reuse Technology In terms of Process Change, not Just on Reuse Potential." (nonconformist)

Numerous reuse technologies – application generators, patterns, high-level languages and cookbooks – have been described in the literature (see [4] and [30] for more details). However, rather than 'leaping into' the technology, it is important to assess the likely impact of the technology. This involves being clear about:

- *Current Practice* — how requirements are currently engineered.
- *Reuse Strategy* — how one envisages reuse will be 'implemented' in the current requirements engineering process.
- *Effects on Current Practice* — how the reuse strategy will change current practice in terms of methods, organisation structure, finance and other facets.

We have identified a number of evaluation criteria within each of these three areas, and used these as the basis for assessing different reuse technologies (described in Table 3 at the end of the paper). The framework acts as a 'checklist', encouraging one to think more deeply about the way in which requirements reuse is to be achieved and sustained in a commercial setting, emphasising the mix of technical, organisational and financial issues. It should be noted, however, that the technology offering the highest reuse potential is not necessarily the most suitable for the organisation (it may be seen as too costly or risky for example).

Contribution to RoSEC: the UTC has produced an assessment document for RoSEC which prescribes a set procedure for evaluating different kinds of reuse technologies, which takes into account both technical and non-technical concerns.

3.4 "Domain Issues act as Requirements Focalpoints, and Can be Used to Organise and Structure Reuse Products and Processes." (nonconformist)

The process of creating reusable requirements is aided by having a road-map for structuring the domain and organising reusable requirements knowledge. In this respect, we have found the notion of "issues" a useful structuring mechanism. We view an issue as an area where requirements, in a particular domain, are typically focused. Table 1 describes issues for thrust reverser systems and the key questions, which we call trigger questions, pertaining to each issue.

Issues can be compared to the notion of 'touchstones' proposed in [31] as a way of structuring and controlling the process of knowledge elicitation. The trigger questions in an issue hint or point to individual requirements, for which there may be a corresponding template requirement. At RoSEC, we have explicitly recorded issues and trigger questions, and used them as a basis for developing reusable, domain-specific requirements engineering processes (described in more detail in sub-section 3.8).

Contribution to RoSEC: For the domain of aero-engine starting, we have identified 16 different issues and their associated questions. These are recorded in a 'Domain issues' document forming part of a wider reusable document set which the UTC is developing aimed primarily for the BR700 engine family.

3.5 "Reasoned abstraction is effective for developing template requirements." (orthodox)

Template (or parametised) requirements encourage reuse by factoring out system-specific details as parameters of the requirement. We have found that template requirements provide a quick and cost-effective route to reuse – the notion of a template is easy to comprehend and does not require a change in the way requirements are described. The method for creating template requirements is one of reasoned abstraction, and is described as:

- Identify commonly re-occurring issues and trigger questions between similar projects.
- Use the trigger questions to locate equivalent 'concrete' requirements in each project.
- Use the similarity between concrete requirements to formalise the 'constant' part of the template requirement.
- Use the differences between concrete requirements to formalise the 'variable' part of the template requirement as parameters.
- Validate the template requirement with an expert.
- Re-use the template requirement in future projects, and refine it as necessary.

Table 4 shows an example of the results of this process with respect to a developing template requirements for dry cranking an engine (rotation of the engine without ignition or fuel). Note that the abstraction process is a reasoned one; in any abstraction process, we need to ask a number of important questions:

- Do we have 'equivalent' exemplar requirements?
- What part of the requirement is constant, what part is variable, and how can the two be separated?
- What is the explanation for the separation?
- Is the resulting template requirement meaningful and sufficiently flexible that it can be considered reusable? Is it possible to test this by applying the template requirement to a separate exemplar?
- Do other requirements engineers understand the template requirement and the abstraction process from which is has been derived?

Answering such questions is not straight-forward, which is why we believe abstraction of this nature will be difficult to automate, despite recent work in the area of computational matching [32]. Abstraction is clearly important in reuse, and the ideas here can be compared to work in artificial intelligence – [33] for example, describes the formation of general plans and heuristics from exemplars. However, some decision must be made as to the most appropriate level of abstraction. Over-abstraction will strip away essential parts of a requirement, while under-abstraction will retain system-specific details which will reduce overall reusability. There is a balance here where the most optimal level of abstraction is ultimately the one which requires the minimum amount of effort to re-use the abstract artifact (template requirements). It is likely that this balance will be reached with usage over time rather than as something we can 'calculate' beforehand.

Contribution to RoSEC: The UTC has created 30 template requirements in the domain of aero-engine starting based on the functional requirements documents (FRDs) from four different engine controllers. We believe these 30 template requirements represent the core 30% of a typical FRD. As well as the ongoing work of refining these, the UTC is also involved in the process of developing template requirements in the domain of thrust reverse.

3.6 "Requirements patterns often emerge after working in a particular domain." (nonconformist)

Domain analysis is based on the idea that if one studies the systems in a particular domain, patterns can be identified. Substantial interest in patterns at the levels of design and code has been shown by those in the object-oriented community [34]. However, patterns can be found much earlier in the development process. An analysis of different requirements documents for aero-engine starting systems and signal validation systems revealed patterns of requirements. For example, we noticed a pattern of requirements for engine relight, a common feature of modern FADECs (Figure 2).

In short, engine relight refers to the relighting of the engine when a flameout condition occurs (such as in the case of severe water ingestion). The pattern depicts five different kinds of requirements, which are often found together. For example, the requirements for engine relight will always include a requirement for how flameouts will be detected. The arrows shown in Figure 2 provide additional information about the pattern, and indicate the order in which the requirements are usually addressed. For example, an expert in starting will usually ask questions about the operation of the ignition system before considering any associated timing requirements.

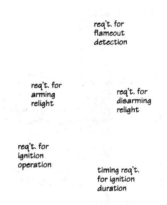

Figure 2 A requirement pattern for engine relight

By capturing patterns, we have moved closer to formalising the structures of requirements knowledge in a particular domain. In doing so, there are a number of issues which need to be addressed:

- *Pattern content.* What knowledge does a pattern impart? For example, the pattern in Figure 2 tells us about 'expected' requirements in a particular area, and something about the order in which they are to be elicited. However, we have often found the existence of dependencies between requirements along the lines of 'requirement B is possible only if requirement A is true'. It is possible therefore that other kinds of patterns might capture this type of knowledge.
- *Representing patterns.* How are patterns represented? In some respects, this is dependent upon the pattern content. However, we have found the use of simple diagrams (as in Figure 2) supplemented with more detailed explanations in natural English, sufficient for our purposes here. It should be noted it is not the

formality of the pattern representation which is important, but the knowledge which the pattern communicates to the pattern reader.

- *Dealing with exceptions.* Are there exceptions to the pattern? Encountering exceptions may indicate the need to revise a pattern, or to further delineate the context in which the pattern can be applied.

In practice, recognising a pattern is the most difficult part of the process, which will only be possible after studying several similar systems. The important point, however, is to aware that patterns exist and to document them so that they can reused to guide future systems.

Contribution to RoSEC: A Windows-based prototype tool, known as COMPASS (COMPonent ASSistant) [8], has been developed which records patterns in the area of aero-engine starting and links them with template requirements stored in a local reuse database. Using COMPASS, a user (a RoSEC engineer) is able to browse and select patterns and instantiate template requirements into project requirments. At present, COMPASS has 13 patterns including patterns for cranking, various starting modes, continuous ignition and engine relight. RoSEC is currently evaluating COMPASS.

3.7 "Make explicit the context of reuse to prevent misuse." (nonconformist)

One of the most striking cases of reuse misuse, reported by [35], concerned the traffic control system used by the Civil Aviation Authority (CAA) in the UK. The software, designed by IBM's Federal Systems Division, contained a model of the airspace it controls. However, the software was designed for air traffic control centres in the US, and the CAA had not taken account of a zero longitude when reusing the software in the UK. This oversight caused the computer to fold its map of Britain in two at the Greenwich meridian.

The case clearly demonstrates the need to reuse with care, especially in the case of high-integrity systems. Work at the UTC suggests that the explicit documentation of context is a step towards the prevention of reuse misuse. In doing so, one is forced to think about the (often hidden) assumptions behind a reuse artefact. For example, in the domain of aero-engine starting systems, we have explicitly defined three typical contexts, shown in Figure 3.

Each context depicts an abstract design model of the (often physical) components in a starting system. Civil aircraft often have an air turbine starter, which provides a high torque-to-weight ratio. Military aircraft on the other hand, will usually have a solid propellant starter which provides rapid starting. Small, light aircraft find the elec-

tric motor starter more suitable because of the ease of maintenance.

We argue that it is important to associate reusable requirements with a particular context, i.e. a particular abstract design model, and to ensure that the *intended* context that a set of reusable requirements is created for matches the *actual* context in which they are going to be reused. For example, requirements which concern the ignition of a solid propellant cartridge are only 'valid' or meant to be reused in the context of a solid propellant starter – attempting to reuse them in the context of an air turbine starter or electric motor starter is inappropriate, potentially dangerous and is likely to lead to poor levels of reuse

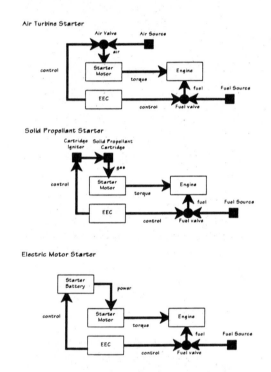

Figure 3 Three different contexts for aero-engine starting

Contribution to RoSEC: There is now raised awareness within RoSEC with respect to the problems of inappropriate reuse. The UTC is currently in the process of writing a 'safe reuse' guidebook for RoSEC which includes a description of safe reuse guidelines.

3.8 "Parts of the requirements engineering process is also reusable." (nonconformist)

It was Osterweil [36] who first suggested that "processes are programs too". Since then, there has been an increasing interest in the explicit modelling of software processes [37], [38]. In the context of requirements engineering at RoSEC, we observed similar sequences of questions and routines being followed by domain experts working on the requirements of systems in the same domain. We believe that if parts of this process is modelled in an abstract manner, it can be reused to guide future requirements engineering exercises. For example, we have modelled the process by which requirements are elicited for 'aborting a start' (Figure 4) using our own variant of Role Activity Diagrams (RAD) [39].

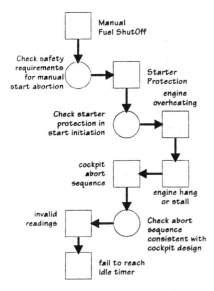

Figure 4 The requirements process for 'aborting a start'

In short, an aircraft engine in the process of being started can be aborted for safety reasons such as engine overheating. In our process description, a square box represents a questioning or elicitation activity, described as a set of trigger questions for the RoSEC requirements engineer to consider. For example, the questioning activity labelled 'Manual fuel shutoff' includes the trigger question:"is there a facility in the cockpit for the pilot to abort the start by switching off the fuel supply to the engine?". A circle represents a checking activity, i.e. some analysis of the information gathered from a questioning activity. For example, the checking activity labelled 'Check safety requirements for manual start abortion' involves following a procedure for finding out the safety and certification requirements for safe manual start aborting and ensuring that they are met in this case. The arrows in our process description suggest a logical order for elicitation.

Our experience shows that process modelling in the manner shown here has a number of benefits:

- A reused process will facilitate a reused product (in this case, a reusable requirements specification).
- The process model can act as a checklist for ensuring all the requirements for a particular area are elicited.

- The 'dynamic nature' of requirements information can often be more clearly seen in the context of a process model (for example, a process model can indicate at what point in requirements engineering a particular piece of information should be elicited, cross-checked with other information, expanded upon etc.).
- Process models can be matured and refined to a point where they become key educational aids for novice engineers learning new areas.

Preliminary process models can be developed by 'watching' the expert at work first, and then refining the model with the expert as a form of validation. In addition, a relationship was often observed between requirement patterns (cf. section 3.6) and process reuse: as patterns 'matured' it became possible to derive the associated process. The UTC is currently investigating the use of process modelling in establishing domain-specific requirements guide-books for RoSEC engineers. More details of our work on process reuse can be found in [40].

Contribution to RoSEC: So far the UTC has informally modeled 7 domain-specific requirements engineering processes in the domain of aero-engine starting using the RAD notation. We have documented these processes as a kind of 'workplan' that an engineer can pickup and follow. These processes are documented in a requirements guide-book for the aero-engine starting domain.

3.9 "Factor requirements variance into pluggable requirement parts." (nonconformist)

Generalisation is often used to capture the commonality between all systems in a domain. As such, there is always a danger that the resulting generalisation is so devoid of detail that it becomes of little practical value (cf. section 3.1: "Beware of seductive generalisations"). However, reuse can be increased if we consider the isolation of requirements which, although are not common to all systems in a domain, re-occur on a frequent basis.

One approach we have used at RoSEC is that of the pluggable' requirement part. To illustrate, consider the following template requirement:

Fuel and Ignition Template Requirement:
'Fuel and ignition will only be switched on when a start has been requested by the pilot and [fuel and ignition conditions] are true'

The square brackets indicates the variable part of the generic requirement. However, although the fuel and ignition conditions for different aircraft systems may differ, it is likely that the same conditions will re-occur across many different aircraft. Therefore, we captured the variability of the generic requirement in the form of pluggable requirement parts (this is something we feel that the domain expert should do because actually recognising variability can be difficult). Table 2 shows a list of six pluggable requirement parts for our generic requirement.

The use of pluggable requirement parts enables engineers to construct requirements quickly, with a degree of flexibility, by choosing relevant requirement parts and plugging them into the generic requirement. The requirement below illustrates this:

'Fuel and ignition will only be switched on when a start has been requested by the pilot and the engine speed > minimum fuel-on speed and the turbine gas temperature < starting TGT and the fuel-on timer has not expired.

Here, we have two reuse concepts working together. The first is that of the template requirement mentioned earlier in the paper. The second is the formalisation of typical parameter values associated with a template requirement, i.e. a set of pluggable requirement parts.

Contribution to RoSEC: As mentioned earlier, we have developed a tool called COMPASS which enables RoSEC engineers to select template requirements from a reuse database and 'instantiate' them to form project requirements. For many instantiations, the user can select from a parts database which has been built into COMPASS.

3.10 "Assess beforehand the impact of requirements reuse on the development 'food chain'." (orthodox)

It is fallacious to think of requirements engineering as a process which is performed in a vacuum. In truth, it is impractical to take a 'purist' view of requirements engineering, as external circumstances such as costs, design implications and even politics will inevitably affect the way in which requirements are shaped. In this respect, an analogy can be made with the food chain – changing a requirement is likely to have profound repercussions down the chain, causing reassessment of the original design, and forcing significant re-testing.

Under the same analogy, changing the way in which requirements engineering is performed by increasing levels of reuse is likely to induce repercussions, good and bad, along the development chain. We need to assess the impact, taking into account a number of viewpoints:

- *Design Options.* Do reusable requirements 'home in' on a standard design or set of designs? Would it be acceptable and worthwhile formalising standard designs based around the reusable requirements? (Work

on the ADAGE project [24] seems to have adopted a reusable design approach)
- *Testing Strategies.* Does reusing requirements lead to more economical testing? If not, can the existing testing strategy be changed to reap such benefits? Does the introduction of novel requirements nullify the potential savings possible with reuse?
- *Certification Pitfalls.* Certification is a crucial hurdle in the development of safety-critical systems such as those often found in avionics. We therefore need to make some honest judgements: is reuse likely to affect the quality of the delivered system? If so, in what way and will this lead to pitfalls during certification? Is the integrity of the delivered system in any way comprised, and if not, how sure can we be?

Attempting to localise reuse, without examining changes along the development food chain, can at best only lead to localised benefits. Our advice here is to establish a small working group, consisting of individuals representing different areas of the software development process. The objective of the working group should be to critically examine how the impact of reuse on other aspects of systems development, and explore potentially advantageous and/or harmful reuse 'spin-offs'.

Contribution to RoSEC: Under the guidance of the UTC, RoSEC has established a 4 man reuse group within the company comprised of individuals from different engine projects. The objective of the reuse group is to identify and exploit opportunities for reuse across the different projects. One achievement of the group has been recent work on the reuse and automatic generation of test scripts.

4 Conclusions

The main contribution of this paper is a description of ten practical steps towards systematic requirements reuse, distilled from three years work at the UTC in institutionalising reuse within RoSEC. These ten steps have helped considerably in taking RoSEC to a point where they are able to begin populating a requirements reuse library, and where requirements reuse is becoming an integral part of their development process. An overall figure of 50% reuse between engine controllers within the BR700 family has been quoted by the senior management at RoSEC (even though the exact breakdown of this figure for requirements reuse is not yet available to the UTC, we expect it to be close to the 50% mark). Although most of our steps have a distinct technical focus, we can not over-emphasise the importance of organisational and managerial factors in institutionalising reuse [41], [42].

Perhaps the most encouraging lesson from our work is that an increase in the level of requirements reuse can be achieved using a number of relatively cheap and simple measures which do not require radical organisational changes. While we believe that each of the steps described in this paper can stand alone in its own right, it is the synergy between steps working in parallel which is likely to bring the most significant benefits of reuse.

References

[1] Lim, W.C. (1994) Effects of Reuse on Quality, Productivity, and Economics, IEEE Software, 11(5):23-30, 1994.

[2] Sommerville, I. (1996) Software Engineering (5th Edition), Addison Wesley, ISBN 0-201-42765-6.

[3] SPC (1992) Software Productivity Consortium Reuse Adoption Guidebook, Version 01.00.03, SPC-92051-CMC, November, 1992.

[4] Biggerstaff, T. and Ritcher, C. (1987) Reusability framework, assessment and directions, IEEE Software, 41(3), March, 1987.

[5] Guerrieri, E. (1994) Case study: Digital's application generator, IEEE Software, 11(5):95-96.

[6] Joos, R. (1994) Software reuse at Motorola, IEEE Software, 11(5):42-47.

[7] Kelly, T.P., Lam, W. and Whittle, B.R. (1996) Diary of a domain analyst: a domain analysis case-study from avionics, In Proceedings of IFIP Working Groups 8.1/13.2 Conference, Domain Knowledge for Interactive System Design, Geneva, May 8-10, 1996.

[8] Lam, W., Whittle, B.R., McDermid, J. and Wilson, S. (1996a), An Integrated Approach to Domain Analysis and Reuse for Engineering Complex Systems, In Proceedings of International IEEE Symposium and Workshop on Engineering of Computer-Based Systems (ECBS '96), Friedrichshafen, Germany, March 11-15, 1996.

[9] Lam, W. and Whittle, B.R. (1996) A Taxonomy of Domain-Specific Reuse Problems and their Resolutions - Version 1.0, Software Engineering Notes (To appear)

[10] Whittle, B.R., Lam, W. and Kelly, T.P. (1996) A Pragmatic Approach to Reuse Introduction in an Industrial Setting, In Proceedings of the International Workshop on Systemmatic Reuse, Liverpool John-Moores University, Springer-Verlag.

[11] Whittle, B.R., Vickers, A.J., Lam, W., McDermid, J., Hill, J.A., Rimmer, R. and Essam, P. (1995), Structuring Requirements Specifications for Reuse, International Journal on Applied Software Technology (To appear)

[12] Finkelstein, A. (1988) Reuse of formatted requirements specifications, Software Engineering Journal, 3(5):186-197, 1988.

[13] Maiden, N. and Sutcliffe, A. (1993) Exploiting reusable specification through analogy, Communications of the ACM, 35(4):55-64, 1993.

[14] Lam, W. (1994) Reasoning about requirements from past cases, PhD thesis, Kings College, University of London, 1994.

[15] Bolton, D., Jones, S., Till, D., Furber, D. and Green, S. (1994) Using domain knowledge in requirements capture and formal specification construction, In Jirotka, M. and Goguen, J. (Eds.), "Requirements Engineering: Social and Technical Issues", Academic Press, London, 1994.

[16] Reubenstein, H.B. (1990) Automated Acquisition of Evolving Informal Descriptions, Report No. AI-TR 1205, Artificial Intelligence Laboratory, Massachusetts Institute of Technology, 545 Technology Square, Cambridge, MA 02139, 1990.

[17] Miriyala, K. and Harandi, T.H. (1991) Automatic derivation of formal software specifications from informal descriptions, IEEE Transactions on Software Engineering, 17(10):1126-1142, 1991.

[18] Ryan, K. and Mathews, B. (1993) Matching conceptual graphs as an aid to requirements reuse, In Proceedings of the IEEE International Symposium on Requirements Engineering, ISBN 0-8186-3120-1, page 112-120, 1993.

[19] Griss, M.L., Favaro, J. and Walton, P. (1994) Managerial and Organisational Issues - Starting and Running a Software Reuse Program, In Software Reusability, eds. W. Schaefer, R. Prieto-Diaz and M. Matsumoto, Ellis Horwood, Chichester, GB, 1994 pp.51-78, 1994.

[20] WISR (1995), Proceedings of the 7th Annual Workshop on Software Reuse, St. Charles, Illinois, August 28-30, 1995.

[21] Prieto-Diaz, R. (1990) Domain analysis: an introduction, ACM Software Engineering Notes, 15(2):47-54, 1990

[22] Wartik S. and Prieto-Diaz P. (1992), Criteria for comparing reuse-oriented domain analysis approaches, International Journal of Software Engineering and Knowledge Engineering, 2(3):403-431, 1992.

[23] Gomaa, H. (1995) Reusable software requirements and architectures for families of systems, Journal of Systems and Software, 28:189-202, 1995.

[24] Tracz, W. (1995), DSSA (Domain-Specific Software Architecture) pedagogical example, ACM SIGSOFT Software Engineering Notes, 20(3):49-62, 1995.

[25] Tracz, W., Coglianese, L. and Young, P. (1993) A domain-specific software architecture engineering process outline, ACM SIGSOFT Software Engineering Notes, 18(2):40-49, 1993.

[26] WSRET (1995) Proceedings of the 4th International Workshop on Software Reuse Education and Training, Morgantown, West Virginia, 14-18th August, 1995.

[27] Bate, I.J. et al. (1996), Technology Transfer: An Integrated 'Culture-Friendly' Approach, In Proceedings of Workshop on Technology Transfer, 18th International Conference on Software Engineering, Berlin, Germany, 25-29 March, 1996.

[28] Vickers, A.J., Whittle, B.R. and McDermid, J.A. (1996), Technology Transfer by Case Study: An Experience Report, Presented at 3rd International Conference on Concurrent Engineering & Electronic Design Automation, Poole, 1996.

[29] Kramer, J. (1993) Generalisations are false?, In Proceedings of the IEEE International Symposium on Requirements Engineering, ISBN 0-8186-3120-1, 1993.

[30] Mili, H., Mili, F. and Mili, A. (1995) Reusing Software: issues and research directions, IEEE Transactions on Software Engineering, 21(6):528-561, 1995.

[31] Littman, D.C. (1987) Modeling human expertise in knowledge engineering: some preliminary observations, International Journal of Man-machine Studies, 26:81-92, 1987.

[32] Spanoudakis G. and Constantopoulos P. (1996) Analogical Reuse of Requirements Specifications: A Computational Model, Applied Artificial Intelligence (to appear)

[33] Carbonell, J.G. (1983) Learning by analogy: Formulating and generalising plans from past experience, In R.S. Michalski, J.G. Carbonell and T.M. Mitchell (Eds.) Machine Learning: an Artificial Intelligence approach, Los Altos, CA, Kaufmann.

[34] Coplien, J.O. and Schmidt, D.C (1995) Eds. Pattern Languages of Program Design, Addison-Wesley, ISBN 0-201-6073-4, 1995.

[35] Wray, T. (1988) The everyday risks of playing it safe New Scientist, September 8, pp61-65, 1988.

[36] Osterweil, L. (1987) Software processes are software too, In Proceedings of the International Conference on Software Engineering, 1987.

[37] Curtis, B., Kellner, M.I. and Over, J. (1992) Process modelling, Communications of the ACM, 35(9), 1992.

[38] McChesney, I.R. (1995) Towards a classification scheme for software process modeling approaches Information and Software Technology, 37(7):363-374, 1995.

[39] Holt, A.W., Ramsey, H.R. and Grimes, J.D. (1983) Co-ordination System Technology as the Basis for a programming environment, Electrical Communication, 57(4), 1983.

[40] Lam, W. (1996) Process reuse using a template approach: a case-study from Avionics, Software Engineering Notes (To appear).

[41] Fafchamps, D. (1994) Organisational factors and reuse, IEEE Software, 11(5):31-41.

[42] Frakes, W.B. and Isoda, S. (1994) Success Factors for systematic reuse, IEEE Software, 11(5):15-19, 1994.

Table 1: Issues for thrust reverser systems

Issue	Trigger Questions
Deploy thrust reverser	How does a pilot activate the thrust reverser system? Is the activation related to the position of the thrust reverser doors? What safety provisions are made if the doors are jammed or inhibited (such as automatic thrust limitation)?
Stow thrust reverser	How does a pilot deactivate the thrust reverser system? Is the deactivation related to the position of the thrust reverser doors? What safety provisions are made if the doors are jammed or inhibited?
Thrust reverser maintenance	How is the thrust reverser deployed and stowed under aircraft maintenance? What safety measures are in place with respect to the operation of the thrust reverser under maintenance?
Thrust reverser interlock	Is interlock provided to give the pilot a tactile indication of thrust deployment? If so, during which period does the interlock take place, and at what point is it released?

Table 2: A list of pluggable requirement parts

Pluggable Requirement Part
1. engine speed > preferred fuel-on speed
2. engine speed > minimum fuel-on speed
3. engine acceleration < maximum engine acceleration
4. turbine gas temperature (TGT) < starting TGT
5. TGT < on-ground pre-start TGT
6. fuel-on timer has not expired

Table 3: Framework for evaluating reuse technology

Evaluation Criteria	Description	RoSEC-related Examples
Current practice		
• document	The type of report	Functional requirements doc., System concept doc.
• document content	The type of requirements included in the document	System, FADEC, hardware design or aircraft interface requirements.
• notation	The representation of the requirements	Structured English and statecharts.
• methods used	Any particular methods or techniques used during the requirements engineering process	Statecharts.
Reuse strategy		
• coverage	An estimate of how much reuse is possible	20% of a typical functional requirements document.
• reuse artefacts	The form of what is to be reused	Generalised structured English requirement statements.
• scope	Any limitations of the reuse artefacts	Only applicable to systems in the BR700 engine series.
• reuse frequency	How often the reusable artefact would be used.	On all engine projects, or just on BR710-related engine projects.
• envisaged process	If reuse was adopted, how the process would look from an engineer's point of view.	The engineer logs onto the reuse library, which is based around a world-wide-web browser. Requirements given as textual statements are cut and paste into a requirements document. Other requirements not taken from the reuse library can be directly added to the document.
• startup actions	What is needed in order to make reuse possible.	A domain analysis of the thrust reverser domain, followed by the setting up a reuse library and its population with generic requirements.
• critical success factors	What are seen to be the most important factors in order for the reuse strategy to succeed.	Engineers actively involved in the domain analysis; the reuse library to be accessible via the engineer's PCs and project manager providing 1 extra week in the project budget to allow engineers familiarisation.
Effects on current practice		
• notation	How a reuse strategy is likely to affect the way in which requirements are represented	None.
• methods used	How a reuse strategy is likely to affect the methods and techniques already in use.	Engineers will be able to select requirements from the reuse database and automatically import them into a RoSEC document.
• organisation	The organisational impact that a reuse strategy will have.	The assignment of a person to maintain a reuse library and to perform a domain analysis.
• financial	The financial consequence that a reuse strategy is likely to entail.	Initial up-front costs in producing a reusable requirements library.
• other	Other consequences a reuse strategy is likely to bring.	General training for engineers on how to use the reuse library.

Table 4: Creating a template requirement using abstraction

Element in Abstraction Process	Example
concrete requirement from system A	When engine not in process of being started, cranked or run, if fuel switch in OFF position and master crank switch in ON position, and engine start switch then turned to ON position then dry crank will be initiated.
equivalent concrete requirement from system B	When engine not in process of being started, cranked or run, if fuel switch in OFF position and engine start switch turned to CRANK position, then dry crank will be initiated.
constant requirement part	When engine not in process of being started, cranked or run, if (X) and then (Y), dry crank will be initiated.
variable requirement part	X and Y are cockpit-specific signals.
abstraction reasoning	Cockpits are specific to a particular system, and not all systems will have the same cockpit layout. Hence, this aspect is a variable requirement part and must be factored out of the generic requirement.
template requirement	When the engine is not in the process of being started, cranked or run, if (cockpit signal 1) and then (cockpit signal 2), a dry crank will be initiated.

Reusing Operational Requirements:
a Process-Oriented Approach

Robert Darimont[1] and Jeanine Souquières[2]

[1] Université catholique de Louvain

Place Ste-Barbe, 2, B-1348 Louvain-la-Neuve (Belgium)

[2] CRIN–CNRS

BP. 239, F-54506 Vandœuvre-les-Nancy Cedex (France)

email: Robert.Darimont@info.ucl.ac.be,Jeanine.Souquieres@loria.fr

Abstract

This paper advocates for a process-oriented approach to reuse operational requirements. In process-oriented approaches, a development (and reuse is just a particular case of development) keeps track of the intermediate states and steps leading to the final artifacts. The paper shows that it is worth recording the reuse process for developing operational requirements and not only the final product artifacts generated by the reuse process. The motivation behind is that syntactical constructs of the specification languages are generally not sufficient to trace the reuse process. Such traces are important for documentation purpose, for maintenance, replay and software evolution.

1 Introduction

Software reuse is the process of using existing software components rather than building them from scratch. Frequently mentioned benefits for reuse are: (i) reduction of development time and effort [20]; (ii) a better global system quality achieved by reusing quality software components, (iii) reduction of the costs for maintaining software systems [9]; (iv) a help to overcome difficulties encountered by inexperienced software engineers during early stages of software development [8].

Those benefits are effective if the reuse approaches succeed to address the two following issues: (i) the components to reuse can be retrieved adequately from a software library; (ii) the selected components must be easily customized to fit the new problem.

Research on specification retrieval and matching aim to address the above issues. [17] investigates the feasibility of using formal specifications to implement a database of software components for the purpose of reuse. [26] defines specification in the relational framework and bases specification matching on the *refines* ordering on relations.

Reusability is promoted as a major quality factor in object-oriented approaches [16]; relationships like inheritance, renaming, and polymorphism incite developers to reuse objects and classes. [7] clusters reuse objects and classes into design patterns.

[13, 14] stress on three processes to cope with specification reuse: problem categorization, selection of candidate specifications belonging to the same category and customization of the selected specification to fit the new problem. They advocate for analogical reasoning to instantiate reusable specifications.

Operational requirements address the objects and actions which the system to be developed has to deal with. Specifying operational requirements formally makes requirements unambiguous. It also allows inconsistencies to be detected easier [15]. Our goal is to help specifiers reuse *pieces of formal specifications* to formalize new operational requirements. The kind of formal specifications on which we are focusing are abstract data type specifications [25].

This paper advocates for a process-oriented approach to reuse. In process-oriented approaches, a development [5] (and reuse is just a particular case of development) keeps track of the intermediate states and steps leading to the final artifacts [4, 3]. The paper shows that it is worth recording the reuse process for developing operational requirements and not only the final product artifacts generated by the reuse process. As originally stated in AI [1], the motivation behind is that syntactical constructs of the specification languages (such as IsA relationships over types) are generally not sufficient to trace the reuse process. Such traces are important for documentation purpose, maintenance, replay and software evolution.

The paper is organized as follows. Section 2 presents an example of reuse for a type specification.

16

Section 3 presents the Proplane process model, which is dedicated to the formalization of specification development processes. Section 4 formalizes the reuse process for operational requirements.

2 An example of reuse process

Consider the specification of a Flight Departure Board (FDB in short) giving information at the main entrance of the airport to customers about the scheduled flights: current time, flight numbers, destination, departure time, gate where to check-in and status of the flight. For instance, a *FDB* display could be the following one:

Flight Number	Destination	Departure Time	Gate	Status
AA784	Los Angeles	10:05	C5	Delayed
BA 54	Toronto	10:20	D3	Closing gate
...	...	...	...	...

Suppose the analyst has browsed the specification library, and has *selected* the specification of a timetable as a promising candidate for reuse. The timetable records the connections starting from a given location; each connection is associated with a unique identifier, has a destination location and schedule. In Glider [2], the Timetable specification looks like the following specification :

type cluster TIMETABLE
    * This cluster describes, for a given location,
    * all its connections (remote location, schedule, ...)
    is CP [Location : LOCATION,
            Connections : SEQ [BINDING]]
    ...
end TIMETABLE
type cluster BINDING
    is CP [Identifier : KEY,
            Connected-Location : LOCATION,
            Schedule : TIME]
end BINDING

Figure 1 shows a graphical view of the *TIMETABLE* type structure. Nodes stand for attribute type names and edges for attribute names. An arrow indicates that a generic type instantiation has taken place; the edges below a node together stands for a Cartesian product.

The starting point, we advocate for reuse, is a type structure mapping because the specification of operations on data types depends on the structure definition of these types. The probability of a better matching between the specification, the analyst has in mind, and the one to be reused is therefore higher if their

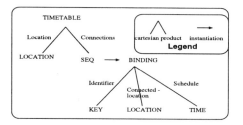

Figure 1: The TIMETABLE data structure

structures are "closed to one another", that is, there is an analogy allowing one structure to be mapped on to the other. Figure 2 results from a mapping from the *TIMETABLE* type structure onto the *FDB* data structure.

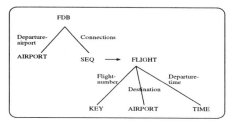

Figure 2: The FDB data structure

Observe that the mapping is only partial: (i) the *Gate* and the *Flight-status* are missing, (ii) no clock is provided, (iii) the *Departure-airport* component is not required.

Therefore, besides some *renaming* operations, the *TIMETABLE* type needs to be *customized* to fit the analyst's needs. Considering each difference in turn, the *process* of customizing the *TIMETABLE* type is as follows.

2.1 Adding gate and flight status

Introducing a gate and a flight status requires to modify the FDB data structure.

1. Focus on the *FLIGHT* component as the gate and flight status are specific to flights.

2. Introduce the new attributes *Gate* and *Flight-status* to the *FLIGHT* type leading to Figure 3.

3. Consider new invariants which could be set with respect to the new components. A new invariant related with gates can be : *"the same gate can be used for two flights provided that there is at least half an hour between their departure time"*. Zoom out to the specification root and update the FDB invariant as follows:

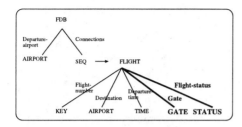

Figure 3: New attributes for a flight

$(\forall$ f: FDB   i, j: NATURAL$)$
$(1 \leq i < j \leq$ Length(Connections(f)) $\wedge$
Gate(Ith(Connections(f), i)) =
    Gate(Ith(Connections(f), j)))
$\Longrightarrow$ Departure-time(Ith(Connections(f), j)) —
Departure-time(Ith(Connections(f), i)) ≥ 30

4. Consider new operations related to the new attributes. For instance, the *Change-Gate* operation can be specified to describe the activity of assigning a new gate to a flight. Zoom out to the specification root and develop the following specification:

    operation   Change-Gate
      arity   FDB $\times$ KEY $\times$ GATE $\longrightarrow$ FDB
      asserts   Change-Gate(f, k, g) = f'
      with
      Gate(Ith(Connections(f), LookUp(f, k))) = g
      pre $(\exists$ i: NATURAL$)$
         $1 \leq i \leq$ Length(Connections(f)) $\wedge$
         Flight-number(Ith(Connections(f), i)) = k
    end Change-Gate

5. Consider existing operations and extend them if needed. For instance, the *Display-Flight-Info* which outputs the destination, time, gate and status of a flight can be extended as follows:

    operation   Display-Flight-Info
      arity   FDB $\times$ KEY $\longrightarrow$
         CP [AIRPORT, TIME, GATE, STATUS]
      asserts   Display-Flight-Info(f, k) = <d, t, g, s >
      with
      Ith(Connections(f), LookUp(f, k)) =
         <k, d, t, g, s >
    end Display-Flight-Info

2.2   Providing a clock

Information about current time is independent of any flight and is global to the board. Therefore, the introduction of a clock addresses the root of the *FDB*:

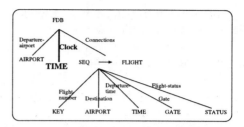

Figure 4: A new attribute for FDB

1. Focus on the root of *FDB*.

2. Introduce the *Clock* attribute (Figure 4).

3. Define a new operation, *Tick*, modeling time elapsing.

    operation   Tick
      arity   FDB $\longrightarrow$ FDB
      asserts   Tick (f) = < —, Inc(Clock(f)), —>
    end Tick

2.3   Drop the departure airport

It is clear that each *FDB* is physically bound to a unique airport. This component can thus be removed from the *FDB* definition.

1. Focus on the root of *FDB*.

2. Remove the component *Departure-airport* (Figure 5).

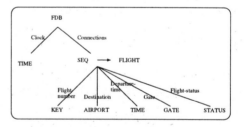

Figure 5: A new definition for *FDB*

3. Consider the assertions in which the *Departure-airport* attribute is used. In our case, the following assertion has to be deleted:

    * (2) A timetable location differs from all its
    * connected locations
    $(\forall$ t: FDB, i: NATURAL$)$
      $1 \leq i \leq$ Length(Connections(t)) $\Longrightarrow$
      Departure-airport(t) $\neq$
         Destination(Ith(Connections(t), i))

4. Consider operations in which the *Departure-airport* attribute is used. In this case, constructors and modifiers have to be modified, and the corresponding observer removed. For instance, the Tick operation defined above should be updated as follows:

operation Tick
    arity FDB ⟶ FDB
    asserts   Tick (f) = < Inc(Clock(f)), —>
end Tick

3 The Proplane Process Model

The Proplane model defines a development process as composed of development states, steps, operators [22, 23, 12] and tactics [3]. More precisely, the development of a specification is defined as a sequence of steps. Each step maps some development state to the next one in the sequence by applying some operator. A development state consists of two consistent viewpoints, a workplan and a product, together related by links. Figure 6 shows an example of a development state for the *FDB* case study developed from the *TIMETABLE* given specification.

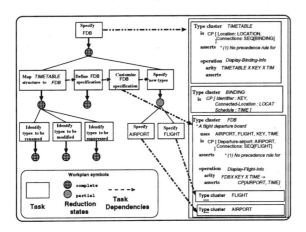

Figure 6: A development state

Workplan. The left–hand side of Figure 6 represents a workplan state. A workplan is a collection of tasks linked through a *reduction relation*. A *task* denotes a work to be done in order to produce a piece of specification. A *reduction* represents a way of accomplishing a task by introducing new subtasks.

As in the classical top-down reduction approach to problem solving [18], workplans are modeled by *and/or* –graphs with two kinds of nodes: *task* nodes

and *reduction* nodes. *And* –links are used to represent the reduction of tasks into subtasks. *Or* –links are used to represent alternative reductions. At each step of the development, only one reduction of a task is *current.* and each reduction has a state. Tasks are connected by *dependency relationships*.

Product. No specification can be developed by setting up some workplan alone. The product, which consists of both formal and informal texts, has to be set up simultaneously. The right-hand side of Figure 6 represents a product state with two kinds of nodes: *and* –node denotes a product version, i.e. instances of abstract-syntax trees in which leaves can be meta-variables; *or* –nodes stand for the product versions.

Workplan–Product Links. The workplan must be related to the product. We require that each task must be related to at least one meta-variable and each reduction must be related to a product version. Conversely, any meta-variable is associated with a unique task. Links are represented in Figure 6 by dotted-dashed arrows between tasks and product components.

Development Operators. Development operators work in parallel on the workplan and the product to reduce tasks and construct or modify the product text. Parameters of the operators are acquired interactively from the specifier and from the current development state. The operators consist of (i) a language-independent section which describes the action on the workplan, and (ii) a language-dependent section which concerns the product definition. The workplan section expresses the decomposition of the task into subtasks possibly with dependency relationships on the subtasks. The product section proposes a product version in terms of abstract-syntax trees; each meta-variable in this product is linked to some subtask of the workplan section.

Development Step. At each step of the development, the specifier chooses a task among those to be reduced and an operator among those that could be applied to the task. Next, the operator parameters are instantiated and the operator is applied. A step description keeps also track of the different operators that could have been used at this step and an informal rationale for the choice (in the justification field).

Development Tactics. Tactics aim at guiding the specifier for the selection of specific operators according to the current development state and heuristics

[3]. They are formalized in terms of (i) the states for which they are relevant, (ii) heuristics explaining when and why to apply specific operators, and (iii) a process guidance describing how to apply the selected operator.

Prototype Tool. A prototype tool consisting of a plan editor connected to a product manager [24] has been implemented. The main activity of the specifier no longer consists in writing down the specification, but in applying predefined operators to evolve the specification. This approach relieves developers of many humdrum activities and allows them to focus on the reasoning underlying the elaboration of the specification. The tool, equipped with a graphical interface, is parameterized with specification languages and libraries of development operators.

4 Reuse process in Proplane

The reuse process identified in Section 2 has been formalized in terms of development states, steps, operators and tactics in Proplane. The most interesting steps are outlined below.

Starting point. The starting point is the informal requirements document and the availability of a library of specifications.

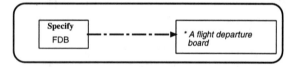

Figure 7: Initial development state

4.1 Reusing pieces of formal specifications

Instead of developing the *FDB* system from scratch, the *TIMETABLE* component is selected from a library of specifications as a candidate to be reused.

The **ReuseSpec** operator is applied on the *Specify FDB* task. The operator takes the name of the type to be reused, namely *TIMETABLE*, and the name of the new type, namely *FDB*, as parameters.

Resulting state. The *TIMETABLE* type and a skeleton for the new type are displayed in the product view (Figure 8). Four new tasks are also introduced in the workplan to help the specifier reuse this piece of formal text. Each task is reviewed in the next subsections.

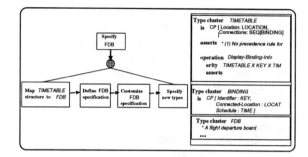

Figure 8: Reusing the TIMETABLE component

4.2 Mapping

The first subtask invites the specifier to map the structure of the reused specification on the target requirements, recording analogies and differences. The aim is to identify (i) unchanged types, (ii) new types to be introduced, (iii) types to be updated, and (iv) types to be suppressed.

Operators. Identification operators are used to set up lists of types to be renamed, modified or deleted.

Resulting state. Figure 9 shows the development state resulting from applying the identification operators. Three new tasks have been generated. The resolution of each of them produces a list of type names. These lists are intermediate product artifacts not directly visible in the product view. They can be retrieved by operators to solve other tasks in the workplan.

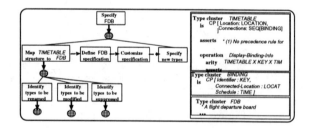

Figure 9: Mapping the TIMETABLE component onto FDB

4.3 Type Definition

Consider now the *Define FDB Specification* task. Having produced the mapping as explained above, the specifier has to choose whether to build the new type as a specialization of the reused one or to define it as a clone.

Tactics. The following heuristics is recommended to decide the best choice for each type appearing in the new specification:

> *Compare the original types and those in the new structures. Some are left unchanged; other are just renamed; some others need to be extended or shrunk. For types that have been renamed or extended, consider the opportunity of defining them as specializations of the types, which they come from. If this reveals to be unadequate for semantic reasons, or if the specification language does not support inheritance and/or renaming, define a* **clone** *of the orignal type. Shrunk types cannot be specializations of the type from which they come. Examine whether they can be specializations of some generalization of the type from which they come. If yes, proceed as above. If not, define a clone of the original type.*

In the example, the *FDB* specification has to be defined as a clone of *TIMETABLE* because some components of *TIMETABLE* have to be dropped.

Operators. The **SpecializeType** operator allows a new type to be defined on top of an existing one. The **CloneType** operator allows a type specification to be cloned into a new, completely separate type specification. The latter operator takes the specification of the type to be cloned, the name of the new type as arguments, and the mapping list between the "cloner" and "clonee" types. It generates the clone specification at the product level propagating attribute and function renaming automatically. New tasks reduce the *Specify new types* task to cope with the specification of the types that have been renamed.

Observation. Each time the clone technique is used, the relationship between the original type and the new type is lost at the product level. This is due to the fact that the relationship can only be recorded at the process level. It is important to keep track of this relationship for documenting the product and for maintenance. Maintenance issues occur when the original type specification needs to be changed for instance to fix a "bug"; in such cases, it can be interesting to support the propagation of the changes to all clones.

Resulting State. Figure 6 and the next specification show the result of applying the **CloneType** operator on the TIMETABLE specification for defining the

FDB specification. Observe that tasks have been introduced to cope with the specification of the *FLIGHT* and *AIRPORT* types. The idea is then to consider such tasks exactly as the *Specify FDB* task. The same process can be replayed for the *FLIGHT* type, but in this case, the candidate type specification for reuse is already selected: it is the *BINDING* specification.

type cluster FDB
* *A flight departure board*
provides all
uses AIRPORT, FLIGHT, KEY, TIME, NATURAL
is CP [Departure-airport: AIRPORT,
        Connections : SEQ [FLIGHT]]
asserts
  * *(1) No precedence rule for 2 simultaneous schedules . . .*
  * *(2) A timetable location differs from all its*
  * *connected locations*
  ($\forall$ t: FDB, i: NATURAL)
  $1 \leq i \leq$ Length(Connections(t)) $\Longrightarrow$
    Departure-airport(t) $\neq$
      Connected-Locations(Ith(Connections(t),i))
operation Display-Binding-Info
  arity FDB $\times$ KEY $\longrightarrow$ CP [AIRPORT, TIME]
  asserts Display-Binding-Info(t,k) = < l, t >
  pre ($\exists$ i: NATURAL)
    $1 \leq i \leq$Length(Connections(t)) $\wedge$
    Identifier(Ith(Connections(t), i)) = k
end Display-Binding-Info
$\cdots$
end FDB

4.4 Customizing the specification

Further customization is often required to reuse type specifications. The specification structure can require to be altered or new operations can be defined. The *Customize FDB Specification* task aims to capture such activities.

4.4.1 Process overview

The process to incorporate changes can be stated informally as follows:

(a) Zoom in the appropriate level of the specification structure (where the difference has to be incorporated).

(b) Apply a transformation on the structure, e.g. introduce a new attribute, rework the type structure, replace a type name with its structure, and so on (see below).

(c) Constrain the acceptable states of the specification by introducing, modifying or deleting assertions on the specification state.

(d) Update the specification behaviour by introducing new operations, modifying or deleting existing operations.

Some transformation operators are reviewed in Section 4.4.2. Tactics to help specifiers to zoom in the appropriate level and to introduce new attributes are presented in Section 4.4.3.

4.4.2 Operators

The most current operators to perform the transformations are the following ones:

- **AddAttribute.** Given a specification structure and a new attribute, this operator grafts the new attribute into the structure [23]. It (i) modifies the **is** clause of the type cluster by adding the new attribute name and type, (ii) propagates the modification to everywhere the cluster is used inside the cluster specification or in other specifications by adding parameters whose value is "...".

- **Unfold.** Given a type name or an operation name and one occurrence of this name in the specification, replace the name occurrence with its definition.

  Example. Consider the following type definitions:

        type cluster T
            is SEQ[V]
        type cluster V
            is CP [Field1: T1, Field2: T2]
  They can be replaced with the following definition:
        Type cluster T
            is SEQ [CP [Field1: T1, Field2: T2]]

- **Restructure.** This operator transforms the specification structure according to rewriting rules [6], e.g.:

  $$CP [CP [T1, T2], T3] \approx CP [T1, T2, T3]$$
  $$SEQ [CP [T1, T2]] \approx CP [SEQ [T1], SEQ [T2]]$$

- **RemoveAttribute.** Removing an attribute can be done by the **Remove-attribute** operator applied to a type cluster whose expression is a cartesian product. This operator removes the attribute name and type from the **is** clause. If the cartesian product is composed of only one attribute after the modification, the expression is restructured. Next, the operator propagates the modification to everywhere the cluster is used inside the specification or in other specifications by deleting the corresponding attribute occurrences.

Observation. The **Unfold** transformation is another example of operation, for which the resulting specification cannot be related with the one from which it comes, at the product level. Maintenance issues will occur again when the unfolded type has to be updated. Such propagations can be performed at the process level by *replaying* the *Unfold* transformation.

4.4.3 Tactics

The tactics described below aim to guide the specifier customizing a reused specification. In the case of adding a new attribute in particular, the tactics helps to select the appropriate structure zooming and to select which transformation to apply.

(a) Zoom

The selection of appropriate zoom levels is guided by heuristics telling when to stop going down the hierarchy:

> *To select the appropriate level of zooming, visit nodes of the specification structure in turn starting from root nodes and going down the hierarchy. Decide at each node whether the attribute to be added is (i) local to the node, (ii) specific to one subcomponent, (iii) common or unrelated to all of them.*
>
> *In case (ii), go down the hierarchy into the subcomponent and repeat the step. In case (iii), consider the current level as the level in which the structure transformation has to be performed. Stop going down in case (i) or if the subcomponent is a black box into which no further access is allowed.*

Example. In the example of Section 2.1, the *Gate* and *Flight-status* attribute names are local attributes of flights. The transformation will take place at level 3 in Figure 10.

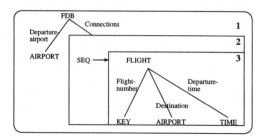

Figure 10: Levels in the data structure

Example. The *Clock* attribute is a piece of global information unrelated to any flight. The level 1 (Figure 10) is the appropriate level for the transformation.

(b) Transformations

Transformations are applied on types at the current zoom level. Four heuristics are introduced below to determine how to incorporate a new attribute into a specification structure depending (i) on the semantics of the attribute, (ii) on whether the structure is composed of components which can be observed and worked on (*open box*) or of components that are to be considered as *black boxes*. The distinction between open and black boxes is important because specifiers may want to hide information in the context of multi-specifier environments.

Case 1. *If* the new attribute is common to all subcomponents, *then* consider the cartesian product (*CP*) that groups the existing subcomponents and the new attribute (if the *CP* does not exist in the specification, create a new one).

Example 1. This transformation has been performed to add the *Clock* attribute in Section 2.2 (Figure 11).

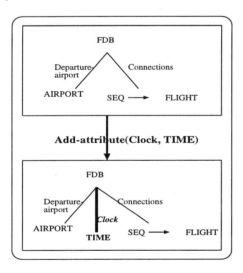

Figure 11: Add the *Clock* attribute

Case 2. *If* the new attribute is a feature local to the root node of the current zoom level, and if this node is an open box, *then* consider the cartesian product that extends the node fan out with the new attribute.

Exemple 2. Consider the *FLIGHT* component as an open box, i.e., with the possibility of changing its

definition. New attributes *Gate* and *Status* can be added for the *FLIGHT* component as described in Figure 2 and 3.

Case 3. *If* the new attribute is a feature about members of a collection, each member of which is a black box, *then* replace each black box with a *CP* composed of the black box and the new attribute.

Exemple 3. Consider now the *FLIGHT* component as a black box, i.e., without any way or permission to modify its definition. One way to add the new attributes *Gate* and *Flight-status* is to replace the *FLIGHT* parameter with a new one defined as *CP [Flight: FLIGHT, Gate: GATE, Flight-status: STATUS]*, *FLIGHT* been unchanged (Figure 12).

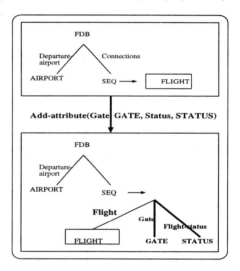

Figure 12: Add *Gate* and *Flight-status* attributes

Case 4. *If* the new attribute is a feature about members of a collection that is to be considered as a whole as a black box, *then* replace the black box with a *CP* composed of the original black box and an isomorphic structure for the new attribute values.

Exemple 4. Consider that *SEQ [FLIGHT]* is a black box. As new attributes deal with specific flights, the *SEQ [FLIGHT]* component should be replaced with *CP [SEQ [FLIGHT], SEQ [CP [Gate: GATE, Flight-status: STATUS]]* in which *SEQ [FLIGHT]* remains unchanged (Figure 13).

4.4.4 Resulting state

Figure 14 and the next specification show the result of applying the **AddAttribute** operator for the *Clock*

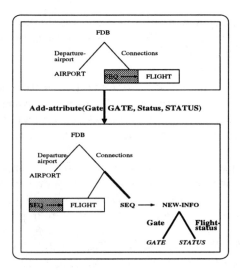

Figure 13: Add *Gate* and *Flight-status* attributes

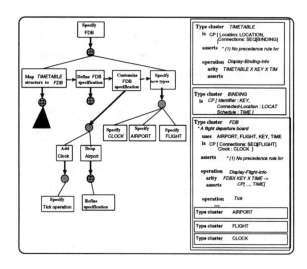

Figure 14: Customizing the resulting specification

attribute and the **RemoveAttribute** operator for the *Departure-airport* attribute on the *FDB* specification.

type cluster FDB
provides all
uses AIRPORT, FLIGHT, KEY, TIME, NATURAL
* A flight departure board
is CP [Connections : SEQ [FLIGHT],
     Clock: CLOCK]
asserts
   * (1) No precedence rule for 2 simultaneous schedules
   * schedules
   ...
   * (2) The condition about connected locations
   * is removed
operation Display-binding-Info
   arity FDB × KEY ⟶ CP [..., TIME]
   asserts Display-Binding-Info(t,k) = < l, t >
   pre (∃ i: NATURAL)
     $1 \leq i \leq$ Length(Connections(t)) ∧
     Identifier(Ith(Connections(t), i)) = k
end Display-binding-Info
...
end FDB

4.5 New types

The mapping and customization activities may identify new types which have to be defined. A similar process can be followed to specify each of these types.

5 Conclusions

A development process to support reuse of operational requirements specification has been presented in this paper. It is based on a structure mapping between abstract data types and on customization. Development operators are used to capture process activities that make a workplan and product evolve. Tactics help developers select the right operators. Keeping process information allows links between a product and its various reuse instances to be kept. The paper has shown cases in which such links cannot be recorded at the product level. Keeping track of these links improve the understanding of a specification development and makes the product easier to maintain for the long term.

The benefits of the process model adopted in this paper can be summed up as follows:

- language-independency. Experiments with the languages Glider, Pluss [22, 23], and Z [4, 11], [21] have been carried out and are in progress with Lotos in the domain of communication protocols [10];

- incremental construction of the specification. Operators enable the specifier to focus on methodological issues before addressing technical details related to the specification language [12];

- documentation [19]. The steps and the workplan record the design decisions taken during the specification development. They make the specification easier to understand and help the developers to verify and refine the system;

- support for backtracking. Backtracking on decisions with undoing of task side-effects is a basic construct in design processes which is rarely supported in existing process meta-models. Backtracking is supported by starting new reductions at appropriate places in the workplan;

- replay. Development decisions are recorded and thus a given development can be replayed.

Further Work. A domain that is worth being investigated is the way non-operational requirements interfere with process development. Then it is interesting to study whether the non-operational requirements can be reused in the same way operational ones are. On the other hand, the supporting tool should also be extended to integrate tactics.

Acknowledgment. The work reported herein was partially funded by the Commission of the European Communities (ESPRIT Project 2537 "ICARUS").

References

[1] J.-C. Carbonell. *DERIVATIONAL ANALOGY: A Theory of Reconstructive Problem Solving and Expertise Acquisition*, volume 2. Machine Learning, An Artificial Intelligence Approach, 1986. Chap. 14.

[2] S. Clerici, R. Jimenez, and F. Orejas. *Semantic Constructions in the GLIDER Specification Language*, volume Recent Trends in Data Type Specification of *LNCS 785*. Springer Verlag, 1994.

[3] R. Darimont. Process support for requirements elaboration. PhD thesis, UCL, Département d'Ingénierie Informatique, June 1995.

[4] R. Darimont and J. Souquières. A Development Model: Application to Z Specifications. In *Proc. of the WG 8.1 Working Conference on Information System Development Process*, Como (It), 1993. North Holland.

[5] E. Dubois, J.-P. Finance, J. Hagelstein, A. van Lamsweerde, F. Orejas, J. Souquières, and P. Wodon. Icarus: Overview of the Project. REQUIREMENTS ENGINEERING NEWSLETTER (36), April 1995.

[6] E. Dubois, N. Lévy, and J. Souquières. Formalising restructuring operators in s specification process. *Proc, of First European Conference on Software Engineering*, pages 173–184, 1987.

[7] E. Gamma, R. Helm, R. Johnson, and J. Vlissides. *Design Patterns - Elements of Reusable Object-Oriented Software*. Addison-Wesley, 1995.

[8] D. Jackson and M. A. Jackson. Problem decomposition for reuse. *Software Engineering Journal*, pages 19–30, January 1996.

[9] C. W. Krueger. Software Reuse. *ACM Computing Surveys*, 24(2):131–183, June 1992.

[10] T. Lambolais. Development of Automata with Lotos. In *Method Integration Workshop'96*. Leeds Metropolitan University, March 1996.

[11] N. Lévy and G. Smith. A language-independent approach to specification construction. In *Proceedings SIGSOFT'94: Symposium on the Foundations of Software Engineering, New Orleans, USA.*, 1994.

[12] N. Lévy and J. Souquières. A "Coming and Going" Approach to Specification Construction : a Scenario. In *Proc. 8th Int. Workshop on Software Specification and Design*, Paderborn (G), March 1996.

[13] N. A. Maiden. Analogy as aparadigm for specification reuse. *Software Engineering journal*, January 1991.

[14] N. A. Maiden and A. C. Sutcliffe. Exploiting Reusable Specifications Through Analogy. *Communication of the ACM*, 35(4):55–64, April 1992.

[15] B. Meyer. On Formalism in Specifications. *IEEE Software*, 1(8):6–26, 1985.

[16] B. Meyer. *Object oriented software construction*. International Series in Computer Science. Prentice Hall, New York, C.A.R. Hoare edition, 1988.

[17] A. Mili, R. Mili, and R. Mittermeir. Storing and retrieving software components: a refinement–based approach. In *Proc. of the 16th ICSE*, 1994.

[18] N.-J. Nilsson. *Problem Solving Methods in Artificial Intelligence*. Computer Sciences Series. MacGraw-Hill, 1971.

[19] D. L. Parnas. Software Aging. In *Proc. of ICSE-16*, pages 279–287. IEEE Press, May 1994.

[20] H. Partsch. Generalize and reuse: An exercise in reusing transformational developments. Technical Report MIP-8915, Passau, University of Nijmegen, 1989.

[21] G. Smith. A Development Framework for Object-Oriented Specification and Refinement. In *Proc. TOOLS EUROPE'94*, March 1994.

[22] J. Souquières. Aides au développement de spécifications. Thèse d'état, CRIN/Université de Nancy 1, Janvier 1993.

[23] J. Souquières and N. Lévy. Description of Specification Developments. In *Proceeding of RE'93*, pages 216–223. I.E.E.E. Press, January 1993.

[24] J. Souquières and N. Lévy. PROPLANE : A Specification Development Environment. In *5th International Conference on Algebraic Methodology and Software Methodology Amast'96*, volume 1101, Munich (G), July 1996. LNCS.

[25] M. Wirsing, R. Hennicker, and R. Stabl. MENU: An example for the systematic reuse of specifications. Technical Report MIP-8930, University of Passau, 1989.

[26] P. Wolff. Related Specifications Reuse. In *Proceedings International Workshop on Software Specification and Design*, Los Angeles (CA, USA), 1993.

Analogical Reuse of Requirements Frameworks

Philippe Massonet and Axel van Lamsweerde

Université catholique de Louvain, Département d'Ingénierie Informatique
B-1348 Louvain-la-Neuve (Belgium)
{phm,avl}@info.ucl.ac.be

Abstract. *Reusing similar requirements fragments is among the promising ways to reduce elaboration time and increase requirements quality. This paper investigates the application of analogical reasoning techniques to complete partial requirements specifications. A case base is assumed to be available; it contains requirements frameworks involving goals, constraints, objects, actions, and agents from systems already specified. We show how a rich requirements meta-model coupled with an expressive formal assertion language may increase the effectiveness of analogical reuse. An acquisition problem is first specified by the requirements engineer as a query formulated in the vocabulary of the specification fragments built so far. Source cases and partial mappings are found by query generalization followed by search through the case base. Once analogies have been confirmed, mappings are completed by use of relevance rules that distinguish in the formal assertions what is relevant to the analogy from what is irrelevant. Best analogies are then selected and extended in such a way that logical properties of the answers to the query may be verified, thus increasing confidence in the analogy. The approach is illustrated by analogical acquisition of specifications of a meeting scheduler in the KAOS goal-oriented specification language.*

Keywords: Specification reuse, analogical reasoning, goal-driven requirements engineering, formal methods.

1. Introduction

Software reuse is intended to decrease software development costs and increase software quality [Kru92]. Specification reuse has in particular been given some attention recently as it provides support for the difficult, critical task of requirements elaboration; moreover one may expect that the reuse of a specification entails the reuse of the software components implementing the specification. In [Fin88], a technique based on matching of simple patterns and causal chains is suggested for reusing specifications expressed in the CORE notation. The Requirements Apprentice allows specifiers to build and check specifications written in a frame-based language by instantiation and specialization of reusable domain-specific clichés [Reu91]. In the same spirit, [Rya93] introduces a graph matching technique as a basis for reusing previously developed specifications expressed by conceptual graphs.

Specification reuse raises a number of technical issues, notably, the representation of specifications for reuse, the organization of the library of reusable components, the retrieval of appropriate specifications through component matching, and the customization of the specification obtained. Various approaches have been proposed to address those issues, e.g., specialization graphs with multiple inheritance [Reu91], faceted classification [Dia89, Ost92], similarity-based retrieval with various similarity measures [Spa93, Ost92, Lau94], and signature matching [Zar93]. The use of analogical reasoning technology [Hal89] for specification reuse and validation has also been suggested [Mai91], [Sut93]. Requirements analysts do rarely start from scratch when they model and specify a new system; they tend to reuse past experience with similar cases to fill in the holes. Analogical reasoning techniques seem therefore quite appropriate in this context.

An analogy is defined as a *mapping* from elements of a *source* domain to elements of a *target* domain, where each domain contains related knowledge and facts. Analogical inference consists of mapping well defined source elements to partially defined or unknown target elements so that additional knowledge about the latter can be obtained by source-target transposition. The source and target elements are called source and target analogs, respectively. The following (overlapping) steps can be identified in an ideal analogical inferencing process [Lam91]. *Presentation:* the problem to be solved is first defined which results in the identification of target elements; *Recognition:* source analogs are identified from target elements; *Mapping:* an analogical match is established between the source and target analogs [Gen88], [Fal89], [Hol89]; *Transfer:* relevant knowledge about the source is transposed to the target according to the match found; *Validation:* the transferred features are checked for appropriateness in the new target situation [Fal87], and revised if necessary; *Consolidation:* the analogy may be analyzed against utility criteria and generalized for later reuse in new situations [Min88].

In the various approaches to specification reuse we are aware of, the aspects of a specification that are formalized and used for retrieval are limited to a global, syntactic "declaration" level --e.g., frame slot names/values, facet names/values, names of nodes and edges in a graph, signature of an abstract data type, etc. The actual requirements on the declared concepts/attributes/links/operations are not formally represented nor used; they are considered as informal assertions annotating the declared objects.

In this context the paper has three objectives:

• to show how a richer ontology for declaring requirements, combined with formal assertions to express them, may result in a more accurate and meaningful level of specification matching and reuse,

• to propose techniques for adapting the various steps of the analogical process above to the problem of reusing *formal* requirements specifications,

• to extend the scope and granularity of reuse from single concept to a collection of related concepts (called framework) though incremental analogy extension.

The perspective in this paper is more on using analogies to complete and/or validate partial specifications already elaborated, rather than starting from almost empty targets. Source and target specifications are written in the KAOS language, a multi-paradigm specification formalism which combines semantic nets [Bra85] for the conceptual modelling of goals, consraints, agents, objects and operations in the system; temporal logic [Man92], [Koy92] for the specification of goals, constraints and objects; and state-based specifications [Pot91] for the specification of operations. The language has a rich ontology which is explicitly defined and accessible at the meta-level [Dar93].

The paper is organized as follows. Section 2 introduces some minimal background on the KAOS framework. Section 3 provides an overview of our approach by outlining the different steps of the analogical reuse process. Section 4 then explains how a reuse problem is defined as an analogical query. Section 5 shows how source cases are identified from such a query; partial, declaration-level target-source mappings are elaborated therefrom. Section 6 then shows how the formal assertion level is used to complete such mappings on more semantic grounds; a completed mapping produces an answer to the original query and provides the basis for analogical transfer. Section 7 summarizes our results and outlines ongoing work on formal verification, extension, and consolidation of analogies.

2. Requirements specification in Kaos

The KAOS methodology provides a specification language to capture *why*, *who*, and *when* aspects in addition to the usual *what* requirements; a goal-driven elaboration method; and meta-level knowledge used for local guidance during method enactment. Hereafter we introduce some features of the language and the meta level that will be used later in the paper; see [Dar93], [Lam95], [Dar96] for more details.

2.1 The Kaos language

The specification language provides constructs for capturing a rich variety of *concepts* involved in the requirements engineering lifecycle, namely, goals, constraints, agents, entities, relationships, events, actions, views, and scenarios. There is one construct for each type of concept. The following types of concepts will be used in the sequel.

• *Object:* an object is a thing of interest in the domain whose instances may evolve from state to state. It is in general specified in a more specialized way -as an *entity*, *relationship*, or *event* according as the object is autonomous, subordinate, or instantaneous, respectively. Objects are described formally by invariant assertions.

• *Action:* an action is an input-output relation over objects; action applications define state transitions. Actions may be *caused/stopped* by events. They are characterized by pre-, post- and trigger conditions.

• *Agent:* an agent is an object acting as a processor for some actions. An agent *performs* an action if it is effectively allocated to it; the agent *knows* an object if the states of the object are made observable to it. Agents can be humans, devices, programs, etc.

• *Goal:* a goal is an objective the system should meet. *Refinement* links relate a goal to a set of subgoals. The goal refinement structure for a given system is in general an AND/OR directed acyclic graph. Goals often *conflict* with others. Goals *concern* the objects they refer to.

• *Constraint:* a constraint is an implementable goal, that is, a goal that can be assigned to some individual agent in the system. Goals must be AND/OR *refined* into constraints. Constraints in turn are AND/OR *operationalized* by actions and objects through strengthenings of their pre-, post-, trigger conditions and invariants, respectively. Alternative ways of assigning responsibilities for a constraint are captured through AND/OR *responsibility* links; the actual assignment of agents to the actions that operationalize the constraint is captured in the corresponding *performance* links.

• *Case:* a case is an aggregation of interrelated instances of the various concept types above. Cases capture well understood areas, abstract or real, with distinguishable boundaries. They may be organized into taxonomies; concepts from a more specific case are linked to corresponding concepts of the same type in a more general case through *IsA* inheritance links.

Each construct in the KAOS language has a two-level generic structure: an outer semantic net layer for *declaring* the concept, its attributes and its various links to other concepts [Bra85]; an inner formal assertion layer for *formally defining* the concept. Goals, constraints and object invariants are formalized in a real-time temporal logic borrowed from [Koy92]. The following operators will be used for temporal referencing: $\bigcirc/\bullet$ (in the next/previous state), $\Diamond/\blacklozenge$ (some time in the future/past), $\square/\blacksquare$ (always in the future/past). Real-time restrictions are indicated by subscripts (e.g., $\leq_{Cd}$ means within C days). Actions are formalized by first-order pre-/postconditions (in the state-based tradition [Pot91]) and trigger conditions.

Figure 1 illustrates the KAOS declaration level for a fragment of a conference organization system. (This fragment will be used as source case later in the paper.)

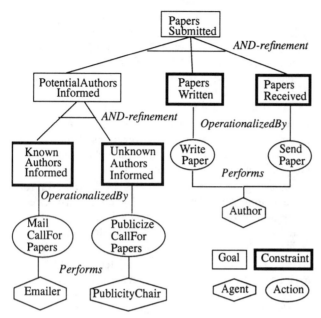

Fig. 1 - Conference organization

The goal PapersSubmitted requires collaboration between authors and conference organizers to be achieved. It may be refined into goals such as PotentialAuthorsInformed, PapersWritten and PapersReceived; the latter is a constraint as it can be can be under responsibility of a single type of agent, namely, Author; authors can perform the SendPaper action under restricted time conditions to ensure that the constraint PapersReceived is met. The specification below gives an alternative textual declaration of the goal PapersSubmitted together with its inner formal assertion. The latter says that if an author is a PotentialAuthor for a CallForPaper, then a paper will be submitted to the ProgramChair by her some time in the future before the Deadline in the call for papers. (Clearly this goal is too ideal and should be weakened.)

Goal Achieve [PapersSubmitted]
    InstanceOf SatisfactionGoal
    Concerns Author, CallForPaper, Paper, ...
    RefinedTo PotentialAuthorsInformed, PapersWritten, ...
    FormalDef $\forall$ a: Author, c: CallForPaper, pc: ProgramChair
        PotentialAuthor (a,c) $\wedge$ Chairing (pc,c)
        $\Rightarrow \Diamond_{\leq c.Deadline}$ ($\exists$p: Paper) Submission (a,p,pc)

In the formal assertion above, the predicate Potential-Author(a,c) means that an instance of the PotentialAuthor relationship links, in the current state, variables a and c of sort Author and CallForPaper, respectively. The Potential-Author relationship type, Author agent type and CallForPaper entity type are declared in other sections of the specification, e.g.,

Agent Author
    CapableOf WritePaper, ...
    Has Expertise: **SetOf** [*SubjectArea*]
    ...

Relationship PotentialAuthor
    Links Author {**card**: 1:N}, CallForPaper {**card**: 1:N}
    DomInvar $\forall$ a:Author, c:CallForPaper
        PotentialAuthor (a,c) $\Leftrightarrow$ a.Expertise $\cap$ c.Scope $\neq \varnothing$

In the declarations above, Expertise is declared as an attribute of Author and used in the invariant defining Potential-Author. Object inheritance is also supported, e.g., an agent type PC-Author might be introduced as a specialization of Author with some extra attributes and invariants.

As mentioned earlier, operations are specified formally by pre- and postconditions in the state-based tradition, for example,

Action WritePaper
    Input SubjectArea {**Arg**: s}; **Output** Paper {**Res**: p}
    PerformedBy Author {**Arg**: a}
    DomPre $\neg$ WrittenBy (p,a)
    DomPost WrittenBy (p,a) $\wedge$ p.SubjectArea = s

2.2 Meta-level knowledge

Domain-independent knowledge is used for local guidance and validation during goal-driven elaboration. Such knowledge will be used in this paper to constrain analogical mappings.

In particular, a rich *taxonomy* of goals, constraints, objects and actions is defined at the meta level together with rules for specifying concepts of the corresponding sub-type. Here are a few examples.

- Goals are classified by pattern of temporal behaviour they require:
    Achieve: P $\Rightarrow \Diamond$ Q or *Cease:* P $\Rightarrow \Diamond \neg$ Q
    Maintain: P $\Rightarrow \Box$ Q or *Avoid:* P $\Rightarrow \Box \neg$ Q
- Goals are also classified by type of requirements they will drive with respect to the agents concerned (e.g., *SatisfactionGoal, InformationGoal, ConsistencyGoal, SafetyGoal, PrivacyGoal,* etc.)
- Constraints are in the *HardConstraint* category if they may never be violated, or in the *SoftConstraint* category if they are likely to be temporarily violated.
- Actions are *Modify* or *Inspect* actions according as they modify some object state or not.

Such taxonomies are constrained by rules, e.g.,
- SafetyGoals are AvoidGoals to be refined in Hard-Constraints;
- PrivacyGoals are AvoidGoals on *Knows* predicates;
- SoftConstraints must have associated ModifyActions to restore them.

3. Overview of the approach

Reusing specifications requires a library of reusable components to be available. Such a library in our context consists of a structured collection of *cases* that have been specified in the past. As seen before, each case is a domain-specific aggregation of interrelated goals, constraints, agents, entities, relationships, events, actions, and scenarios. The collection is structured by *IsA* specialization links that may relate concepts of the same type from different cases; multiple inheritance between different cases is thus possible. Such a library will hereafter be called *case base*.

The analogical reuse problem we consider in this paper is formulated as follows.

GIVEN - a partial specification already elaborated, called *target* specification,
- the identification of some missing part,

FIND - a set of relevant *source* specifications from the case base, *translated* into the vocabulary of the target specification, as candidates for filling in the missing part.

The retrieved source specifications are called solutions to the reuse problem. The analyst then has to select among solutions and evaluate how to adapt and integrate the preferred alternative.

3.1 Example

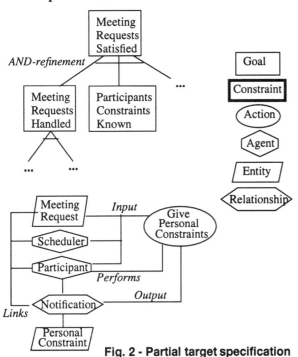

Fig. 2 - Partial target specification

Figure 2 shows two unrelated fragments already elaborated for a target specification of a meeting scheduler. The goal MeetingRequestsSatisfied has been refined into subgoals such as MeetingRequestsHandled, ParticipantsConstraintsKnown, etc. The latter has not been refined yet. On another hand, the action GivePersonalConstraints has been identified and partially defined through an independent process --e.g., from usage scenarios. This action allows a participant to provide his personal constraints, in the time frame prescribed by the request, to the scheduler.

Figure 3 shows one solution obtained by analogical reuse. Graphical symbols in gray highlight the concepts and links which have been found. The formal definitions of those concepts are transferred as well --e.g., there is a strengthened postcondition on the GivePersonalConstraints action to ensure the ParticipantsConstraintsProvided con-

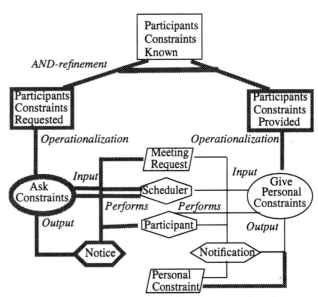

Fig. 3 - Target specification completed

straint, namely, that the participant's constraints delivered fit in the time frame prescribed by the meeting request.

3.2 Constraints on the analogical mapping

Analogies are often seen as mappings from source to target cases such that relations holding in the source also hold in the target [Gen88, Fal89]. The corresponding sources and targets do not necessarily have to be "similar"; they are placed into correspondence according to the *role* they play in the relational structure. The intuition is that analogies are more about relations than about individual features.

This view is customized and extended to the declaration and assertion levels of our KAOS framework. The following constraints are required to hold on any analogical mapping between specifications.

- *One-to-one concept mapping:* a source (target) concept may only be mapped to a single target (source) concept.

- *Same meta-level type:* analogs must be of the same meta-type --e.g., if the source concept is an agent then the target must be an agent as well; if the source concept is a constraint, then the corresponding target concept must be a constraint as well.

- *Same relationship/predicate arity:* source and target relationships/predicates may be mapped one onto the other only if they have the same number of linked objects/arguments.

- *Role preservation:* source and target relationships/predicates may be mapped one onto the other only if their arguments can be mapped pairwise in their order of appearance. Consider, for example, the predicates *Notification*(Participant,PersonalConstraint,MeetingRequest) and *Submission*(Author,Paper,CallForPapers). The Notification relationship may be mapped to Submission only if Participant can be mapped to Author, PersonalConstraint to Paper, and MeetingRequest to CallForPapers, respectively.

- *Same meta-level category:* source and target concepts may be mapped one onto the other only if they are in the same category. For example, a SatisfactionGoal may not be mapped to an InformationGoal; a HardConstraintl may not be mapped to a SoftConstraint. When formal definitions are compared, they must be in the same category as well; this constraint prevents an Information assertion from being compared to a Privacy assertion even though they may be structurally similar.

- *Same assertion pattern:* source and target concepts may be mapped one onto the other only if they have the same assertion pattern. For example, an Achieve constraint may not be mapped to a Maintain constraint. When formal definitions are compared, they must have the same pattern as well; concepts involved in a source (target) assertion may only be mapped to concepts in a target (source) assertion if the assertions have the same temporal pattern.

The above mapping constraints must be satisfied both at declaration and assertion levels. They drive the entire analogical reuse process. It should be clear at this point that *ontological richness of the language directly impacts on the accuracy of the analogy.*

3.3 Steps of the analogical reuse process

Analogies are built incrementally. First a *structural* match between declarations is sought; the match is then refined through *semantic* matching on assertions; the analogy is finally extended so that KAOS-specific properties may be verified, thus increasing confidence in it. The following steps are distinguished.

1. *Specification of the reuse problem:* a query is formulated to identify missing parts in the target specification for which candidate specification fragments should be found by analogy.

2. *Classification of target concepts:* relevant target concepts in the query are classified with respect to concepts in the case base; the relevant target concepts are the ones being related to the missing ones. Conceptual classification is needed for relating the current situation to past cases.

3. *Query reformulation and satisfaction:* the query is generalized using the classification found; concept taxonomies in the case base are then traversed to retrieve candidate source analogs that satisfy the generalized query. Partial mappings are derived on the basis of *structural* correspondences between target and source declarations.

4. *Refinement of analogical mappings:* the declaration-level mappings are refined on the basis of *semantic* correspondences between source and target assertions. Refined mappings allow relevant assertions to be transferred from source to target.

5. *Source case selection:* this step is necessary when multiple candidate source cases are found. Source cases are compared on the basis of a similarity metrics that measures the number of similar relevant assertions with respect to the total number of relevant assertions.

6. *Verification-driven analogy extension:* a commonly accepted heuristic in analogical reasoning is that things agreeing in one respect will probably agree in yet another. This step aims to extend the original analogy so that more than the initial reuse problem is solved. Extension rules based on KAOS-specific features are used for that purpose. For example, if a partial target AND-refinement is involved in the analogy, an attempt is made to complete the AND-refinement so that its correctness can be verified [Dar96].

7. *Transfer and integration:* analogical declarations and assertions are transferred after translation according to the refined mapping, and integrated into the target specification so that the analyst may assess the result and adapt it.

8. *Consolidation:* effective reuse requires a broad and up to date case base. This interactive step identifies potentially reusable target concepts that are worth being included in the case base; it uses a measure of newness and distance with respect to closest generalizations in the case base.

In the sequel we will focus on steps 1-4, and briefly discuss the other steps in order to provide a complete picture of our approach.

4. Specifying the reuse problem

The analyst must indicate some missing parts in her specification which she would like to fill in by analogical reuse. This is done by submitting a query in the vocabulary of the specification being elaborated. The answer to the query will be a set of candidate source specifications reformulated in that vocabulary.

The query defines the *scope* of the analogy; all concepts mentioned in it are considered *relevant.* The whole process of analyzing partial specifications to find out missing parts is thereby shortcut. Moreover, retrieving a source case that would match a big partial specification is highly unlikely in real situations. A precise, restricted scope that points out concepts relevant to missing parts has much more chance to result in source analogs being retrieved at lower cost.

Consider, for example, the analogical reuse problem suggested in Fig. 2. This problem may be more precisely specified by the query

set result = **Constraint**
    which Refines ParticipantsConstraintsKnown
    and which OperationalizedBy GivePersonalConstraints

We use the generic entity-relationship query language described in [Lam87], customized to the KAOS meta-model. Names of meta-level types, links and attributes become language keywords. The language has a set-theoretic framework. The effect of a **set**-statement is to define a *group* of one or more elements (denoted here by variable result) from types or other groups by means of classical set operations and **and/or/not** logical compositions of predicates on meta-level links (**which**-clause) and attributes (**with**-clause).

Little work has been done on specifying analogical

reuse problems that must be solved within larger contexts. Most of work has been devoted to single-concept reuse. Queries seem a natural way to focus the analogical scope within a larger framework. Queries were used in [Lau94] to specify user-defined similarity criteria for locating source analogs; this provides much more flexibility over arbitrary syntactic similarity measures that have been proposed in the past, based usually on weighed counts of similar attributes. Queries have also been used to locate library components by providing desired component features and comparing them to library component features [Dev90], [Fea93], [Mil94]. Signature information can be seen as one such feature [Zar93].

5. Query generalization and satisfaction

Exhaustive search through the case base with structural and semantic matching is clearly unfeasible for a sufficiently rich cases base. Our approach takes advantage of the structure of the case base and the two-level structure of the KAOS language. The basic idea is to first retrieve "promising" source cases through declaration-level *structural* matching, and then determine whether the mapping may be refined through assertion-level *semantic* matching. Sources are retrieved by (i) *reformulating* the query into more abstract terms using conceptual taxonomies in the case base, (ii) *satisfying* the more abstract query using the case base query system, and (iii) deriving a partial mapping for each answer to the query, on the basis of the *structural* match between the query and the answer. Once candidate sources have been found and a partial mapping established, the formal assertions are analyzed pairwise for similarity. This more expensive step is performed only for refining the partial mapping.

5.1 Classification of target concepts

Query satisfaction requires that the concepts involved in the query be related to concepts in the case base. This is done by generalizing the query so as to reach a vocabulary of concepts found in the case base.

We define the *conceptual neighbourhood* of a concept C in a specification S as the set of concepts related to C in S through meta-level links. For example, the conceptual neighboorhood of the action GivePersonalConstraints in Fig. 2 is the set {MeetingRequest, Scheduler, Participant, Notification}.

The *scope* of a query is then defined as the set of domain-level concepts appearing in the query together with their conceptual neighbourhood in the target specification.

The preliminary substep for query generalization consists of *classifying* the various concepts in the scope of the query with respect to the contents of the case base. The classification establishes temporary *IsA* specialization links between target concepts and case base concepts to allow comparisons to be made --e.g., to determine whether two concepts have a common ancestor or common features. (Temporary *IsA* links may be frozen later during the consolidation step if the case base is enriched with the target.)

Figure 4 gives one possible result of the classification substep for the sample query above. Note that parent concepts in the classification may come from different cases; the entity MeetingRequest is a specialization of an abstract entity Request from the ReservableResourceManagement case; the action GivePersonalConstraints is a specialization of the Send action from the MessageDelivery case; and the goal ParticipantsConstraintsKnown is a specialization of the abstract goal ConstrainedAnswersGathered from the QuestionAnswering case. The Notification relationship specializes the Limitation relationship (linking Supplier, Request and Requestor not shown there) by specializing the linked entity Request to MeetingRequest. (Lack of

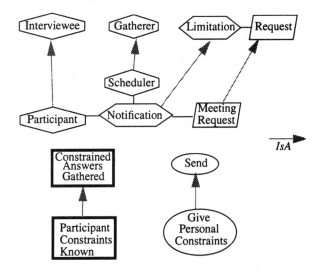

Fig. 4 - Conceptual classification in the case base

space prevents us from describing these cases here; a detailed specification of the ReservableResourceManagement case can be found in [Dar95].)

The classification of target concepts relies on standard classification technology [Dev90], [Bor94]; the classification is further constrained by the meta-level type, category and role preservation constraints introduced in Section 3.2. For example, a human agent must be classified in a human agent taxonomy.

Graphical browsing tools may sometimes be used as a substitute to automatic classification. In particular, if classifications are made to parents from a same case, it may be worth browsing through the whole parent case. Suppose, for example, that the goal ParticipantsConstraintsKnown *and* the action GivePersonalConstraints have been classified under ConstrainedAnswersGathered and ExpressConstrainedAnswer, respectively. One would then expect some constraint from the QuestionAnswering case to refine ConstrainedAnswersGathered and to be operationalized by ExpressConstrainedAnswer; such a constraint might be specialized to fill in the missing part directly. Use of the wordnet software [Mil88] was very helpful in finding some of the generalizations in Fig. 4.

5.2 Query generalization

The vocabulary of the initial query must be translated into a sufficiently abstract version to allow source cases to be retrieved by query satisfaction.

A query *Q2* is said to be a *generalization* of query *Q1* in a classified case base *C* if *Q2* is the result of substituting one or more concepts in the scope of *Q1* by some corresponding more general concept in *C*.

For the classification given in Fig. 4, the following query is a generalization of the sample query above:

set result = **Constraint**
    which Refines ConstrainedAnswersGathered
    and which OperationalizedBy Send

The search for query generalizations is integrated in the query satisfaction algorithm, see below.

5.3 Structural matching

Elements in the target and source cases are considered structurally similar if there is a correspondence at declaration-level between the source, target, and generalized cases. This correspondence is more precisely defined as follows.

Let *S* and *T* be fragments of a source and target case, respectively; *S* and *T* are said to *match structurally* if there exists a generalized case fragment *G* such that one can decompose *S*, *T*, and *G* into corresponding pairs of elements that satisfy the following conditions: (i) the corresponding elements in the corresponding pairs of *S*, *T*, and *G* are of the same meta-level type and category; (ii) the elements in the corresponding pairs of *S*, *T*, and *G* are linked together by links of the same meta-level type; (iii) the corresponding elements in the corresponding pairs of *S* and *T* are both specializations of the corresponding element in the corresponding pair of *G*.

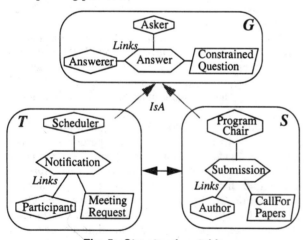

Fig. 5 - Structural matching

Figure 5 shows a structural matching between fragments of the MeetingScheduler and ConferenceOrganization cases. The pair (Participant, Notification) in the MeetingScheduler fragment corresponds to the pairs (Author, Submission) in the ConferenceOrganization fragment

and (Answerer, Answer) in the common generalization. It is easy to see that conditions (i)-(iii) hold, e.g., (i) Participant, Author, Answerer are all of type Agent whereas Notification, Submission, Answer are of type Relationship; (ii) Participant and Notification, Author and Submission, Answerer and Answer, are pairwise connected by links of the same meta-level type --namely, Links (other types would be Performs, OperationalizedBy, etc.); (iii) Participant and Author are both specializations of Answerer; Notification and Submission are both specializations of Answer; etc.

A *structural mapping* between *T* and *S* is directly derived from a structural match of *T* and *S*; it is defined by the set of corresponding elements in the corresponding pairs. For Fig. 5 this mapping is:

    { (Notification, Submission), (Participant, Author),
      (Scheduler, ProgramChair), (MeetingRequest, CallForPapers) }

The first four mapping constraints in Section 3.2 are clearly satisfied by structural mappings. Such mappings are partial because the formal assertions in *S* and *T* are not taken into account.

5.4 Query satisfaction and structural mapping

The objective of the query satisfaction algorithm is to retrieve source case fragments that structurally match the target case fragment defined by the scope of the query submitted, and to derive structural mappings therefrom. Each concept in the scope of the query is assumed to be classified in the case base (see Section 5.1).

The general principle is to search for query generalizations and source fragments satisfying them in parallel. Generalized scopes are found by climbing *IsA* taxonomies in the case base, starting from the parent concepts determined by the classification. Alternative ways of climbing taxonomies are explored; alternative generalizations are encountered when a concept has multiple parents. In order to explore more promising source fragments first, the search is guided by the following two heuristics:

- *Specificity*: favor scope generalizations that involve more specialized concepts rather than more general ones;
- *Coherence*: favor scope generalizations that involve concepts from a same case rather than from multiple ones.

The specificity heuristics is based on the intuition that two concepts that have a common generalization lower in the taxonomy have more common features by inheritance; the corresponding source is therefore more likely to be relevant. The coherence heuristics is based on the intuition that if the generalized query is specific to a case, then the answers to the query will be specific to specialized cases of it.

The specificity heuristics is implemented by a *breadth-first* traversal of taxonomies; more concrete generalizations are thereby explored before more abstract ones. The coherence heuristics is taken into account through a coherence metrics defined by s / c, where s denotes the number of elements in the scope of the query and c denotes the number of concepts in the smallest case covering the entire scope of the query.

The algorithm dynamically expands a so-called *reformulation graph*. Nodes in this graph represent target-parent concept substitution pairs to be used for generalizing the query; arrows represent ways of moving to more general target-parent concept pairs, by replacing parent concepts by their own parents in the case base taxonomy. *Expanding* a node means computing its successors by making such replacements. A node is said *open* if it is expandable but has not been expanded yet. The set of open nodes in a reformulation graph under expansion is denoted by OPEN.

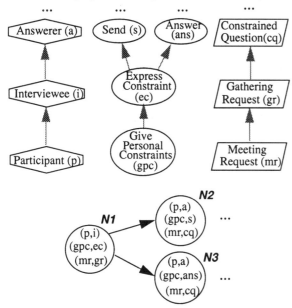

Fig. 6 - Reformulation graph

Fig. 6 shows a reformulation graph for the taxonomies appearing on the upper side. The action GivePersonalConstraints (gpc) has one parent, namely, ExpressConstraint (ec); the initial node in the reformulation graph therefore contains the pair (gpc,ec). Since ec has two parents, namely, Send (s) and Answer (ans) from the MessageDelivery and QuestionAnswering cases, respectively, there are two alternative successor nodes which contain the pairs (gpc,s) and (gpc,ans), respectively. According to the specificity heuristics, node N3 will be expanded before node N2 because the latter contains more abstract substitutions. According to the coherence heuristics, the generalized query

set result = **Constraint**
   which Refines ConstrainedAnswersGathered
   and which OperationalizedBy Answer

will be preferred over the generalized query

set result = **Constraint**
   which Refines ConstrainedAnswersGathered
   and which OperationalizedBy Send

that appeared in Section 5.2; the concepts in the former all belong to the QuestionAnswering case whereas in the latter the QuestionAnswering and MessageDelivery cases are involved.

The query satisfaction process can now be described by the following heuristic algorithm.

1. Create root of reformulation graph from classification of target concepts in case base;
2. **while** OPEN is not empty **and** maximum search time not exceeded **do**
  2.1. select a most specific and coherent node in OPEN, say *MSC*;
  2.2. - reformulate query according to substitutions in *MSC*;
      - submit generalized query to case base query system;
      - extend set of answers and derive structural mappings;
  2.3. expand *MSC* and update OPEN accordingly
3. Return set of answers and structural mappings.

The algorithm may in general return a set of alternative answers. In our ongoing example, one possible answer is the constraint PapersReceived from the ConferenceOrganization case introduced in Fig. 1, together with its conceptual neighborhood. Additional relevant source concepts are obtained iteratively through new queries involving the answer returned, e.g.,

set result = Action {**which Isa** Answer
               and operationalizes *PapersReceived* **and** ...}

which in turn yields the SendPaper action.

The coresponding structural mapping obtained is:
{ (ConstrainedAnswersGathered ← [ParticipantConstraintsKnown,
                       PapersSubmitted]),
(Answer ← [GivePersonalConstraints, SendPaper]),
(ConstrainedQuestion ← [MeetingRequest, CallForPapers]),
(Answerer ← [Participant, Author]),
(Asker ← [Scheduler, ProgramChair]),
(Answer ← [Notification, Submission]) }

where the form (CommonGeneralization ← [TargetConcept, SourceConcept]) is used to highlight the common parent.

6. From structural to semantic mappings

The structural mapping resulting from query satisfaction is only partial because the formal assertions attached to the answers are not taken into account; such assertions should be eventually transferred (at least partially) to the target after suitable translation of the concepts involved in them. Moreover source assertions may refer to concepts which were not covered by the structural mapping; finding target analogs of such concepts may result in the identification of missing elements in the target specification.

Consider the constraint PapersReceived that was retrieved as an answer to the analogical query above.

SoftConstraint Achieve [PapersReceived]
  InstanceOf SatisfactionConstraint;
  Concerns Author, CallForPapers, Paper, ...
  Refines PaperSubmitted;
  OperationalizedBy SendPaper
  FormalDef
    $\forall$ a: Author, c: CallForPapers, p: Paper, pc: ProgramChair
    PotentialAuthor (a,c) $\wedge$ Chairing (pc,c) $\wedge$ WrittenBy (p,a)
    $\Rightarrow \Diamond_{\leq c.Deadline}$ Submission (a,p,pc)

Reformulating the above formal definition into the target vocabulary is not possible from the partial mapping obtained so far; the relationships WrittenBy, Chairing and the entity PersonalConstraint are not covered by the structural mapping above.

Our next step is to complete the structural mapping so

that assertions that need to be transferred to the target can be translated. The general principle is to select "relevant" source assertions that involve unmapped concepts, translate them into target assertions with the unmapped concepts being replaced by variables, and then attempt to match them with existing target assertions. A successful matching instantiates variables to target concepts, that become analogs of the source ones.

6.1 Relevance rules

The rules for deciding which assertions are relevant for refining the structural mapping are specific to the KAOS meta-model; they depend on the meta-level types, links and attributes of the concepts appearing in the query. Fig. 7 gives a small portion of the KAOS meta-model (see [Dar93] for more details). We give a sample of relevance rules associated with this portion.

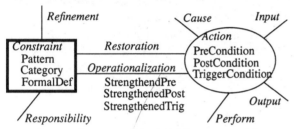

Fig. 7 - Portion of the KAOS meta-model

- If the query is about a target Action and the Input and Output meta-level links are mentioned in it, then the Precondition and Postcondition of the source Action are relevant.

- If the query is about a target Action and the Input meta-level link is mentioned in it, then the corresponding Output link in the source Action and its arguments along with the Precondition and Postcondition are relevant.

- If the query is about a target Action and the Operationalization meta-level link is mentioned in it, then the StrenghtenedPrecondition and StrenghtenedPostcondition for Operationalization of the source Action are relevant.

Consider, for example, the action GivePersonalConstraints. Suppose that we only specified its signature so far and want to define it further through analogical reuse. For the query

set result = Action
    which Inputs MeetingRequest, Participant, PersonalConstraint
    and which Outputs Notification ,

the only assertions considered relevant for refinement will be the pre- and postconditions of the analog source action. If the above query is extended with "... **and which** Operationalizes ParticipantsConstraintsGathered", the rules above will extend the transfer of the pre/postconditions to the strengthened pre- and postconditions to ensure ParticipantsConstraintsGathered. The principle is to focus the refinement on the parts of the source case that seem relevant to the analogy and ignore all the others for the moment (they will be taken into account later in the extension step).

6.2 Semantic matching

Candidate target analogs of unmapped concepts in source assertions are obtained by matching target assertions with relevant source assertions in which the unmapped concepts appear.

Let *SA*, *TA* and *M* denote a source assertion, target assertion, and structural mapping of source/target case fragments to which SA/TA are attached, respectively; let *GTrans(SA,M)* denote a generalized translation of *SA* according to *M*, in which unmapped source concepts have been replaced by meta-variables of the corresponding meta-level type.

SA and *TA* are said to *match semantically* if there exists a substitution of the source variables in *GTrans(SA,M)* by target concepts that meets the mapping constraints and makes *TA* and *GTrans* logically equivalent. Such a substitution, where source variables are reinstantiated to the corresponding source concepts, is called *semantic mapping extension*. The mapping constraints are those introduced in Section 3.2; logical equivalence may be established using the proof theory of temporal logic [Man92].

To illustrate this notion of semantic matching, consider the following two assertions. The first one captures the fact that, if a call for papers refers to a conference and an author's paper is accepted, then the author will attend the conference; the second one captures the fact that, if a meeting request results in a meeting being planned and a participant's personal constraints have been addressed, then the participant will attend the meeting:

$\forall$a: Author, c: CallForPapers, co: Conference, p: Paper
RefersTo (c,co) $\wedge$ *Accepted* (p, co)
$\Rightarrow \Diamond_{\leq\,co.begin}$ Attends (a, co)]   *[MustAttendConference]*

$\forall$ p: Participant, mr: MeetReq, m:Meeting, *pc: PersonalConstraint*
PlannedInto (mr,m) $\wedge$ Addressed (*pc*, m)
$\Rightarrow \Diamond_{\leq\,m.begin}$ Participates (p, m)   *[MustAttendMeeting]*

Suppose that all concepts in these two assertions are structurally mapped except the Accepted relationship and the PersonalConstraint entity. Assuming the source assertion to be MustAttendConference, the generalized translation *GTrans(SA,M)* yields the following assertion:

$\forall$ p: Participant, mr: MeetReq, m: Meeting, x : <?X: Entity>
PlannedInto (mr,m) $\wedge$ <?Y: Relationship> (x, m)
$\Rightarrow \Diamond_{\leq\,m.begin}$ Participates (p, m)

The substitution {(<?X:Entity>, PersonalConstraint) , (<?Y:Relationship>, Addressed)} makes *TA* and *GTrans* logically equivalent. This substitution meets the various mapping constraints, viz. one-to-one concept mapping, same meta-level type and category, same relationship/predicate arity, role preservation, same assertion pattern (Achieve), and same assertion category (Satisfaction). The semantic matching thus extends the analogical mapping with the pairs (PersonalConstraint, Paper) and (Addressed, Accepted).

Attempting to semantically match a source and target assertion makes sense only if the assertions are comparable, that is, if they have a same pattern and same category.

Compound assertions may therefore need to be decomposed into basic assertions so that semantic matching may be achieved. Basic assertions are assertions that have a pattern and a category; compound assertions are ones that are made up of basic and compound assertions using logical connectors such as ∧ and ∨.

6.3 Completing the mapping

The principle is to extend the structural mapping until all relevant source assertions that need to be transferred are translated into the target vocabulary. A mapping *M* is said to be a *completed mapping* if (1) it extends a structural mapping, (2) all concepts from relevant source assertions attached to the answer to the query are in the range of *M*, (3) their analogs are defined in the target case, and (4) *M* meets the pattern and category mapping constraints.

In our ongoing example, the mapping is completed when all concepts from the formal definition of the constraint PapersReceived are in the range of the mapping.

A mapping completion algorithm based on the above definition is sketched below.

1. Apply relevance rules to select relevant assertions *A*
2. Identify in *A* concepts unmapped by structural mapping
3. **while** concepts still unmapped **do**
   3.1. select one, say *C*
   3.2. identify relevant source assertions *SA* about *C*
   3.3 **while** *C* is unmapped **and** there are still *SA* about *C* **do**
   - select one *SA*
   - **if** semantic match with relevant target assertion **then** add analogs to mapping
4. Return completed mapping or failure

For the PapersReceived answer in our ongoing example, step 1 applies the rule "if the query is about a Constraint and the Operationalization meta-level link is mentioned in it, then the formal definition of the constraint is relevant"; therefore the formal definition of PapersReceived needs to be transferred. Step 2 identifies the relationships PotentialAuthor, WrittenBy, Chairing and the entity Paper as unmapped by the structural mapping (see the mapping at the end of Section 5.4 and the formal definition at the beginning of Section 6). Step 3.1 selects an unmapped concept such as Paper. Step 3.2 identifies relevant assertions about Paper, such as the invariant attached to Submission, and decomposes it into basic assertions like MustAttendConference. An assertion is then selected (e.g. MustAttendConference) and a semantic match with relevant target assertions is attempted. As shown earlier MustAttendConference can be matched with the target assertion MustAttendMeeting; as a result the pairs (PersonalConstraint, Paper) and (Addressed, Accepted) are added to the mapping. Failure to return a completed mapping indicates that no analogy can be found due to dead end; in this case the source case should be dropped and another one should be taken from the answer set. A completed mapping allows the formal definition of the source constraint PapersReceived to be translated to a candidate formal definition for the missing target constraint (the name PersonalConstraintsGathered is supplied by the analyst), like the following.

HardConstraint Achieve [PersonalConstraintsGathered]
  InstanceOf SatisfactionConstraint;
  Concerns Participant, PersonalConstraint, ...
  OperationalizedBy GivePersonalConstraints
  FormalDef
  ∀ p:Participant, mr:MeetReq, pc:PersonalConstraint, s:Scheduler
  Invited (p, mr) ∧ Scheduling (s, mr) ∧ ConstraintProduced (pc, p)
  ⇒ ◊$_{\leq \text{mr.Deadline}}$ Notification (p, pc, s, mr)

Compared with other methods for declaration-level reuse through structural matching, analogical transfer through semantic matching results in a much richer, meaningful requirements fragment. The transposed requirements have of course to be checked against adequacy by the analyst and adapted if necessary.

7. Discussion

Analogical reasoning techniques are aimed to mimic the frequent situations where requirements engineers rely on past experience to elaborate and validate new requirements. With respect to existing work on analogical reasoning techniques [Gen88], [Fal89], [Hal89], [Lam91] and their application to specification reuse [Mai91], [Sut93], the focus of the paper was on showing how a *rich ontology* for capturing requirements combined with *formal assertions* to express them may yield more meaningful matching for specification reuse. Richer meta-level taxonomies result in more accurate structural matching (thanks to the constraints on same meta-level type, category and assertion pattern imposed in addition to the one-to-one and relational consistency constraints from [Gen88], [Fal89], see Section 3.2); formal assertions provide additional semantic accuracy. Analogical reuse is thereby extended from semi-formal to formal specifications. Such reuse may contribute to decreasing the cost of applying formal methods in requirements engineering while retaining their benefits.

Missing parts in a partial specification are identified by queries. The scope of the query may involve a wide variety of concepts whose type is defined at the meta-level: goals, constraints, agents, entities, relationships, events, actions, or scenarios. The target concepts in this scope are first classified with respect to concepts in the case base. Generalized queries and specialized answers are then found by climbing case base taxonomies up and down, respectively, under structural and semantic mapping constraints. The answers produce source cases and structural mappings; the latter are refined through semantic matching of relevant formal assertions. Analogical transfer is achieved by applying completed mappings to the source declarations and their relevant assertions. The process can be iterated step by step so that entire requirements frameworks can be reused.

The search-intensive phases of retrieving source cases and elaborating structural analogies do not require the analyst's intervention. Heuristic search for structural matches is aimed at guaranteeing reasonable performance. Semantic matching may raise more serious problems as a powerful theorem prover would be needed to fully automate the process. Semantic matching is currently performed manually; our experience suggests that manual assertion matching is already quite helpful for elaborating and/or validating requirements specifications. Automating semantic matching at reasonable cost would probably require some compromise between formalism expressiveness and efficient assertion matching (as already shown by classification technology). Our principle of introducing specialized assertion patterns and categories that constrain the mapping may be seen as a preliminary step in that direction.

Two important aspects of the analogical reuse process were left aside from this paper: the validation of transferred requirements and the consolidation of the case base.

Once candidate source specifications have been transposed, the analyst has to assess whether the resulting specifications are appropriate to the target situation and, if so, adapt them if necessary.

Besides, an extra benefit of integrating a formal assertion layer is that formal verification technology may be used to increase the level of confidence in reused specifications. We are currently investigating techniques for verification-based analogy extension. For example, when a partial goal refinement has been transferred, the analogy may sometimes be extended from the source refinement towards a provably complete refinement; the correctness of the transposed specification may be verified formally using techniques described in [Dar96].

The approach to reuse described in this paper relies on a rich, well-structured case base. The increasing popularity of design patterns [Gam95], [Bec96] makes us confident that this assumption is realistic for the future. The incremental update of the case base with new and "interesting" cases remains a big issue though. As a last step of the analogical reuse process, consolidation is aimed at determining whether parts of the target specification obtained should be integrated into the case base due to their potential reusability. This is known as the utility problem in explanation-based learning [Min88]. Our consolidation step is based on three utility criteria. *Generality* estimates the variety of concepts that could be defined as specializations of the concept being considered for inclusion in the case base. *Newness* measures how many new features the concept introduces with respect to its closest parent in the case base. Declaration-level newness is the number of new attributes and links the concept adds; assertion-level newness is based on a count of the number of basic assertions the concept adds with respect to the closest parent. *Conceptual distance* is aimed to identify conceptual holes in case base taxonomies. Defining *Distance(c,c')* as the number of intermediate concepts on the shortest path between concept c and his ancestor c', one may use the ratio *Distance(s,cg) / Distance(t,cg)*, where s is a source concept, t the target analog and cg the common generalization found during structural match, to identify conceptual holes on the path from t to cg. Generalized versions of the target concept may then be introduced to fill in such holes. Case base consolidation is thus an interactive process that involves the evaluation of utility criteria, the integration of useful target concepts according to the classification links established during query satisfaction (see Section 5.1), and the generalization of these concepts when utility needs to be improved.

Experience with our analogical reuse approach has been limited so far to small requirements fragments from different domains including various specializations of resource management systems, distributed communication systems, and simple transportation systems. Real, large scale experimentation would require a realistic size case base. The building of such a case base using the KAOS requirements engineering environment under development is among our high-priority tasks. On another hand, recent experience with the use of the KAOS methodology in two industrial projects has confirmed our feeling that requirements elaboration should be more viewed as a reuse-based activity. Beside shortcutting elaborations and supporting validation, specification reuse guarantees better uniformity and comparability.

Acknowledgement. This work was partially supported by the Belgian Ministry of Science (SPPS Project IT/IF/10 "Oasis") and the "Communauté Française de Belgique" (FRISCO project, Actions de Recherche Concertées Nr. 95/00-187 - Direction générale de la Recherche). Thanks are due to the RE'97 reviewers for helpful feedback.

References

[Bec96] K. Beck, R. Crocker, G. Meszaros, J.O. Coplien, L. Dominick, F. Paulisch and J. Vlissides, "Industrial Experience with Design Patterns", *Proc. ICSE-18 - 18th International Conference on Software Engineering*, Berlin, 103-114.

[Bor94] A. Borgida and P.. Patel-Schneider, "A Semantic and Complete Algorithm for Subsumption in the CLASSIC Description Logic", *Journal of Artificial Intelligence Research*, vol. 1, June 1994, 207-308.

[Bra85] R.J. Brachman and H.J. Levesque (eds.), Readings in Knowledge Representation, Morgan Kaufmann, 1985.

[Dar93] A. Dardenne, A. van Lamsweerde and S. Fickas, "Goal-Directed Requirements Acquisition", *Science of Computer Programming*, vol. 20, 1993, 3-50.

[Dar95] R. Darimont, "Process Support for Requirements Elaboration", PhD Thesis, Department of Computer Science, University of Louvain (Belgium), 1995.

[Dar96] R. Darimont and A. van Lamsweerde, "Formal Refinement Patterns for Goal-Driven Requirements Elaboration", *Proc. FSE-4: Fourth ACM Symposium on the Foundations of Software Engineering*, October 1996.

[Dev90] P. Devambu, "LASSIE: A Knowledge-Based Software Information System", *Proc. 12th Intl. Conf. on Software Engineering*, Nice, 1990, 249-261.

[Dia89] R. Prieto Diaz, "Classification of reusable modules", in: *Software Reusability, Volume 2: Applications and Experience*, T.J. Biggerstaff, AJ. Perlis (Eds.), Vol. 2, ACM Press, 1989, 99-123.

[Fal87] B. Falkenhaimer, "An Examination of the third stage in the Analogy Process : Verification-Based Analogical Learning", *Proc. 10th International Joint Conference on Artificial Intelligence*, 1987, 260-263.

[Fal89] B. Falkenhaimer, K. D. Forbus and D. Gentner, "The Structure-Mapping Engine: Algorithm and Examples", *Artificial Intelligence*, vol. 41, 1989, 1-63.

[Fea93] M. Feather, "Requirements Reconnoitering at the Juncture of Domain and Instance", *Proc. RE'93 - First International IEEE Symposium on Requirements Engineering*, San Diego (Ca), January 1993, 73-77.

[Fin88] A. Finkelstein, "Re-use of formatted requirements specifications", Software Engineering Journal, September 1988, 186-197.

[Gam95] Gamma, E., Helm, R., Johnson, R., Vlissides, J., *Design Patterns — Elements of Reusable Object-Oriented Software*, Addison-Wesley, 1995.

[Gen88] D. Gentner, "Analogical Inference and Analogical Access", in: *Analogica*, A. Prieditis (Ed.), Pitman, 1988, 63-88.

[Hal89] R.P. Hall, "Computational Approaches to Analogical Reasoning: A Comparative Analysis", *Artificial Intelligence*, vol. 39, 1989, 39-120.

[Hol89] K.J. Holyoak and P. Thagard, "Analogical Mapping by Constraint Satisfaction", *Cognitive Science,* vol. 13, 1989, 295-355.

[Koy92] R. Koymans, *Specifying message passing and time-critical systems with temporal logic*, LNCS 651, Springer-Verlag, 1992.

[Kru92] C. W. Krueger, "Software Reuse", *ACM Computing Surveys*, vol. 24, no. 2, June 1992, 131-183.

[Lam87] A. van Lamsweerde, B. Delcourt, E. Delor, M.-C. Schayes and R. Champagne, "Generic Lifecycle Support in the ALMA Environment", *IEEE Transactions on Software Engineering*, vol. 14, no. 6, June 1988, 720-741.

[Lam91] A. van Lamsweerde, "Learning Machine Learning", in: *Introducing a Logic Based Approach to Artificial Intelligence*, A. Thayse (Ed.), Vol 3, Wiley, 1991, 263-356.

[Lam95] A. van Lamsweerde, R. Darimont and P., "Goal-Directed Elaboration of Requirements for a Meeting Scheduler: Problems and Lessons Learnt", *Proc. RE'95 - 2nd International Symposium on Requirements Engineering*, York (UK), March 1995, 194-203.

[Lau94] D. Lauzon and T. Rose, "Task-Oriented and Similarity-Based Retrieval", *Proc. 9th Knowledge-based Software Engineering Conference*, Monterey, September 1994, pp. 98-107.

[Mai91] N. Maiden and A. Sutcliffe, "Analogical Matching for Specification Reuse", *Proc. 6th Knowledge-based Software Engineering Conference*, Syracuse (NY), September 1991, 101-112.

[Man92] Z. Manna and A. Pnueli, *The Temporal Logic of Reactive and Concurrent Systems*, Springer Verlag, 1992.

[Mil88] G. Miller. C. Fellbaum, J. Kegl and K. Miller, "WordNet: An electronic lexical reference system based on theories of lexical memory", Princeton University Cognitive Science Laboratory Technical Report 11, 1988.

[Mil94] A. Mili, R. Mili and R. Mittermeir, "Storing and Retrieving Software Components: A Refinement Based System", *Proc. ICSE-16 - 16th International Conference on Software Engineering*, Sorrento (Italy), 91-100.

[Min88] S. Minton, "Quantitative results concerning the utility of Explanation-Based Learning", Proc. AAAI'88, 564-569.

[Ost92] E. Ostertag, J. Hendler, C. Braun and R. Prieto Diaz, "Computing Similarity in a Reuse Library System: an AI-Based Approach", *ACM Transactions on Software Engineering and Methodology*, vol. 1, no. 3, July 1992.

[Pot91] B. Potter, J. Sinclair and D. Till, "An Introduction to Formal Specification and Z". Prentice Hall, 1991.

[Reu91] H. Reubenstein and R. Waters , "The Requirements Apprentice: Automated Assistance for Requirements Acquisition", *IEEE Transactions on Software Engineering*, vol. 17, no. 3, March 1991, 226-240.

[Rya92] K. Ryan, B. Mathews, "Matching conceptual Graphs as an aid to Requirements Reuse", *Proc. RE'93 - First IEEE International Symposium on Requirements Engineering*, San Diego,

January 1993, 112-120.

[Spa93] G. Spanoudakis and P. Constantopoulos, "Similarity for Analogical Software Reuse: A Conceptual Modelling Approach", *Proc. 5th International Conference on Advanced Information Systems Engineering*, LNCS 685, June 1993, 483-503.

[Sut93] A.G. Sutcliffe and N.A.M. Maiden, "Use of Domain Knowledge for Requirements Validation", *Proc. IFIP WG8.1 Conference on Information System Development Process*, 1993, 99-115.

[Zar93] A.M. Zaremski and J. M. Wing, "Signature Matching: A Key to Reuse", *ACM SIGSOFT Software Engineerig Notes*, vol. 18, no. 5, December 1993, 182-190.

Session 2B

Applications and Tools 1

Groupware-Assisted Requirements Assessment

Ahti Salo
Nokia Research Center
Heikkilantie 7
P.O. Box 45
FIN-00211 Finland
ahti.salo@research.nokia.com

This report presents experiences in deploying groupware applications as a platform for facilitating the collection and analysis of requirements in large, geographically distributed organizations.

Specifically, evidence from ongoing product development projects indicates that groupware-based requirement repositories contribute significantly to the efficiency of the earliest phases of requirements engineering. On one hand, distributed repositories permit the solicitation of requirements from broad cross-departmental audiences. On the other hand, these repositories enhance the visibility of product requirements and allow requirements documentation to be analyzed in relation to the originating source information. Furthermore, by enhancing requirement repositories with sufficiently rich domain models of the organizational context, it becomes possible to automate the processes of seeking expert feedback to bear on the requirements under-evaluation.

Requirement repositories are also analyzed from the perspective of organizational memory, as these repositories have been found central in 1) imparting product knowledge to new personnel and 2) preserving the rationale behind product decisions.

Lessons Learned from Applying the Spiral Model in the Software Requirements Analysis Phase

Dr. Chonchanok Viravan
Head of Software Technology Laboratory
National Electronics and Computer Technology Center (NECTEC)
Bangkok Thai Tower, 12th floor, Room 1202
108 Rangnam Road, Rajthevi
Bangkok, Thailand 10400
viravan@nwg.nectec.or.th

Boehm's Spiral model is currently gaining popularity over the traditional Waterfall software development model. The Spiral model is a risk-driven approach. The process steps are determined by the need to resolve the high risk situations – ones that have greatest chance to ruin the project. This approach contrasts with traditional document-driven approach where the process step is determined by the type of document that has to be submitted.

Over a period of two years, the author applied the spiral model approach in requirement analysis phase for four projects. These projects were with different sectors of the Thai government. The author was the head of the requirement analysis team for three projects and was responsible for quality assurance in one project.

The author's experience reveals factors that foster the spiral approach's success in requirement analysis as well as factors that inhibit its effectiveness. Some of these factors include the Thai culture, governmental regulations, education background of requirement engineers, the governmental employees' understanding of requirement analysis, terms in the contract, etc. The experience also reveals the risks in conducting software requirement analysis in a country that endures shortages of software engineers.

Industrial Workshop on
Requirements for R & D in Requirements Engineering

Philip Morris, Marcelo Masera, and Marc Wilikens
TP 210
21020 Ispra (VA) Italy
Philip.Morris@jrc.it

Introduction. This paper describes one of the activities of the Dependable Software Applications (DSA) sector at the Joint Research Centre in Italy. The activity is part of their support work to DGIII of the European Commission (EC). It reflects an aspect of DSA's continuous support process and mirrors our beliefs in the process of enhancing European industrial competitiveness by defining research needs based on technology observation. This presentation summarizes a report written for the EC describing industrial needs in Requirements Engineering as well as the mechanisms used to generate and validate the report. The report will be used by DGIII to define future research calls and for assessing research proposals in Requirements Engineering.

Defining R & D Needs. We briefly describe the industrially-driven workshop, used as the mechanism, for defining the industrial needs in RE. We outline the participant selection criteria, number of participants, their application domains, and their roles in those domains. We then discuss the seven domain areas identified by the industrial participants as requiring further research in Requirements Engineering: Requirement Elicitation, Requirements Specification, Requirements Management, Application Process, Requirements Validation and Verification, Process Improvement, and Framework Issues.

For each domain area the report provides the following elaboration:

- *Domain Description:* Description of the research area as defined by the industrial workshop participants
- *Rationale:* Identified problems in the area
- *Needs:* Research goals for the area addressing the identified problems
- *Drivers:* Industrial drivers are for these needs
- *Implementation mechanisms* What type of research and development is required to satisfy the needs
- *Success criteria:* How the research progress can be measured.
- *Downstream activities:* What other activities could be done upon completion of this research

This section concludes with the priority ratings assigned by the industrialists to each of the above research domains.

We then provide a description of the ways in which the DSA will promote industrial involvement in the Requirements Engineering domain and the perceived benefits to industrialists, researchers, and the EC research administration.

SOFL: A Formal Engineering Methodology
for Industrial Applications

Shaoying Liu
Hiroshima City University
shaoying@cs.hiroshima-cu.ac.jp

A major challenge for formal methods is to effectively address the needs of industry and achieve wide acceptance. This challenge remains unmet, as formal methods are difficult to use and their application consumes prohibitive amounts of resource. Much research on the integration of available formal methods (e.g. Z, VDM, B-Method) and either Structured Methodology or Object-Oriented Methodology has been conducted to make formal methods more practical, but with limited success. No attempt has yet been made to integrate the three approaches of formal methods, structured methodology and object-oriented methodology to take advantage of the desirable features of the three approaches.

As one approach to the solution of these problems, we propose a language called SOFL (Structured Object-Oriented Formal Language) for system development. It supports the concept that a system can be constructed using the structured methodology in the early stages of its development, and by using object-oriented methodology at later, more detailed levels. During the complete system development process, formal methods are applied in a manner that best uses their capabilities.

Session 3A

Scenarios and Use Cases

Enhancing a Requirements Baseline with Scenarios
Julio Cesar Leite, Gustavo Rossi, Federico Balaguer, Vanesa Maiorana,
Gladys Kaplan, Graciela Hadad, and Alejandro Oliveros

You want to know how things *really* work around here? Have I got some stories for you!" The early capture of requirements often starts with a set of "salient scenarios" — the stories that tell about the environment in which a new system is to operate; stories that attempt to capture the essence of the role that the system is to play in its environment. This paper examines various aspects of scenarios including a grammar for scenarios, some useful properties of scenarios, and hypertext treatment of scenarios, and advances a simple but compelling "scenario evolution" example that demonstrates how a new system will impact "how things really work."
— *Mark Feblowitz*

Producing Object-Oriented Dynamic Specifications: An Approach Based on the Concept of 'Use Case'
Benedicte Dano, Henri Briand, and Franck Barbier

Recently, use case or scenarios have been attracting the attention of researchers and practitioners. There are reports from practitioners that the deployment of use case enhances the quality of object oriented specifications, but there are still a lot of misconceptions and problems in the use of such a strategy. This paper describes a promising approach, that departing from a tabular representation of a use case and using a systematic process, produces object type state transition diagrams. This process uses Petri nets as the basic representation scheme in order to provide more rigour to the use case descriptions. An important side effect of the process is the possibility of using Petri nets tools to detect problems in a set of use case descriptions.
— *Julio Cesar Leite*

A Technique Combination Approach to Requirements Engineering
Alistair Sutcliffe

As more new techniques are developed to address specific problems in RE, it is not surprising that any single technique may not fully address the needs of a practical RE situation. The practitioner may choose to use a combination of several techniques to do the job. This paper reports on an empirical study which combines three important techniques: early prototyping, scenario-based analysis, and design rationale. The findings should be of interest to practitioners and researchers alike.
— *Eric Yu*

Enhancing a Requirements Baseline with Scenarios

Julio Cesar Sampaio do Prado Leite*
Gustavo Rossi†
Departamento de Informática, PUC-Rio
R. Marquês de S. Vicente 225
Rio de Janeiro 22453-900 Brasil
julio, rossi@inf.puc-rio.br

Federico Balaguer, Vanesa Maiorana
Gladys Kaplan, Graciela Hadad, Alejandro Oliveros
Dep. de Investigación, Universidad de Belgrano
Zabala y Villanueva Piso 12,
(1426) Buenos Aires, Argentina

Abstract

Scenarios are well reconized as an important strategy towards understanding the interface between the environment and the system as well as a means to elicit and specify software behavior. We have a broader understanding of scenarios. For us a scenario is an evolving description of situations in the environment. Our proposal is framed by Leite's work on a client oriented requirements baseline, which aims to model the external requirements of a software system and its evolution. Scenarios starts as describing the environment situations, according to the main actions performed outside the software system. Scenarios do also help the clarification of the interrelation between functional and non-functional requirements. We have validated our strategy and the related representations based on case studies of a real situation.

1 Introduction

Traceability, a major issue in software engineering, is seldom present at the initial requirements engineering process. This paper reports on a proposal for adding a scenario view to the requirements baseline model [11], in which evolution is the major driving factor. The scenario view should evolve, as part of the requirements baseline, along with the software construction process. The baseline is perennial, so scenarios will change as the software development progresses. The baseline can not be considered as the requirements specification. The specification will be built based on the information contained in the baseline.

Our understanding of scenarios is a combination of a series of ideas presented in the literature [2] [22] [19] [8] [18] to which we added four main concepts:

- a scenario starts by describing situations in the macrosystem, and its relation with the outer system. That is, we first consider the interfaces of the macrosystem, and then describe the interfaces of the software with its macrosystem.

- a scenario evolves as we progress in the software construction process.

- scenarios are naturally linked to the LEL (Language Extended Lexicon) [10] and to the basic model view (BMV) of the requirements baseline [11].

- a scenario describes situations, with an emphasis on the behavior description. Scenarios, similarly as the BMV, uses natural language description as its basic representation.

The Language Extended Lexicon is a meta model designed to help the elicitation of the language used in the macrosystem. This model is centered on the idea that a circular description of language terms improves the comprehension of the environment (the macrosystem), see Figures 1 to 4 as an example.

The basic model view (BMV) is a structure which incorporates sentences about the desired system. These sentences are written in natural language following defined patterns. Figure 5, using an entity relationship notation, shows the basic conceptual model behind the basic model.

Our scenario model view is a structure composed of goal, context, resources, actors and episodes. Goal, context, resources and actors are declarative sentences. Episodes are a set of sentences according to a very simple language that makes it possible the operational description of behaviors.

The addition of a scenario view to the requirements baseline made it possible to uncover a series of aspects to which neither the LEL nor the basic model view gave particular attention. The explicit introduction of goals, the relationship between resources and the identification of actors enriched the requirements baseline extending its *static* amplitude. The episodes gave the requirements baseline a representation capable of dealing with behaviors aspects. For instance, in the basic model view we had actions and external events linked to this action in a sequential fashion, however it was not possible to represent conditions, exceptions or parallel events. A language able to express such situations makes it possible that the episodes better mirror what does happen in the macrosystem.

*This work is supported in part by CNPq grant n. 510845/93-2 and Universidad de Belgrano

†This work is supported partially by Conicet (Argentina) and Universidad de Belgrano (Argentina).

If for one side the addition of a scenario view augments the expression power of the requirements baseline, it is also the case that it improves the traceability capability of the baseline. For instance having links to actors, resources and episodes of the macrosystem helps to anchor the origins of the requirements. Gotel[5] has pointed out that most of the research and use of traceability methods and tools does happen after the availability of a software specification. We understand that part of our work addresses this issue. Unlike other proponents of pre-traceability models [17], we did not make explicit the link to a requirements specification. We understand that the construction of a requirements specification is a process that comes after the availability of a requirements baseline.

Our text organization follows Parnas advice[16], that is, we present the proposal as it should be, which is not exactly how we used in our case studies. As such, the proposal described and exemplified in Section 2 is the result of cleaning up several aspects observed during the process of using the proposal in a real case. The same applies to Section 3, where we describe how scenarios evolve. It is important to note as well, that our description focuses on the representation, its navigation (presentation) and how it can be managed from the point of view of traceability. We do not focus on the process of producing the scenarios, nor the LEL, nor the basic model, this is treated elsewhere [1] [9] [10] [14]. At Section 4 we briefly describe observations about the use of scenarios and how we reached the result presented at the previous sections. We conclude stressing our contribution and linking our work to the work of others.

2 The Requirements Baseline Conceptual Model

As a consequence of adding scenarios to our baseline we now have it structured as follows:

- the lexicon model view,
- a basic model view,
- a scenario model view
- a hypertext view, and
- a configuration view.

It is mister to note that the configuration view and the hypertext view are orthogonal to the other three. These views may be seem as indispensable support services to be provided by the baseline in order to guarantee traceability (configuration view) and access to the stored information (hypertext view). The hypertext support works as an integrator of the lexicon, the basic model and the scenario views, enabling the definition of links between these views.

2.1 LEL, the Lexicon View

The Language Extended Lexicon is a representation of the symbols in the problem domain language. The LEL is anchored on a very simple idea: *understand the language of the problem, without worrying about understanding the problem* [10]. It is a natural language

representation that aims to capture the vocabulary of an application. The Lexicon main goal is to register signs (words or phrases), which are peculiar to the domain. Each entry on the lexicon has two types of descriptions, opposed to the usual dictionary which has just one type. The first type, called Notion, is the usual type and its goal is to describe the denotation of the word or the phrase. The second type, called Behavioral Response, goal is to describe the connotation of the word or the phrase, that is, it describes extra information on the context at hand. Besides using this extra information, LEL construction heuristics forces the use of links between the entries in the lexicon.

Over this basic representation, we established two major principles.

- When describing a notion or a behavioral response **maximize** the use of the signs of the language extended lexicon. We call this the *principle of circularity*.

- When describing a notion or a behavioral response **minimize** the use of signs exterior to the target domain. When using external signs make sure that they belong to the basic vocabulary of the natural language in use, as well as, as much as possible, have a clear mathematical representation (eg. set, belongs, intersection, function). We call this the *principle of minimal vocabulary*.

By imposing the principle of minimal vocabulary and the principle of circularity we are forming a self-contained set with several links between its elements, thus forming a graph. This graph is in reality a hypertext document, where the authoring rules are the structure of LEL and the two principles. We designed a customized hypertext system [10], HyperLex, which implements the structure of LEL and gives support to its use. HyperLex not only provides support for the acquisition of lexicons but also provides reports on several statistics regarding the hyperdocument.

Figures 1 to 4 exemplify the use of the lexicon and are based on the passport emission domain[1] [9].

2.2 Basic Model View

The basic model uses the entity relationship framework[4] as a representation language, and is pictured at Figure 5. Figure 6 provides an example in the passport domain.

2.3 The Scenario Model View

In the same style of the basic model view [11] we describe the scenario model using the entity relationship framework [4]. Figure 7 shows the entity relationship diagram of the scenario model behind the baseline. Note in the entity relationship diagram, that an episode[2] can be explained as a scenario itself, thus enabling the possibility of decomposition of a scenario in sub-scenarios.

---

[1]Please note that all the examples were written in Spanish and were *freely* translated to English.

[2]Episodes are a set of particular actions or situations that describe the behavior of a scenario. A scenario may have more than one episode. An episode can also be described as a scenario.

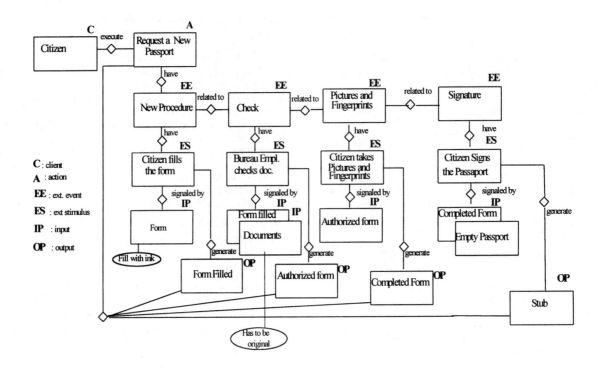

Figure 6: The BMV instantiated for the action REQUEST A NEW PASSPORT

Picture Cabin

- **Notion**:
  - It is a sector of the **Documents and Certificates Division**.
  - It is where the **citizen** picture is taken and charged.
- **Behavioral Response**:
  - The **form** is stamped with the same number as the picture.
  - The **citizen** receives two pictures.
  - The **Picture cabin** clerk archives the third picture.

Figure 1: LEL Entry: Picture Cabin

Check fingerprints

- **Notion**:
  - An action taken by the **Dactyloscopy Division** to verify if the citizen is the one he/she says he/she is.
- **Behavioral Response**:
  - In the case of **request new passport** the dactyloscopy **form** is archived.
  - In the case of **request passport renewal** the **Dactyloscopy Divison** checks the fingerprints in the **form** with the dactyloscopy **form**.
  - In the case of problems with the fingerprints the **citizen** has to be directed to the **Revision Division**.

Figure 2: LEL entry: Check Fingerprints

Form

- **Notion**:
  - It is a preprinted paper a **citizen** uses to **request a new passport** or **request passport renewal**.
  - It registers the **citizen** data.
- **Behavioral Response**:
  - It is filled by the **citizen**.
  - It is stamped by the **cashier** clerk.
  - It is stamped by the **picture cabin** clerk.
  - It is filled, signed and stamped by a **General Index Division** clerk.
  - It is filled, signed and stamped by a **Dactyloscopy Division** clerk.
  - It is appended to the **folder**.
  - The **stub** is teared from the **form**.

Figure 3: LEL entry: Form

46

Stub

- **Notion:**
  - It is the bottom part of the **form** needed to receive the **passport**.
  - It has the **citizen** identification.
- **Behavioral Response:**
  - It is signed, stamped and delivered to the **citizen** at the **Reception Cabin**.
  - The **citizen** presents it to the **Delivery Sector** in order to receive the **passport**.

Figure 4: LEL entry: Stub

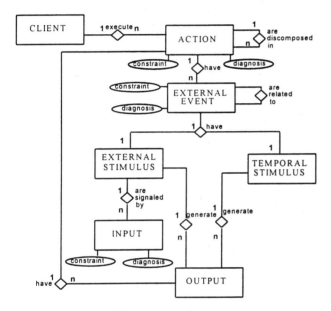

Figure 5: The ER Diagram for the Baseline Basic Model

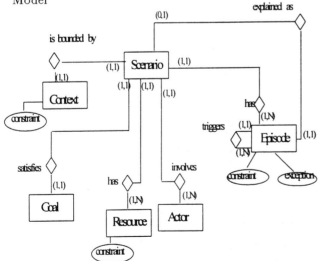

Figure 7: The ER Diagram for the Scenario Model

Below, we describe the entities pertaining to the ER model[3]

- **Title** - The title of the scenario. In the case of a sub-scenario the title is the same as the episode sentence (see below), without the exceptions and/or constraints. It has the following structure:

```
[ Phrase | ([Actor | Resource] + Verb + Predicate)].

Ex: Request a new passport.
        (Phrase)
```

- **Goal** - Is an objective to be achieved in the macrosystem. The scenario describes the achievement of the goal. It is described by the following structure.

```
[Subject] + Verb + Predicate.

Ex. Charge the citizen the passport fee.
        (Verb)       (Predicate)
```

- **Context** - Describes the geographical location (setting) of the scenario as well as an important initial state. Context is represented by a sentence with the following structure:

```
Location + State

Where Location is:
    Name.
Where State is:
    [Actor | Resource] + Verb + Predicate + {Constraint}

Ex: Cashier.  (Name)
    Citizen has received the forms and
    (Actor)  (Verb)       (Predicate)
    must be in cashier's line.
        (Constraint)
```

- **Resource** - Means of support, devices, necessary to be available in the scenario. Resource is represented by a sentence with the following structure:

```
Name + {Constraint}

Ex: Stub,
    (Name)
which must match the name of the requesting citizen
        (Constraint)
```

---

[3] + means composition, { x } means zero or more occurrence of x, () is used for grouping, | stands for logical or, and [x] denotes the optionality of x.

- **Actor** - A person or an organization structure that has a role in the scenario. It has the following structure:

```
Name

Ex.  Cashier's clerk
```

- **Episodes** - A series of episode sentences which detail the scenario and provides its behavior. The following partial BNF description gives an idea of how episodes are structured.

```
<episodes> ::= <series>
<series> ::= <sentence> | <series>
<sentence>
<sentence> ::= <sequential sentence> |
<non-sequential sentece> | <conditional
sentence>
<sequential-sentence> ::= <episode
sentence> CR
<conditional
sentence> :: = If <condition> Then
<episode sentence> CR
<non-sequential   sentence>  ::=  #
<series> #
```

Where <episode sentence> is described by the following structure:

```
[Actor | Resource] + Verb +
+ Predicate + {Constraint} + {Exception}

Ex: Picture Cabin clerk takes Citizen picture.
    (Actor)         (Verb)    (Predicate)
Exception: Camera does not work
```

The most important attributes of our ER model are the Constraint and the Exception attributes. A Constraint is a scope or quality requirement referring to a given entity. It is represented by a short sentence with the following structure: MUST + Verb + Predicate. **An exception causes serious disruption in the scenario, asking for a different set of actions, described in separate as exception scenarios.** It is represented by a short sentence, or by a small paragraph and usually reflects the lack or malfunction of a necessary resource.

The terms: Name, Location, Subject, Verb, Predicate, Actor and Resource can be chosen from the Language Extended Lexicon table, a structure that makes it possible the usage of a controlled vocabulary.

Below we give a scenario description related to the BMV of Figure 6 and one sub-scenario of the REQUEST A NEW PASSPORT scenario. This sub-scenario is described in graphical form (Figure 8).

TITLE:
    Request a New Passport.
GOAL:
    Satisfy the initial requirements for a new passport.
CONTEXT:

Documents and Certificates Division. The citizen does not have a passport.

ACTORS:
    Citizen
    Cashier's clerk
    Fingerprint Cabin clerk
    Picture Cabin
    Documents and Certificates Division clerk
RESOURCES:
    Camera
    Form
    Empty Passport
    Stub
    Citizen Documents
EPISODES:
Citizen fills the form. *Constraint:* Fill with ink
Documents and Certificates Division clerk checks the form. *Constraint* Form must be correctly filled. *Exception:* Missing documents
\# Picture Cabin clerk takes citizen picture. *Exception:* Camera does not work
Fingerprint Cabin clerk takes Citizen's fingerprints \#
Citizen pays the passport fee.
Citizen signs the passport.
Citizen receives stub.

2.4 The Hypertext View

The hypertext view is orthogonal to the BMV, LEL and SMV. The use of hypertext for the BMV is described in [11] and the LEL, itself a hypertext, is described in [10]. Here we will focus on the hypertext view from the scenario standpoint.

There are many relationships to be explored among scenarios and among scenarios and other components of the requirements baseline. From each scenario we derive one or more hypertext nodes which are related by hypertext links. We may have one node class for each type of entity defined in the scenario model (Figure 7). Links in turn are derived by three different forms:

- from existing structural relationships among scenarios, for example the relationship sub-scenario,

- from information in the LEL table and

- defined opportunistically

The reason why hypertext nodes are derived from scenarios is that we can build different views of the same scenario according to the user profile or task, e.g. given the scenario REQUEST A NEW PASSPORT we could define a view that is useful for a user, another view for a software engineer, and so on. Each view defines a hypertext node and the set of nodes and links define the hypertext. This approach is now usual in modern hypermedia design approaches [20] [21] and allows us, for example, to concentrate all information we want from an scenario in a composite node containing: Title, Goal, Context, and Resources as attributes

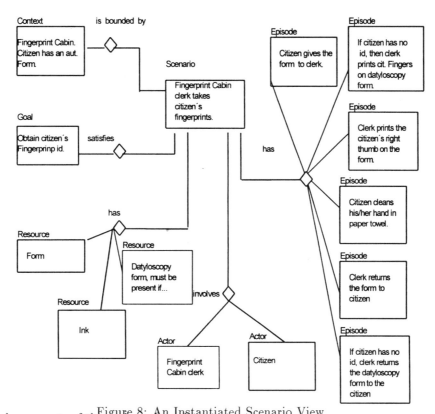

Figure 8: An Instantiated Scenario View

and the set of episodes as parts of the node and reachable by following structural links. When episodes are themselves scenarios we have an interesting navigation pattern that will be described below.

As we may have a large number of nodes (scenarios, sub-scenarios) and relationships, the hypertext model is organized in navigational contexts, i.e. set of nodes that are closely related with each other and that we intend to navigate in a straigthtforward way. Some navigational contexts are derived from the compositional structure of scenarios as defined in the scenario meta model; for example we will have the navigational context formed out with all sub-scenarios of a given one and will have corresponding links allowing to reach the first of this set and each member of the set (even recursively, see Figure 9). Note that we have added links among the components of of scenario A. When we are navigationg this context, the episodes of A, we understand that all components are connected by the *next* link, induced by the compositional structure of A, and the same is true for A.1.

Another interesting context is derived from information contained both in the scenario an in the LEL. For example we could select an entry in the LEL table representing a resource in the scenario model and find all scenarios using that resource. These nodes may not be related with a hierarchical relationship existing in their conceptual counterparts as in the previous case, though their corresponding nodes will be organized in a similar way, i.e as a set with a naviagtion strategy allowing to navigate the whole set easily. Others interesting relationships arise when we take into account the configuration view, such that we could build a set formed out with the different versions of a scenario. Note that the same scenario may be navigated in dif-

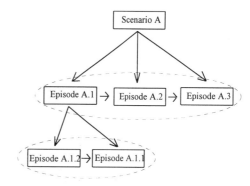

Figure 9: Navigational Contexts

ferent contexts and that the meaning of *next* is always the next scenario in this context.

It is important to note that by separating scenarios from scenarios views (the hypertext nodes) we can provide a different appearance to different user's profiles and can organize the hypertext navigation according to each user's need. For instance, we would not define a navigational context: *versions of a given scenario* for some kind of users, meanwhile we will surely define this context for members of the software engineering team.

We have enriched the concept of LEL table as described in [11] including information about scenarios, see Figure 10; in this way, for example, two scenarios that are not related structurally but have some term in common may be automatically linked. How-

LEL TABLE / DELTA TABLE

TERM	ENTITIES / SCENARIOS - ADDRESSES		
Fingerprint Cabin	scenario: Fing. Cab clerk takes citizen fingerprints	actor: Fing. Cabin clerk	...
Form	output: Form	episode: Citizen gives form to clerk	...
Stub	output: Stub	episode: Citizen receives stub	...
...	...	...	...

Figure 10: Enriched LEL table

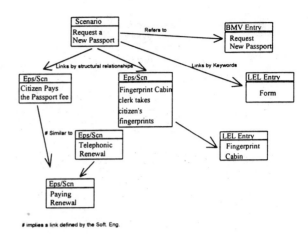

Figure 11: An Example of Hypertext links

ever, as this approach would cause a **link explosion** in the hypertext view, the software engineer may use the LEL-table to select links that did not result evident.

We may also want to link two scenarios that are not related structurally, i.e. the sub-scenario of conceptual relationship, or such that they do not share some item in the LEL. This kind of link is not generated automatically in our model and must be explicitly described. It may happen when we discover a new term that may be considered for further inclusion in the LEL-talbe. We use an approach similar to [11] by using a Delta-table and thus reducing this kind of navigation to the formely described.

Finally, as we said before, scenarios are linked to nodes representing entries in the LEL table and to nodes reflecting information in the Basic Model View. Entries in the LEL table appear in a scenario node as hotwords, for instance, **Form** in the scenario REQUEST A NEW PASSPORT is defined by the LEL, Figure 3, and is used at the BMV described at Figure 6. Figure 11 gives an example of the navigational possibilities of our baseline. It is interesting to note that by providing a rich set of navigational contexts and links we are somehow replacing the need to define specific queries to the baseline meta models.

2.5 The Configuration View

The configuration view is essential in order to maintain the traceability of the products and their revisions. The LEL, the BMV (Basic Model View) and SMV (Scenario Model View) are all subject to a configuration and version control. At a given time a view from the baseline may be requested based on the actual configuration or in past configurations. Each version of each model keeps the following information: date, time, person making the change (user), reasons for the change (change trigger, date of trigger, authorization) and type of change (input, modification, exclusion).

The consistency of the configuration is warranted by consistency constraints determined by each model. As such, change operations trigger a process for consistency checking responsible for the consistency of a given configuration. If, in the SMV, an episode is ex-

cluded, and if it is described as a sub-scenario, the sub-scenario must to be excluded as well. The example of the configuration view is given below at Section 3, where we stress the evolving aspect of SMV.

3 Scenario Evolution

We have taken the episode *Fingerprint Cabin clerk takes citizen's fingerprints* from the scenario REQUEST A NEW PASSPORT to show how a scenario evolves from a pure macrosystem view to the one that deals with the interface with the future computer system, to which we are defining the requirements. After the two versions we show the use case [8] representation for describing version 2 (Figure 12). Our work regarding scenario evolution is at its early stages, but we believe that the usage of a hypertext and a configuration management approach will be the a solid ground to study aspects of interconnection between evolving scenarios.

Below we have Version 1, with the following information:

- **Date:** 15/2/96
- **Time:** 14:00 hs
- **User:** Federico
- **Trigger:** The passport case study
- **Date of Trigger:** 20/10/95
- **Type:** Inclusion

TITLE:

Fingerprint Cabin clerk takes citizen's fingerprints.

GOAL:

Obtain citizen's fingerprint identification.

CONTEXT:

Fingerprint Cabin. Citizen has an authorized form.

ACTORS:

Fingerprint Cabin clerk.
Citizen.

RESOURCES:
  Form, which must have been previously checked.
  Ink.
  Dactyloscopy form, must be present if citizen has no prior identification.

EPISODES:
Citizen gives the form to clerk.
If citizen has no prior identification, **Then** clerk prints the citizen's fingers on the dactyloscopy form.
Clerk prints the citizen's right thumb on the form.
Citizen cleans his/her hand in paper towel.
Clerk returns the form to the citizen.
If citizen has no prior identification, **Then** clerk returns the dactyloscopy form to the citizen.

Below we have Version 2, with the following information:

- **Date:** 10/3/96
- **Time:** 15:30 hs
- **User:** Federico
- **Trigger:** The discussion about how the Fingerprint Cabin would work with the fingerprint reading machine.
- **Date of Trigger:** 1/3/96
- **Type:** Modification

TITLE:
  Fingerprint Cabin clerk takes citizen's fingerprints.
GOAL:
  Obtain citizen's fingerprint identification.
CONTEXT:
  Fingerprint Cabin. Citizen has an authorized form.
ACTORS:
  Fingerprint Cabin clerk.
  Citizen.
RESOURCES:
  Form, which must have been previously checked.
  Fingerprint reading machine.
  Fingerprint database.

EPISODES:
Citizen gives the form to clerk.
Clerk gives instructions of how to position the citizen's hand.
Clerk positions the form for printing and starts the machine.
If Fingerprint reading machine signals red, **Then** clerk make sure that the citizen's finger is properly positioned and restart the machine.
\# Machine prints the fingerprint on the form.
Machine saves the information on the database. *Constraint*: Save must be done in less than four seconds.
\#

4 Observations on the Use of Scenarios
The work reported here is the first result from a joint project of PUC-Rio and Universidad de Bel-

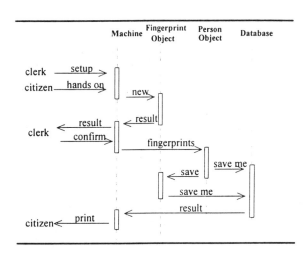

Figure 12: The Use Case for Fingerprint Cabin Clerk takes citizen's fingerprints

grano. Our project [15] is exploring the use of scenarios through the development process, looking into how to evolve scenarios towards testing strategies. Our first step was to investigate an initial representation for scenarios. We decided that Leite's previous work on natural language based representations for requirements could be used a starting point. So we went on and started to look of how scenarios could be defined at the same level of abstractions as the LEL and the BMV.

Once decided to integrate the scenario into the requirements baseline, we established a plan with the following steps:

- revision of the literature on scenarios,
- definition of a basic scenario representation,
- use in a restricted case study,
- use in a real case by two different teams and
- evaluation of case studies.

The revision of the literature confirmed our hypothesis that we had an innovative view of scenarios. The basic representation we have sketched did not have a standard for episode description, neither constraints nor exceptions. We also did not have a very clear idea of how to deal with scenario decomposition. We used the scenario in the restricted elevator problem [7] [1] and then decided to use the Argentina passport emission as our domain. A first team elicited the scenarios departing from the LEL without using the BMV [9], and the other team constructed the BMV for the domain and anchored the construction of the scenarios on the BMV [12].

The case study reported in [9] used unstructured and structured interviews to have a first version of the LEL, which was validated with the user (employees of the Argentina Passport Bureau, *Policia Federal*). From observations, structured interviews and the LEL scenarios were constructed. The construction of scenarios led to some changes in the LEL. The LEL has 34 entries. The group decided to give emphasis on the entities (objects, actors and states) and less on process (actions) on the LEL entries, so several verbal phrases pertaining to the problem were not included in the LEL, but were present in the scenarios. In describing the scenarios the group listed 24 scenarios, being 20 sub-scenarios, with an average of 5 episodes each. The group also has found out the necessity of using special comments, not previously planned, to register alternative cases and restrictions, which we then renamed as exceptions and constraints. The group also pointed out the necessity of having the different types of sentences, since we have only thinked on sequential phrases, so we have included the *non-sequential sentence* and the *conditional sentence* in the grammar for episodes. Another finding was that the hierarchical structure of scenarios was an important aspect in organizing scenarios. The overall impression of the team conducting the case study was that the scenarios substantially helped the comprehension and description of the domain and that the integration with the LEL was straightforward.

The case study reported in [12] had the other group as their main source of information about the domain, but have also used direct observation in order to have a better understanding of the domain. This group did not use the LEL directly. They built the BMV from the information available to them. The resulting BMV had 8 actions and 26 external events. The group decided to focus only on the actor *solicitante*, which in the english version we called `citizen`, because doing so, they would concentrate on the interface between the external world and the macrosystem. From the BMV the group described 8 scenarios. They have not organized the scenarios in a hierarchical way, all the scenarios described were atomic, with an average of 4 episodes each. In this exercise the group came out with a general set of heuristics to compose the skeleton of a scenario. These heuristics help the definition of title, goal, resources and actors, so the elicitation of episodes can occur in a bounded context. For instance, an action gives the title of the scenario, the input can provide hints for resources, as the client points to one of the actors. The group also helped the definition of CONTEXT, noting that besides the geographical location it was necessary to point to an important initial state previous to the happening of a scenario. This group centered their scenario on the workflow of the passport emission, and did not dealt explicitly with constraints and exceptions. This group also produced an evolution of half of the scenarios detected, thus changing the focus to the interface of the macrosystem and the computer system that would help the process of passport emission. This evolution was performed using common sense, that is it does not represent a real situation.

Overall we confirmed that scenarios are an important tool for the requirements engineer, not only because of the natural way in which they describe behavior, but because they force the software engineer to map the macrosystem in which the future software

will work. The case studies confirmed our hypothesis that scenarios must be in the requirements baseline, and that their link with the BMV and the LEL can be achieved without a lot of effort. We have also learned some on how to evolve scenarios and that we must have a hypertext system to make this requirements baseline usable. The case studies were not conducted with the hypertext support, which is under construction. Regarding the elicitation aspect, our strategy of maintaining the application vocabulary has helped the task of elicitors as well as the task of validation.

5 Conclusion

We have presented a proposal to integrate scenarios into a requirements baseline, making possible their evolution as well as the traceability of the different views of the requirements baseline. Our proposal is innovative in three important aspects: the use of scenarios as means for evolution, the vision that scenarios start from situations in the macrosystem, and the integration of its representation in an environment oriented towards hypertext navigation and configuration management. Our proposal is an evolution of work being performed at PUC-Rio, where the main focus is using natural language descriptions to help the elicitation and modeling of requirements. The case studies performed helped us tunning the scenario view as well as confirmed the properness of the LEL and the BMV as requirements representation models.

Our work was influenced by several authors, in particular; [2], [22], [19], [8] and [18]. From Rubin and Jacobson we have been convinced of the importance of scenarios to better describe object behavior, and the importance of tracing this behavior to interfaces aspects of the software system. Carrol gaves us a better understanding of the cognitives aspects of scenario based development, as well as a confirmation of our idea that scenarios should be born in the macrosystem (Universe of Discourse) and not only at the interface of the software system. This idea was also re-enforced by the work of Zorman, that well defines scenarios as situations. From Zorman we also used her survey of scenarios representations. Potts showed us the importance of relating scenarios to goals, by the same token it was important our previous knowledge of the goal oriented meta model proposed by Dardenne [3].

Unlike Potts, we organize goals from the point of view of scenarios, so that a scenario hierarchy will be similar to a goal hierarchy. Like Potts et al, we give special attention to constraints and exceptions, which they named obstacles. We made a differentiation between constraints and exceptions, because exceptions require a special treatment and constraints signalize to important aspects of non-functional requirements [13]. It is also important to note that the BMV also has special treatment for non-functional requirements (see Figure 5).

Future work will focus on continuing the use of the requirements baseline in real cases. Now besides Petrobrás who used it [11], there is a pilot experience going on in a major brazilian bank. We still do not have a solid prototype, the one used at the Petrobrás case did not evolve to integrate the configuration and the hypertext view, we continue working on the support for the original baseline and are now incorporating the scenario representation as well. We also plan to explore the hypermedia [6] capability of our model, which already has all the infrastructure to support such extension. We hope that a consolidation of the baseline, by experiences with real cases, will provide a more solid ground to the proposal of a requirements baseline as a knowledge base, and as such be amenable to automated analysis of different types.

6 Acknowledgement

We wish to thank the referees for the questions and comments made on the submitted version.

References

[1] Balaguer, F., Leite, J.C.S.P., Rossi G. Using Scenarios in Real Time Systems' Requirements Elicitation, submitted to Canela'96.

[2] Carrol, J. (ed.) *Scenario-Based Design: Envisoning Work and Technology in System Development*, Wiley, New York, 1995.

[3] Dardenne, A., van Lamsweerde, Fickas, S., Goal Directed Requirements Acquisiton, *Science of Computer Programming*, Vol. 20, Apr. 1993.

[4] Elmasri and Navathe, S. *Fundamentals of Data Base Systems*, Benjamin/Cummings Publishing Comp. Inc, 1989.

[5] Gotel, O.C.Z. and Finkelstein, A.C.W., An Analysis of the Requirements Traceability Problem, *In Proceedings of the First International Conference on Requirements Engineering*, Colorado Springs, IEEE Computer Society Press, 1994, pp. 94-101.

[6] Gough, P, Fodemski, F.T., Higgins, S.A., Ray, S.J. Scenarios - An Industrial Case Study and Hypermedia Enchancements, *In Proceedings of the Scnd IEEE International Symposium on Requirements Engineering*, IEEE Computer Society Press, 1995 pp. 10-17.

[7] Jackson, M.A. *Systems Development*, Prentice-Hall, 1983.

[8] Jacobson, I. Christerson M., Jonsson P., Overgaard, G., *Object-Oriented Software Engineering – A Use Case Driven Approach*, Reading, MA: Addison-Wesley; New York: Acm Press, 1992.

[9] Kaplan, G., Hadad, G. Oliveros, A., Uso de Lexico Extendido del Lenguaje (LEL) y de Escenarios para la Elicitacion de Requerimientos. Aplicacion a un Caso Real, *Informe de Investigación* Departamento de Inverstigación, Universidad de Belgrano, Buenos Aires, 1996.

[10] Leite, J.C.S.P. and Franco, A. P. M., , Languages, *In Proceedings of the First IEEE International Symposium on Requirements Engineering*, San Diego, Ca, IEEE Computer Society Press, 1994 pp. 243-246.

[11] Leite, J.C.S.P. and Oliveira, A.P.A., A Client Oriented Requirements Baseline, *In Proceedings of the Scnd IEEE International Symposium on Requirements Engineering*, IEEE Computer Society Press, 1995 pp. 108–115.

[12] Maiorana, V., Balaguer, F., La Relacion Entre el Modelo Baseline y Escenarios, *Informe de Investigación* Departamento de Inverstigación, Universidad de Belgrano, Buenos Aires, 1996.

[13] Mylopoulos J., Chung L., Nixon B., Representing and Using Non-Functional Requirements: A Process Oriented Approach, *IEEE Transactions on Software Engineering*, Vol. 18, No. 6, Jun. 1992.

[14] Oliveira, A. P., Leite, J.C.S.P., SERBAC: Uma Estratégia para a Definição de Requisitos, *In Proceedings of the VIII Simpósio Brasileiro de Engenharia de Software*, Sociedade Brasileira de Computação, Out. 1994, pp. 109-123.

[15] Oliveros, A., Leite, J.C.S.P., Rossi G., Uso de Escenarios en el Desarrollo de Software, *Proyecto de Investigacion*, Departamento de Inverstigación, Universidad de Belgrano, Buenos Aires, 1995.

[16] Parnas, D. L., Clements, P.C., A Rational Design Process: How and Why to Fake it, *IEEE Transactions on Software Engieering*, Vol. SE-12, No. 2, Feb. 1996, pp. 251–257.

[17] Pohl, K. PRO-ART: Enabling Requirements Pre-Traceability, *In Proceedings of the Scnd International Conference on Requirements Engineering*, IEEE Computer Society Press, 1996 pp. 76-84.

[18] Potts, C., Takahashi, K., Antón, A.I, Inquiry-Based Requirements Analysis, *IEEE Software*, Vol. 11, n. 2, Mar. 1994, pp. 21-32.

[19] Rubin K.S., Goldberg J., Object Behavior Analysis, *Communications of the ACM*, Vol. 35, No. 9, Sep. 1992, pp. 48–62.

[20] D. Schwabe, G. Rossi, The Object-Oriented Hypermedia Design Model, *Communications of the ACM*, VOl 38 (8), August 1995, pp45-46.

[21] D. Schwabe, G. Rossi, S. Barbosa, Systematic Hypermedia Design with OOHDM, *Proceedings of the Seventh ACM International Conference on Hypertext, Hypertext'96*, pp. 116-128.

[22] Zorman, L. *Requirements Envisaging by Utilizing Scenarios (Rebus)*, Ph.D. Dissertation, University of Southern California, 1995.

An Approach Based on the Concept of Use Case to Produce Dynamic Object-Oriented Specifications

Bénédicte DANO
TELIS / IRIN
18, impasse des Jades
44088 - Nantes Cedex 03
FRANCE
E-mail: dano@irin.univ-nantes.fr

Henri BRIAND
IRIN / IRESTE
University of Nantes
2, rue de la Houssinière
44072 - Nantes Cedex 03
FRANCE

Franck BARBIER
IRIN
University of Nantes
2, rue de la Houssinière
44072 - Nantes Cedex 03
FRANCE

Abstract

This paper presents an approach based on the concept of use case to support the requirements engineering process. The proposed approach assists the analyst in producing dynamic object-oriented requirements specifications. The approach is "domain expert-oriented" in the sense that the domain expert can actively participate during the requirements acquisition activity by identifying and by describing the use cases. A formal technique using Petri nets is also proposed during the requirements acquisition activity in order to make the domain expert's requirements as complete and as unambiguous as possible. During the requirements conceptualization activity, the explanation of how some parts of dynamic requirements specifications can be produced from the formalized use cases is presented.

1: Introduction

This paper describes an approach to support the requirements engineering process. The proposed approach is based on the concept of "use case" to progress towards the object-oriented requirements specifications. This paper essentially focuses on the dynamic aspects, as opposed to the static ones, which we dealt with in our previous work [1]. Although the static specifications (precisely, the object types and the relationship types named according to the terminology defined by the OMG[1]) are used at the first step of our approach (Figure 1), the feedback from the consideration of dynamic aspects will be used in improving these static specifications. Using the static specifications in order to produce the dynamic specifications allows greater consistency with the approach presented in most of the object-oriented methods listed by the OMG [2], including

OMT [3], OOSA [4], and presented in other new methods such as Syntropy [5].

Within the requirements engineering process, our work focuses on the dynamic requirements acquisition and conceptualization activities (Figure 1):

- the dynamic requirements acquisition activity is the activity during which the use cases are collected and described. It is important that both the domain expert and the analyst participate and collaborate during this activity. We therefore propose two ways of describing use cases. The first uses tables and is particularly domain expert-oriented. The second uses Petri nets and so brings the necessary formalism to the analyst. In order to help the analyst generate Petri nets, we have established some rules which are applied to the elements contained in the tables.
- the dynamic requirements conceptualization activity is the activity during which the acquired requirements are expressed in terms of object behavioral concepts defined by the OMG, including the state, the event and the transition concepts. The requirements conceptualization activity consists in producing state transition diagrams of some object types.

The main goal of this paper is to propose an approach to support the requirements engineering process compatible with many object-oriented methods. Each activity of the proposed approach is precisely described and illustrated through the example of a "gas station" application which is presented in [5] and in [6]. Section 2 presents the dynamic requirements acquisition activity. Section 3 describes the dynamic requirements conceptualization activity. The paper ends with a discussion on future research plans in Section 4.

---

[1] Object Management Group

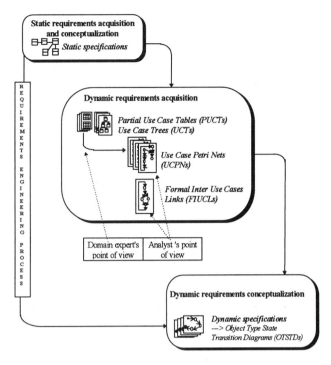

Figure 1 : Overview of the process

2: The dynamic requirements acquisition activity

The activity of dynamic requirements acquisition consists in gathering behavioral elements relevant to an application by capturing the domain expert's needs in a systematic manner. We think that a "use case-driven approach" naturally facilitates the communication between the analyst and the domain expert during the requirements acquisition activity. The concept of use case is originated from Objectory [7] and its use is particularly detailed in [8]. In OBA [9], the authors have used a similar concept named "scenario". Indeed, the terms "scenario" and "use case" have often been used synonymously [10].

The use case concept has recently been well received by many methodologists such as [11], [12], [13]. Although there is a growing acceptance of use cases, the lack of standardized and precise definitions and of formal descriptions of use cases is certainly the main drawback of the use case-driven approaches.
Following these remarks, we propose a rigorous definition of the use case concept in Section 2.1. We also suggest two techniques, including a formal one, to acquire and to describe the use cases in Section 2.2: firstly, an extended tabular notation which is particularly well-adapted for the domain expert is proposed. Secondly, a formalism such as Petri nets is presented as a helpful support for the analyst. In order to obtain a

consistent view from the two techniques and in order to facilitate the passage of the domain expert's description to the analyst's description, some mapping rules are introduced and presented in Section 2.3. A use cases description gives a separate view of the application. The notion of "temporal links" between use cases is thus introduced in Section 2.4 in order to obtain a global description of the application.

2.1: Definition of a use case

A use case is a behaviorally related sequence of transactions performed by an actor. An actor can be a human or a machine and it corresponds to a specific role played in the application. As it happens, we can find the following definition in [8]: "a use case is a specific way of using the system by using some parts of functionality".

Such a definition seems too fuzzy to allow easy identification of the use cases. In particular, the definition allows too free an interpretation of the use case concept so that it is difficult to determine what must be integrated in a use case, and what must be part of another use case.

In our opinion, a use case is an "objective" that an actor should achieve by using one complete functionality of the application. For example, some use cases in a "gas station" application are "take gas from a given pump", "collect payment for a given pump" or "monitor the level of the tank of a given pump". The first use case is the customers' objective, and the last two use cases are the attendant 's objectives.

2.2: Description of a use case

In accordance with our definition of use case, describing a use case uc_i consists in characterizing sets of n functions $f_{i,1},..., f_{i,n}$ the sequence of which either allows an actor to achieve a given objective or not.

Describing a use case using natural language, as recommended in [8] and in [13], presents many drawbacks. Although the use of natural language facilitates the communication between the analyst and the domain expert, it increases the risks of ambiguity, inconsistency and incompleteness of the description. In order to avoid these typical problems with natural language, it is important to use a more structured or a formal technique for the description of use cases.

In the literature, some structured techniques for the use cases description have been proposed [9], [12] and are based on a tabular notation. Some formal techniques such as grammars or conceptual state machines [11] and such as statecharts [14] have recently been introduced to describe use cases.

According to us, the techniques used for the description of the use cases must be chosen from the three following criteria. They must be :

a) understandable and supervisable by the domain expert.

b) semantically rich enough so that all pertinent description of the use case can be taken into account without ambiguity.

c) executable.

A domain expert-centered viewpoint. In our approach, we recommend the use of tables to facilitate the dynamic requirements acquisition activity from the domain expert's view. Of course, the main goal of using tabular notations is to strongly integrate the use cases description into an object-oriented approach. In particular, in an object-oriented context, a use case is expressed as an interaction between several object types which have been identified during the static requirements conceptualization activity.

As in OBA [9], our tables are defined with elements such as actions, which we call functions. We add some new elements such as states of object types during a function as well as conditions and assumptions. These elements collected in one table describe only one part of a use case. Each table is therefore called a "Partial Use Case Table" (PUCT).

Table A (Appendix) partially describes the "take gas from a given pump" use case. It contains a set of sequential functions $f_{1,1}$, $f_{1,2}$, $f_{1,3}$ and $f_{1,4}$ which allows a customer to take gas from a given pump. Two supplementary dummy functions called $f_{1,begin}$ and $f_{1,end}$ are added to the previous ones in Table A in order to apply some rules which will be presented below.

Table A also contains the states of the object types during each function. The object types are those which have been identified during the static conceptualization activity and those which are concerned by the current use case. The set of states of each object type during a $f_{i,j}$ function is called a configuration of the $f_{i,j}$ function.

There can exist several configurations of a given function: for example, two configurations of the $f_{1,1}$ function are defined in Table A (Appendix) which are {idle, in, not empty, displaying an amount, on, UD[2] } and {idle, in, UK[3] , displaying an amount, off, UD}. In order to distinguish several configurations of a given function, some conditions associated with each configuration of the function are collected. Thus, the first configuration of $f_{1,1}$ is valid when the "the pump is enabled" condition is verified, whereas the second configuration is valid when the "the pump is disabled" condition is verified.

The configurations of a $f_{i,j}$ function are defined from the configuration of the $f_{i,j-1}$ function for which there exists an associated assumption in the table. An assumption associated with a given function is a condition associated with one of the configurations of that function. The other possible configurations of the $f_{i,j-1}$ function can be taken into account by building some new PUCTs, identical to Table A in which the considered assumptions have to be indicated. Table B is thus another "take gas from a given pump" PUCT which has been derived from Table A by assuming that "the pump is disabled" during the $f_{1,1}$ function.

All the PUCTs which describe a use case can be represented by several trees called UCTs (Use Cases Trees). The root of a UCT is one initial PUCT like Table A. The nodes are PUCTs which contain conditions and the leaves are PUCTs which do not contain conditions. The assumptions made to build a given PUCT are indicated in a gray rectangle. In Figure 2, a UCT of the "take gas from a given pump" use case is represented. The initial PUCT is Table A (Appendix). Other PUCTs like Table B (Appendix) and Table C can thus be built from the initial PUCT.

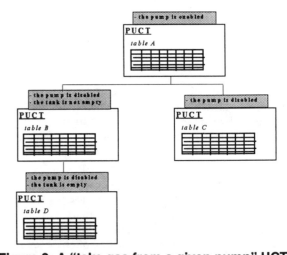

Figure 2: A "take gas from a given pump" UCT

A dotted line notation between two configurations of a given function is introduced in the PUCTs in order to indicate that a given configuration of the function is valid just after the previous configuration of the same function is obtained, without executing any additional function. For example, the second configuration of the $f_{1,2}$ function in Table A (Appendix) is valid just after the first configuration has been obtained whereas no additional function has been executed.

---

[2] UnDefined state
[3] UnKnown state

An analyst-centered viewpoint. Although the description of the use cases by means of tables makes the involvement of the domain expert during the dynamic requirements acquisition activity easier, it is not entirely satisfactory for the following reasons:

- the description is too complex to be completely expressed by means of tables.

- the description is not rigorous enough for the analyst who needs a degree of formalism in order:

    - to make sure that no requirement is misunderstood.

    - to ensure the completeness and the consistency of the requirements.

In order to take these two major observations into account, a description of the use cases by means of Petri nets is proposed. This technique is one of the most common techniques for the dynamic modeling [15] and it allows to deal with concurrency, non-determinism and causal connections between events. In our case, the "synchronized" [16] and "colored" Petri nets is an especially good choice. A synchronized Petri net is a marked Petri net in which an event is associated with each transition. A synchronized and colored Petri net is an eight-tuple PN = <P, T, prev, post, M_0, E, sync, C> where:

- P is the set of places.
- T is the set of transitions.
- prev: P×T→f is the function which returns the weight of the arc from a given place to a given transition. The weight is a function which belongs to f and which returns the tokens to be removed from the place after the firing of the transition.
- post: P×T→f is the function which returns the weight of the arc from a given transition to a given place. The weight is a function which belongs to f and which returns the tokens to be added in the place after the firing of the transition.
- M_0 is the initial marking of places (i.e. the initial distribution of the tokens in the places of the net).
- E is the set of events.
- sync: T→E∪{e'} is the function which returns the event associated with a given transition where e' is the "always occurs" event.
- C is the set of colors of the tokens.

Figure 3 presents the formal description of the "take gas from a given pump" use case by means of a synchronized and a colored Petri net. Places are represented with circles. Places hold tokens. The tokens have associated colors, y for yellow, r for red and w for white, each of them referring to a given pump of the "gas station" application. Transitions are indicated using horizontal lines. Arcs are represented with arrows and have weight. Firing a transition changes the marking in

accordance with the weight of the input and the output arcs of the transition. In our example, the weights of the arcs are not represented in Figure 3 since they are defined with the "identity" function. This means that firing a transition of the net consists in removing one colored token from the transition entry places, and in putting one token of the same color in the transition output places. Events are associated with transitions, so firing a transition is only possible when its associated event occurs.

An important benefit of using a formalism such as Petri nets is that there exists a variety of automated Petri net tools (as stated in [17]) which can be used to verify the nets.

Some of the tools not only allow the analyst to check Petri nets automatically for syntactic correctness, but enable the analysis of the application for properties like :

    a) "liveness": will the application run indefinitely or would it get stuck?.

    b) "boundedness": is the number of places finite or infinite?.

    c) "reachability": are certain desirable/undesirable places reachable/unreachable from an initial marking?.

    d) "T-invariants": does the application have a cyclic behavior?.

    e) "S-invariants": is the number of certain movable parts of the application constant or are there sources and/or drains which may change the number of such parts?.

In our approach, we have used an automated Petri net tool called XPetri tool [18] which deals with all the properties mentioned above.

2.3: Mapping rules

As elaborating Petri nets is not a trivial task, we propose some rules to derive parts of them from the elements contained in the PUCTs. Some parts of the "take gas from a given pump" UCPN can thus be obtained from each "take gas from a given pump" PUCT. In Figure 3, the gray rectangle delimits that part of the "take gas from a given pump" UCPN which is obtained from the elements contained in Table A (Appendix).

The rules we propose below permit the identification of some places, some transitions and some arcs of a net. In order to establish these rules, we have considered the fact that a given distribution of the tokens in some places of the net at a given time expresses a particular state of the application. A particular state of the application is in fact composed of the states of each object type. Our assumptions during the elaboration of the rules are that places correspond to a total or to a partial state of the application and that transitions correspond to events (as it is defined in a synchronized Petri net).

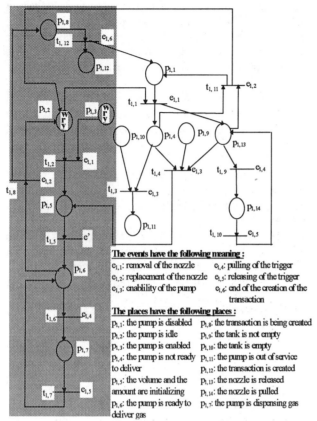

The events have the following meaning :
$e_{1,1}$: removal of the nozzle $e_{1,4}$: pulling of the trigger
$e_{1,2}$: replacement of the nozzle $e_{1,5}$: releasing of the trigger
$e_{1,3}$: enability of the pump $e_{1,6}$: end of the creation of the
         transaction

The places have the following places :
$p_{1,1}$: the pump is disabled $p_{1,8}$: the transaction is being created
$p_{1,2}$: the pump is idle $p_{1,9}$: the tank is not empty
$p_{1,3}$: the pump is enabled $p_{1,10}$: the tank is empty
$p_{1,4}$: the pump is not ready $p_{1,11}$: the pump is out of service
to deliver $p_{1,12}$: the transaction is created
$p_{1,5}$: the volume and the $p_{1,13}$: the nozzle is released
amount are initializing $p_{1,14}$: the nozzle is pulled
$p_{1,6}$: the pump is ready to $p_{1,7}$: the pump is dispensing gas
deliver gas

Figure 3: The "take gas from a given pump" UCPN

The rules which we propose to identify the places, the transitions and the arcs of a net are explained and illustrated below. A more formal description of these rules can be found in [19].

Rule 1. The objective of rule 1 is to identify some places from the states contained in one configuration of a given function. Rule 1 thus applies to each configuration of the functions defined in a PUCT. It consists in grouping together the states contained in one configuration into one or several places by considering the following facts:
- each place must have a label
- the previously identified places have to be taken into account.
- the number of identified places must be minimal.

Definition: a set of states S_k of n object types contained in the configuration of the f_k function $S_k=\{s_{1,k}, s_{2,k}, ..., s_{r,k},..., s_{n,k}\}$ can be grouped together into one place p_j labeled "$s_{r,k}$" if $s_{r,k}$ is a "representative" state of S_k. $s_{r,k}$ (state of the o_r object type during the f_k function) is a representative state of S_k if:
- $s_{r,k}$ is different from $s_{r,k-1}$ (state of the o_r object type during the f_{k-1} function where f_{k-1} is the function which

precedes the f_k function).
- the following expression is verified:

$$\forall s_{j,k} \in S_k, / s_{j,k} \neq s_{r,k} \text{ then } s_{r,k} \Rightarrow s_{j,k}$$

where $s_{r,k} \Rightarrow s_{j,k}$ means the $s_{r,k}$ state implies the $s_{j,k}$ state that is to say that the o_j object type is always in the $s_{j,k}$ state when the o_r object type is in the $s_{r,k}$ state.

Example: the "dispensing gas" state of the "pump" object type, the "pulled" state of the "nozzle" object type, the "not empty" state of the "tank" object type, the "incrementing amount" of the "display" object type, the "on" state of the "motor" object type and the "UD" state of the "transaction" object type which are contained in the configuration of the $f_{1,3}$ function (Table A - Appendix) can be grouped together into place $p_{1,7}$ (labeled "the pump is dispensing gas") since the "dispensing gas" state of the "pump" object type is a representative state of all the other states contained in the configuration of the $f_{1,3}$ function. Indeed:
- the state of the "pump" object type during $f_{1,3}$ is different from the state of the "pump" object type during $f_{1,2}$ ("dispensing gas" $\neq$ " ready to deliver ").
- the "dispensing gas" state of the "pump" object type implies that the nozzle is in the "pulled" state, that the tank is in the "not empty" state, that the display is in the "incrementing amount" state, that the motor is in the "on" state and that the transaction is undefined.

Rule 2. The objective of rule 2 is to identify some places of the net from the assumptions associated with a given function. Rule 2 applies to a given function for which there exists an associated assumption.

Definition: an assumption associated with a f_k function (f_k follows f_{k-1}) represents a new place (with the assumption as label) which has one of the following properties:
- if $f_{k-1} \neq f_{x,begin}$, the identified place will be marked at the same time as the place(s) which have been identified from the application of rule 1 to the configuration of the function which precedes the f_k function.
- if $f_{k-1} = f_{x,begin}$, the identified place will be marked at the same time as the place(s) which have been identified from the application of rule 1 to the configuration of the f_k function.

Example: the "the tank is not empty" assumption, which is associated with the $f_{1,8}$ function (Table B - Appendix), represents a new place ($p_{1,9}$) which will be marked at the same time as the $p_{1,4}$ and the $p_{1,13}$ places which have been identified from the application of rule 1 to the configuration of the $f_{1,7}$ function since $f_{1,7} \neq f_{1,begin}$.

Rule 3. The objective of rule 3 consists in merging some places identified from the application of rule 1 and rule 2 into one place.

Definition: one place p_j (labeled $s_{r,k}$) identified from the application of rule 1 can be merged with one place $p_{j'}$ (labeled $s_{r',k'}$) identified from the application of rule 2 into one place p_m (labeled $s_{r,k}$) if p_j is marked at the same time than $p_{j'}$ and if $s_{r',k'} \Rightarrow s_{r,k}$. The states grouped into the p_j place and the $s_{r',k'}$ state are thus merged into the p_m place.

Example: the application of rule 1 to the first configuration of the $f_{1,1}$ function allows the detection of two places:
- place $p_{1,2}$ (labeled "the pump is idle") groups together the "idle" state of the "pump" object type, the "in" state of the "nozzle" object type and the "displaying an amount" state of the "nozzle" object type.
- place $p_{1,x}$ (labeled "the tank is not empty") groups together the "not empty" state of the "tank" object type, the "on" state of the "motor" object type and the "UD" state of the "transaction" object type.

The application of rule 2 to the $f_{1,1}$ function permits the detection of a place $p_{1,x'}$ (labeled "the pump is enabled") which only contains the "enabled" state of the "pump" object type.

$p_{1,2}$, $p_{1,x}$ and $p_{1,x'}$ are places which are marked at the same time. Moreover, the "enabled" state of the "pump" object type implies the "not empty" state of the "tank" object type so, the states grouped into place $p_{1,x}$ and the state contained in place $p_{1,x'}$ can be merged to be grouped into one place $p_{1,3}$ labeled "the pump is enabled" (Figure 3).

Rule 4. The objective of rule 4 is to identify some transitions and some arcs of the net. The rule connects the places which have been identified from the application of rule 1, rule 2 and rule 3 with transitions and arcs. The event associated with each identified transition has no label since there is no direct correspondence between a UCPN event and the elements contained in the PUCT. The events' labels have to be manually added in the UCPN when a transition between places has been identified.

Definition: if P_{k-1} and P_k are respectively the set of places identified from the application of rule 1, rule 2 and/or rule 3 to the configuration of a f_{k-1} function and the set of places identified from the application of rule 1 to the configuration of a f_k function (f_{k-1} precedes f_k) then there exist a transition, an arc from each place in P_{k-1} to that transition and an arc from that transition to each place in P_k.

Example: $P_{1,4}=\{p_{1,6}\}$ and $P_{1,end} =\{p_{1,2}, p_{1,8}\}$ are respectively the set of places identified from the application of rule 1 to the configuration of the $f_{1,4}$ function and the set of places identified from the application of rule 1 to the configuration of the $f_{1,end}$ function.
Thus, there exist a transition called $t_{1,8}$, an arc from $p_{1,6}$ to $t_{1,8}$ and an arc from $t_{1,8}$ to each place in $P_{1,end}$ (Figure 3).

Rule 5. The objective of rule 5 is to identify some transitions and some arcs of the net. The rule connects the places which have been identified from the application of rule 1, rule 2 and rule 3. For each identified transition, the "e'" event (the "always occurs" event) is systematically associated.

Definition: if P_k and $P_{k'}$ are respectively the set of places identified from the application of rule 1, rule 2 and/or rule 3 to one configuration of a f_k function and the set of places identified from the application of rule 1 to another configuration of the f_k function and if these two configurations are separated with a dotted line in the PUCT then there exist a transition (for which e' is the associated event), an arc from each place in P_k to that transition and an arc from that transition to each place in $P_{k'}$.

Example: $P_{1,8a}=\{p_{1,5}\}$ and $P_{1,8b}=\{p_{1,6}\}$ are respectively the set of places identified from the application of rule 1 to the first configuration of the $f_{1,8}$ function and the set of places identified from the application of rule 1 to the second configuration of the $f_{1,8}$ function. Moreover, the two configurations are separated with a dotted line.
Thus, there exist a transition called $t_{1,5}$ with the e' associated event, an arc from $p_{1,5}$ to $t_{1,5}$ and an arc from $t_{1,5}$ to $p_{1,6}$ (Figure 3).

The non-applicability of rule 1 to a given configuration makes possible the detection of missing object types, non-relevant functions or erroneous states of object types. Indeed, if there does not exist at least one representative state in a configuration of a given function, the analyst has to interpret the reasons rule 1 can not be applied.

The other rules do not make possible to detect incorrect descriptions of use cases since they are always applicable once some preconditions have been verified. Rule 2 is always applicable to a given function once an assumption is associated with that function. Rule 3, rule 4 and rule 5 are always applicable once some sets of places have been first identified from the application of rule 1, rule 2 and rule 3.

2.4: Temporal links between use cases

Definition. The separate description of each use case is insufficient to obtain a unified view of the application: we thus introduce a new type of link between use cases.

The links introduced in our approach are "temporal links", and they express how the use cases are time-dependant. Concurrently with the seven relations which can exist between two activities in [20], we define seven types of links between two use cases uc_i and uc_j (Table 1). The links introduced in our approach are different from those which have previously been defined in the sense that they are not established for the same purpose: indeed, links such as the "extends" and the "uses" links defined in [8] or the similar "adds" link introduced in [13] are composition links between use cases whereas our links are scheduling links between use cases.

uc_i *(B)efore* uc_j	uc_i	██ ▭	uc_j
uc_i *(M)eets* uc_j	uc_i	██ ▭	uc_j
uc_i *(S)tarts* uc_j	uc_i	██ ▭	uc_j
uc_i *(E)nds* uc_j	uc_i	▭ ██	uc_j
uc_i *(EQ)uals* uc_j	uc_i	██	uc_j
uc_i *(D)uring* uc_j	uc_i	▭ ██ ▭	uc_j
uc_i *(O)verlaps* uc_j	uc_i	██ ▭	uc_j

Table 1 : Temporal links between use cases

In the "gas station" application, the temporal links between the "take gas from a given pump" (uc_1), the "collect payment for a given pump" (uc_2) and the "monitor the level of the tank of a given pump" (uc_3) use cases are graphically represented in Figure 4.

Each oriented arc indicates the existence of one or several possible temporal links between two use cases. The oriented arcs are marked B,M,S,E,EQ,D and/or O in accordance with the appropriate temporal links mentioned above (Table 1).

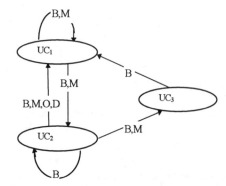

Figure 4: Temporal links between use cases

As an example, let us explain the meaning of the marked arc which is oriented from uc_2 to uc_1:

- the arc is marked B and M since a customer can take gas from a given pump after a certain time or just after the attendant has collected the payment for that pump.

- the arc is marked O since the attendant can start to collect the payment for a given pump when a customer starts to take gas from that pump. Moreover, the attendant can have finished to collect the payment for that pump, whereas the customer has not yet finished taking gas from that pump.

- the arc is marked D since the attendant can start to collect the payment for a given pump whereas a customer has started to take gas from that pump. Moreover, as it has been mentioned above, the attendant can have finished to collect the payment for that pump, whereas the customer has not yet finished taking gas from that pump.

Formal description. A formalization of the links between uc_1, uc_2 and uc_3, called the "gas station" Formal Inter Use Cases Links (FIUCLs), using a colored and a synchronized Petri nets formalism is presented in Figure 5.

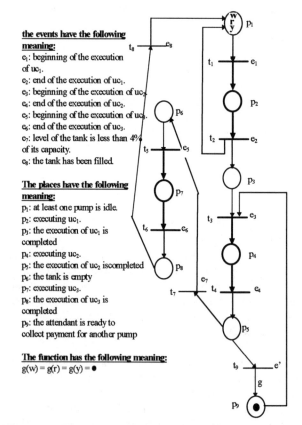

Figure 5: The "gas station" FIUCLs

In comparison with the previous net (Figure 3), some additional notations are used: the black token (i.e. the black dot) can denote any of the y, r or w tokens. An inscription on arcs with a function is represented when the function is different from the "identity" function. For example, the arc between the t_9 transition and the p_9 place carries a function called "g" (Figure 5). Thus, firing the t_9 transition results in removing one colored token from place p_5 and in putting one black token in place p_9. Places, transitions and arcs which are in bold type, refer to one of the three use cases. The other places, transitions and arcs define the links between the use cases.

The main meaning of the net (Figure 5) is the following one:

- a customer is able to take gas if and only if at least one pump is idle. This means that in order to fire the t_1 transition, at least one token must be in place p_1 and e_1 must occur.

- when a customer has taken gas from a given pump, the attendant must collect the payment for that pump. The attendant can collect only one payment at a time. This means that in order to fire the t_3 transition, places p_3 and p_9 must hold one token each and e_3 must occur.

- if there is not enough gas in the tank to serve a given pump, the attendant has to monitor the level of that pump's tank. However, he can decide to collect the payment for some other pumps. This means that in order to fire the t_7 transition, the p_5 place must hold one token and e_7 must occur.

3: The dynamic requirements conceptualization activity

During the dynamic requirements acquisition activity, some elements were gathered by means of tables and by means of Petri nets. The dynamic requirements conceptualization activity consists in mapping the gathered elements to object behavioral concepts (including the state, the event and the transition concepts). During this activity, we propose some rules to help the analyst to produce the state transitions diagrams of each object type. Each state transition diagram of a given object type is called an Object Type State Transition Diagram (OTSTD).

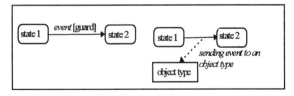

Figure 6: Basic OMT dynamic notation in an OTSTD

The rules we propose automatically build parts of the OTSTDs from parts of the UCPNs and from the static specifications (the relationship types concept). These

rules are listed and graphically illustrated below. In the graphics, the states of some object types grouped together into a given place in a UCPN are represented in brackets within a rectangle. The OTSTDs obtained by application of the rules are expressed by using the OMT [3] notation (Figure 6).

Rule 1. The objective of rule 1 is to identify one transition between two states in an OTSTD.

Definition: if p_i is a place which groups together a set of states including state $s_{i,q}$ (state of the o_q object type when a token is in the p_i place), *if* p_j is a place which groups together a set of states including state $s_{j,q}$ with $s_{i,q} \neq s_{j,q}$ and *if* a transition connects p_i to p_j, *then* a transition between state $s_{i,q}$ and state $s_{j,q}$ in the o_q OTSTD is created.

Example: place $p_{1,6}$ groups together a set of states including the "ready to deliver" state of the "pump" object type. Place $p_{1,7}$ groups together a set of states including the "dispensing gas" state of the "pump" object type. A $t_{1,6}$ transition connects $p_{1,6}$ to $p_{1,7}$.
The preconditions of rule 1 are thus verified. A transition between the "ready to deliver" state and the "dispensing gas" state in the "pump" OTSTD can be created (Figure 7).

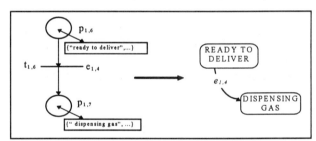

Figure 7: Part of the "pump" OTSTD from rule 1

Rule 2. The objective of rule 2 is to identify some guarded transitions between two states in the OTSTD of a given object type.

Definition: if p_i is a place which groups together a set of states including state $s_{i,q}$, *if* p_j is a place which groups together a set of states including state $s_{j,q}$, *if* p_k is a place which groups together a set of states including state $s_{k,q}$ with $s_{i,q} \neq s_{j,q} \neq s_{k,q}$, *if* a transition with an associated e_p event connects a set of two places $\{p_i, p_x\}$ to p_j and *if* a transition with the associated e_p event connects a set of two places $\{p_i, p_y\}$ to p_k, *then* some guarded transitions, on the one hand between state $s_{i,q}$ and state $s_{j,q}$, and on the other hand, between state $s_{i,q}$ and state $s_{k,q}$, in the o_q OTSTD can be created.

Example: place $p_{1,2}$ groups together a set of states including the "idle" state of the "pump" object type. Place $p_{1,5}$ groups together a set of states including the "initializing" state of the "pump" object type. Place $p_{1,4}$ groups together a set of states including the "not ready to deliver" state of the "pump" object type. Transition $t_{1,2}$ with the $e_{1,1}$ associated event connects the set of places $\{p_{1,2}, p_{1,3}\}$ to $p_{1,5}$. Transition $t_{1,1}$ with the $e_{1,1}$ associated event connects the set of places $\{p_{1,2}, p_{1,1}\}$ to $p_{1,4}$.

The preconditions of rule 2 are thus verified. A guarded transition between the "idle" state and the "initializing" state and a transition between the "idle" state and the "not ready to deliver" state in the "pump" OTSTD are created (Figure 8).

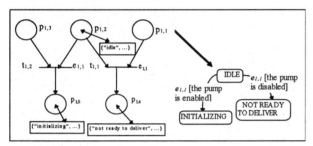

Figure 8: Part of the "pump" OTSTD from rule 2

Rule 3. The objective of rule 3 is to identify some interactions between object types (expressed by means of sending events to object types) within an OTSTD.

Definition: **if** P_a is a set of places which will be marked at the same time, **if** some places in P_a group a set of states including state $s'_{a,q}$ (state of the o_q object type when each place in P_a is marked) and state $s'_{a,r}$, **if** P_b is a set of places which will be marked at the same time, **if** some places in P_b group together a set of states including state $s'_{b,q}$ and state $s'_{b,r}$ with $s'_{a,q} \neq s'_{b,q}$ and $s'_{a,r} \neq s'_{b,r}$, **if** a transition connects the places in P_a to the places in P_b and **if** there exists a relationship type (an association or an aggregation) between the o_q and the o_r object types (indeed, we consider that two object types can interact if there exists a logical link between these two object types), **then** a sending event from o_q to o_r in the o_q OTSTD or a sending event to o_r to o_q in the o_r OTSTD can be created.

Example: place $p_{1,6}$ groups together a set of states including the "not ready to deliver" state of the "pump" object type and the "UD" state of the "transaction" object type. Places $p_{1,8}$ and $p_{1,2}$, which are marked at the same time, group together a set of states including the "idle" state of the "pump" object type and the "being created" state of the "transaction" object type. A transition $t_{1,8}$ connects $p_{1,6}$ to the set of places $\{p_{1,8}, p_{1,2}\}$. There exists an association between the "pump" and the "transaction" object types.

The preconditions of rule 3 are thus verified. A sending event from the "pump" object type to the "transaction" object type in the "pump" OTSTD is created (Figure 9).

Rule 4. The objective of rule 4 is to identify some interactions between object types within an OTSTD.

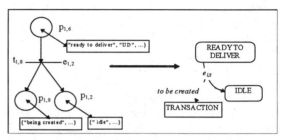

Figure 9: Part of the "pump" OTSTD from rule 3

Definition: **if** P_a is a set of places which will be marked at the same time, **if** some places in P_a group a set of states including state $s'_{a,q}$ and state $s'_{a,r}$, **if** P_b is a set of places which will be marked at the same time, **if** some places in P_b group together a set of states including state $s'_{b,q}$ and state $s'_{b,r}$ with $s'_{a,q} \neq s'_{b,q}$ and $s'_{a,r} = s'_{b,r}$, **if** a transition connects the places in P_a to the places in P_b and **if** there exists a relationship type (an association or an aggregation) between the o_q and the o_r object types, **then** a sending event from o_q to o_r in the o_q OTSTD can be created to ask the o_r object type for its current state.

Example: places $p_{1,4}$ and $p_{1,10}$ group together a set of states including the "not ready to deliver" state of the "pump" object type and the "empty" state of the "tank" object type. Place $p_{1,11}$ groups together a set of states including the "out of service" state of the "pump" object type and the "empty" state of the "tank" object type. Transition $t_{1,3}$ connects the set of places $\{p_{1,4}, p_{1,10}\}$ to $p_{1,11}$. There exists an association between the "pump" and the "tank" object types.

The preconditions of rule 4 are thus verified. A sending event from the "pump" object type to the "tank" object type in the "pump" OTSTD is created (Figure 10) to ask the "tank" object type for its current state.

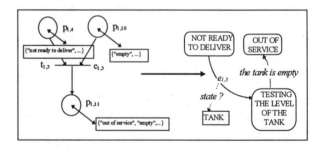

Figure 10: Part of the "pump" OTSTD from rule 1

4: Conclusion

In this paper, we have presented an approach for producing object-oriented requirements specifications. The proposed approach focuses on the dynamic specifications and makes it possible to derive Object Type State Transition Diagrams.

The approach integrates the use case concept. During the requirements acquisition activity, the use cases are collected and described by means of two techniques. The first technique uses tables to facilitate the communication between the analyst and the domain expert. The second technique uses Petri nets and brings the required formalism to the analyst. Some mapping rules have been presented to assist the analyst in building the Petri nets from the tables. Finally, during the requirements conceptualization activity, rules are proposed to produce dynamic object-oriented specifications from the Petri nets.

Current and future research mainly concerns the improvement of the proposed approach:

- during the requirements acquisition activity, we propose to express not only selections and sequential functions in tables as we did in this paper but also iterations and parallel executions of functions. This will enable to deal with more complex applications than the "gas station" application we have presented. Today, more experiments on industrial sized examples have to be carried out to prove our approach efficient.
- our goal is also to improve and to extend the rules we have elaborated during the requirements acquisition and conceptualization activities.

The research work presented in this paper is the highest layer of a CASE tool project called "Conceptor Major" supported by Télis Inc.. This layer has been thought and built in relation to a formal collaboration between Télis and IRIN (Computer Science Research Institute of Nantes). A prototype which supports the static requirements acquisition and conceptualization activities has already been developed in Visual C++ which generates some static specifications using a dialogue-based approach. The dynamic requirements acquisition and conceptualization activities are currently being added to the prototype.

Acknowledgements

The authors would like to acknowledge the support of Professor Julio Cesar Leite in preparing this paper. They also want to thank colleagues Bob Ruediger, Juliana Faur and the anonymous referees for their thoughful remarks.

References

[1] B. Dano, H. Briand and F. Barbier, "A dialogue-based approach for the development of information systems", *Proc. of ADBIS*, Moscow, June 1995.

[2] A.T.F Hutt, *Object analysis and design, Description of methods*, Wiley publications, OMG edition, 1994.

[3] J. Rumbaugh, M. Blaha, W. Premerlani, F. Eddy and W. Lorensen, *Object Oriented Modeling and Design*, Prentice Hall, 1991.

[4] S. Shlaer and S. Mellor, *Object lifecycles, modeling the world with states*, Prentice Hall, 1991.

[5] S. Cook and J. Daniels, *Designing object systems: Object-oriented modelling with Syntropy*, Prentice Hall, 1994.

[6] D. Coleman, P. Jeremaes and C. Dollin, "Fusion: A Systematic Method for Object-Oriented Development", Hewlett Packard Laboratories, Nov. 1992.

[7] I. Jacobson, "Object Oriented Development in an Industrial Environment", *Proc. of OOPSLA*, Oct. 1987.

[8] I. Jacobson, M. Christerson, P. Jonsson and G. Overgaard, *Object-Oriented Software Engineering, A Use Case Driven Approach*, Addison-Wesley, 1992.

[9] K.S Rubin and A. Goldberg, Object Behavior Analysis", *Communications of the ACM*, vol. 35, n°9, Sept. 1992.

[10] P. Gough, F. Fodemski, S. higgins and S. Ray, "Scenarios - an Industrial Case Study and Hypermedia Enhancements", *Proc. of RE*, York, UK, March 1995.

[11] P. Hsia, J. Samuel, J. Gao, D. Kung, Y. Toyoshima and C. Chen, "Formal approach to Scenario Analysis", *IEEE Software*, March 1994.

[12] C. Potts, K. Takahashi and A. I.Anton, "Inquiry-Based Requirements Analysis", *IEEE software*, March 1994.

[13] J. Rumbaugh, "Getting started, Using use cases to capture requirements", *Journal of Object Oriented Programming*, Sept. 1994.

[14] M. Glinz, "An Integrated Formal Model of Scenarios Based on Statecharts", *Proc. of ESEC*, Springer-Verlag, Sept. 1995.

[15] A. Davis, *Software requirements: Objects, functions and states*, Prentice Hall (second edition), 1993.

[16] M. Moalla, J. Pulou and J. Sifakis, "Synchronized Petri nets", *Rairo Automatique journal*, 1978, vol. 12, n°2.

[17] Y. Grude, "Letter to ACM Forum", *Communications of the ACM*, Jan. 1989, vol. 32, n°1.

[18] B. Kahn, R. Noel, S. O'Keefe, *XPETRI*, 1993.

[19] B. Dano, "Dynamic object-oriented requirements specifications: a scenarios-based approach" (in french), *Proc. of INFORSID*, Bordeaux, FRANCE, June 1996.

[20] J. Allen, "Maintaining Knowledge about Temporal Intervals", *Communications of the ACM*, Nov 1983, vol. 26, n°11.

APPENDIX

Table A

FUNCTION Nber	FUNCTION	STATES OF OBJECT TYPES					TRANSACTION	CONDITION	ASSUMPTION
		PUMP	NOZZLE	TANK	PUMP DISPLAY	MOTOR			
$f_{1,begin}$	initiate the execution of the "take gas from a given pump" use case	UD	UD	UD	UD	UD	UD		
$f_{1,1}$	the customer removes the nozzle	idle / idle	in / in	not empty / UK	displaying an amount / displaying an amount	on / off	UD / UD	the pump is enabled / the pump is disabled	the pump is enabled
$f_{1,2}$	the customer pulls the trigger	initializing / ready to deliver	released / released	not empty / not empty	initializing / displaying an amount	on / on	UD / UD		
$f_{1,3}$	the customer releases the trigger	dispensing gas	pulled	not empty	incrementing amount	on	UD		
$f_{1,4}$	the customer replaces the nozzle	ready to deliver	released	not empty	displaying an amount	on	UD		
$f_{1,end}$	terminate the execution of the "take gas from a given pump" use case	idle / idle	in / in	UK / UK	displaying an amount / displaying an amount	on / off	being created / created		

Table A : A "take gas from a given pump" PUCT

Table B

FUNCTION Nber	FUNCTION	STATES OF OBJECT TYPES					TRANSACTION	CONDITION	ASSUMPTION
		PUMP	NOZZLE	TANK	PUMP DISPLAY	MOTOR			
$f_{1,begin}$	initiate the execution of the "take gas from a given pump" use case	UD	UD	UD	UD	UD	UD		
$f_{1,1}$	the customer removes the nozzle	idle	in	UK	displaying an amount	off	UD		the pump is disabled
$f_{1,5}$	the customer pulls the trigger	not ready to deliver	released	UK	displaying an amount	off	UD		
$f_{1,6}$	the customer releases the trigger	not ready to deliver	pulled	UK	displaying an amount	off	UD		
$f_{1,7}$	the attendant enables the pump	not ready to deliver	released	UK	displaying an amount	off	UD		
$f_{1,8}$	the customer pulls the trigger	initializing / ready to deliver / out of service	released / released / released	not empty / not empty / empty	initializing / displaying an amount / displaying "out of service"	on / on / off	UD / UD / UD	the tank is not empty / the tank is empty	the tank is not empty
$f_{1,9}$	the customer releases the trigger	dispensing gas	pulled	not empty	incrementing amount	on	UD		
$f_{1,10}$	the customer replaces the nozzle	ready to deliver	released	not empty	displaying an amount	on	UD		
$f_{1,end}$	terminate the execution of the "take gas from a given pump" use case	idle / idle	in / in	UK / UK	displaying an amount / displaying an amount	on / off	being created / created		

Table B : A "take gas from a given pump" PUCT

A Technique Combination Approach to Requirements Engineering

Alistair Sutcliffe
Centre for HCI Design,
School of Informatics,
City University,
Northampton Square,
London EC1V 0HB, UK
sf328@city.ac.uk
+44-171-477-8411

Abstract

An approach to requirements engineering based on a combination of early prototyping, scenario-based analysis and design rationale is described. Requirements are elicited by presenting users with a prototype-simulation of a prospective design, combined with rationale based techniques for structuring probe questions. Design of analysis sessions and a walkthrough method for requirements elicitation are reported in three layers for linking questions to artefact/scenario demonstrations, follow-up questioning for user-system dependencies and handling user-analyst discourse. An empirical study of the requirements analysis approach is reported. The study used a ship board emergency application. The technique combination approach proved very effective in eliciting requirements but differences in analyst style were an important variable. Recommendations are made for designing and managing requirements capture sessions using scenario-based approaches.

Keywords: requirements analysis, design rationale, scenarios.

1. Introduction

Acquiring and analysing requirements in industrial practice has followed tried and trusted techniques such as interviews, observations, document analysis, etc. [8]. Rapid Application Development RAD/JAD [6] workshops are the current state of the art; however, these offer only a tool box of techniques for user participation and systematic guidance for the requirements engineering process. Requirements engineering researchers have proposed a variety of methods for improving requirements capture such as iterative development and negotiation strategies [2], knowledge acquistion techniques [14], ethnomethodological approaches [11], and reuse of problem templates [15]. Meanwhile HCI researchers have advocated a different range of approaches, e.g. stakeholder analysis methods [12], design rationales [13] and scenario based techniques [5]. However, few studies consider a combination of these approaches.

Requirements analysis poses the problem of developing a mutual understanding of an artefact which does not exist. Prototypes and artefact led elaboration of requirements [4] provide an alternative approach whereby requirements can be clarified by reification. Unfortunately, prototypes incur construction costs and do not guarantee efficient acquisition and validation of requirements as users often uncritically accept what designers offer [1]. Research in explaining complex requirements has demonstrated that a combination of visualisation, examples and simulation is necessary [5, 15]. Scenario based representations and animated simulations help users see the implications of system behaviour and thereby improve validation [7]; Such evidence suggests that a combination of methodical and artefact-based approaches may improve requirements acquisition and validation, and forms the subject matter for this paper.

The approach to be described builds on the Inquiry Cycle of Potts and Anton [16] who propose a method for goal related requirements analysis which uses scenarios of projected system use to discover obstacles. A walkthrough analysis technique is partially articulated explicitly in the Inquiry Cycle [16, 17], although the role of designed artefacts is not clear. We propose an approach of demonstrating a prototype in a scenario context, combined with systematic questioning and exposure of design rationale.

The paper is organised in four sections. First a technique combination approach to requirements engineering is described. This is followed by more detailed description of the analytic methods and process guidance. The next section reports a preliminary evaluation of the proposed approach. The paper concludes with recommendations for future use of requirements analysis techniques and a short discussion.

2. The Technique Combination Approach

The approach is based on the hypothesis that technique integration provides the best avenue for improving requirements engineering. Three techniques are used:

- Use of prototypes or concept demonstrators: a key concept is providing a designed artefact which users can react to.
- Scenarios: the designed artefact is situated in a context of use, thereby helping users relate the design to their work/task context.
- Design rationale: the designers' reasoning is deliberately exposed to the user to encourage user participation in the decision process.

The techniques are combined with a method to provide process guidance for the requirements engineer. The method is composed of advice on setting up sessions, use of the above techniques and more detailed guidance on fact acquisition and requirements validation strategies. The method consists of the following phases:

(i) Initial requirements capture and domain familiarisation. This is conducted by conventional interviewing and fact finding techniques to gain sufficient information to develop a first concept demonstrator. In practice we expect to spend 1-2 client visits on this activity.

(ii) Specification and development of the concept demonstrator. We define a concept demonstrator as a very early prototype with limited functionality and interactivity, so it can only be run as a 'script' to illustrate a typical task undertaken by the user. Scripts illustrate a scenario of typical user actions with effects mimicked by the designer. Typical implementations are in Marcomedia Director, Hypercard, multimedia authoring tools or more ambitiously in Visual Basic. Development time is in the order of 2-4 days for a moderately sized application.

(iii) Requirements analysis-validation session. The users involved in the initial requirements capture are invited to critique the concept demonstrator. The session is recorded for subsequent analysis.

(iv) Session analysis. Data collected during the analysis session is analysed and conclusions reported back to the users. This frequently leads to a further iteration of revising the concept demonstrator and another analysis session.

The end point of the method delivers a requirements specification comprising the concept demonstrator, a set of analysed design rationale diagrams expressing users' preferences for different design options, and specification as text, graphics or more formal notations depending on the requirements engineer's choice. In addition video of the sessions is available for requirements tracability analysis.

2.1 Session design

The physical layout of the analysis session is illustrated in Figure 1. The set up is intended to encourage cooperative requirements capture between two, possibly three, users and two requirements engineers. One acts as a driver of the concept demonstrator and the other fulfils an explainer-rapporteur role. The presence of at least two users helps balance the 'ownership' of the session away from the developers and is productive in producing conversation about the artefact, domain and requirements.

Prior to the session the concept demonstrator is developed and tested. A grounding scenario is developed based on the preliminary domain analysis. This is a short narrative (1/2.. 1 page) describing a situation taken from the users' work context, e.g. "a typical day in the life of"... running through key tasks. This is sent to the user beforehand for comments as are briefing documents for the requirements session.

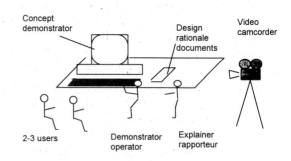

Figure 1. Layout of the requirements analysis session.

The sessions are run according to the schedule:

(i) Introduction and briefing, to put the users at ease, explain the developer roles and emphasise that it is the artefact, and not the users, that is on trial.

(ii) Demonstration and scenario run through. The concept demonstrator is illustrated in a scripted sequence, linked to the grounding scenario. Probe questions are asked at key points in the demonstration script. Design rationale diagrams are used to explain design options at the key points.

(iii) Follow-up phase. The users are encouraged to clarify any points they found ambiguous, go back to any parts of the demonstration, and elaborate further requirements. The requirements engineers may also follow up points raised in the design rationale or user comments during the session. An opportunity for 'hands-on' user interaction with the concept demonstrator is also given.

(iv) Summary phase. The explainer-rapporteur summaries the key facts learned during the session and requests any comments. If the users wish to take copies of the concept demonstrator away with them they are encouraged to do so.

Following the session the video data and audio soundtrack are analysed with notes taken during the session. The depth of analysis depends on resources available. A complete transcription may be undertaken so the users' verbal commentary can be semi-automatically analysed for different requirements and domain facts. Alternatively the video may be 'eye-balled' and facts analysed by playback of key sequences.

2.2. Walkthrough method for requirements analysis

The method provides outline guidance on how to conduct a session as well as heuristics for fact capture and handling the user-analyst discourse.

Operational guidance for analysis sessions. The walkthrough method employs scenario scripts which describe the users work situation and a typical key task. The session is started with an introduction and verbal summary of the situation described in the scenario narrative. One developer operates the concept demonstrator while the explainer-rapporteur role asks questions at key points in the demonstration script. It is important for the developers not to dominate the floor space so users can give their opinions freely. Hence questioning is restricted to a small number of key points in the script. However, when users are shy or not forthcoming the explainer-rapporteur should take the initiative and prompt a user response.

At key points in the sequence a design solution for a requirements is illustrated. This is best explained by reference to the example used in section 4. Figure 2 illustrates a screen dump from a shipboard emergency management system. The user's requirement is for timely

and appropriate information to support their decision making. The operational steps accompanying Figure 2 are:

User: to identify the hazard location
System: shows location of fire
User: sound alarm
User: find location of fire fighting crews
System: displays crew information and location on the diagram.
User: decide appropriate instructions to give to crew
System: displays a checklist of actions.

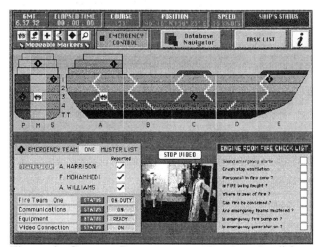

Figure 2. Concept Demonstrator showing the "show emergency teams and hazard location" design option for the Muster Emergency teams task.

The key point in the task is how to instruct the emergency team on where to go and how to deal with the hazard, in this case a fire. The concept demonstrator illustrates one design option. Alternative solutions expressed in design rationale format are illustrated in figure 3. The users' attention is drawn to the design options, in this case providing complete information of decision support. The first option displaying comprehensive information is illustrated with the demonstrator; followed by option 2, provision of more restricted but relevant information for the task, by identifying the team nearest the fire; and then the final option to provide emergency team autonomy and broadcast the loaction of the fire. The users are asked to rate each option and consider the trade-off criteria. The diagram also functions as a recording medium as ranking of options, additional ideas and notes can be scribbled on the diagram. Indeed in many cases discussion may promote re-drawing the diagram.

The design rationale diagram is used as a shared artefact to promote discussion, and features in the concept demonstrator and can focus discussion of design options by pointing to the screen. One obvious problem is bias towards the option implemented in the demonstrator. This can be counteracted by using storyboard sketches of the other options and by more vigorous critiquing by the developers of the implemented version. In particular, use of the criteria, which incidentally capture non- functional requirements, is a powerful way of promoting critical thought. The motivation of using design rationale is to explore the possible solution space with the user in a cost-effective manner. However, should additional resources be available, alternative versions of the concept demonstrator can be implemented and both versions illustrated at the key point.

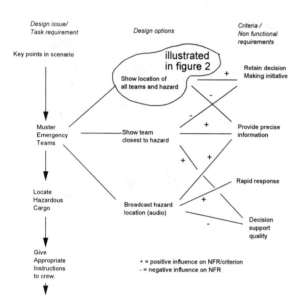

Figure 3. Design rationale diagram used at a key point in the Concept demonstrator script.

The number of key points planned per session depends on the size of the application and number of requirement issues the developers wish to explore. As an approximate guide sessions are advised to last no longer than 60 minutes (e.g. the straight demonstration- script is 20-25 minutes) with 6-7 key points per session.

The next level of the walkthrough method encourages follow-up questioning in response to user comments. Balance of questioning and conversation initiative is ultimately a matter of experience and sensitivity of the requirements analyst.

Questioning strategies. This part of the walkthrough can be regarded as constructing side loops within the main thread of demonstration-led conversation. There are three themes:

(a) Implication exploring questions. The questions amplify concepts drawn from the Inquiry cycle [16] and may be linked into key points in the script. Questions are

prepared beforehand by the requirements engineer and user input is cross checked against the requirements specification (or its reification in the concept demonstrator). If no system function exists to deal with the particular event then a goal must be added to the requirements specification.

For inbound validation, the impact of missing, mis-timed, and inappropriate events is analysed following an Entity Life History approach. Questions encourage the users to explain the range of different event types which might be input, changes in event attributes, likelihood of delayed, missing events, etc. For example: "what happens if the emergency warning is delayed?". When the developer discovers events which the current specification can not deal with, follow up questions are used to ask the user what the system should, or should not, do in this circumstance.

For outbound validation the focus is on the acceptability and impact of system output on users. The grounding scenario used as a basis, and the users are prompted to imagine the effect different contexts (e.g. location) and user roles may have on the acceptability and appropriateness of the system output. Questions focus on whether the information would be appropriate or acceptable to different user groups or if the implication of system decisions will be workable in the real world, e.g. "will detailed instructions be understood by the emergency crews?".

(b) Fact capture discourse. Facts can be elicited by semi-closed questions which contain some information indicating the expected answer, checking statements which are not interrogative but test a user's belief (e.g. ' in normal operation distribute the hazardous cargo in different parts of the ship') and use of gesture to refer to the demonstrator artefact. Fact capture may focus on gathering domain or task related information, or further functional and non-functional requirements. Initiative by the analyst should not dominate the session as users often volunteer facts in response to the concept demonstrator.

(c) Intention capture discourse. This is mainly used in the summarisation phase to elicit user requirements which have been omitted or are outside the scope of the current requirement model. Open questions are asked e.g. are there further information needs?, and the scenario is used to critique the scope of the system, e.g. should the system cover additional areas of operation?

Heuristics are provided to guide the questioning and discourse style in different contexts. For instance closed questions which require a yes/no answer are generally discouraged. Semi-closed questions are recommended for following-up points raised during the demonstration, whereas more open questions are better for intention capture. Exchange structures for user-analyst conversation are suggested as 'tools for thought' for developers to rehearse their questioning skills. Space precludes any detailed treatment, but to briefly illustrate the conversation structuring units with discourse acts:

Key point explanation pattern for the developer:

Inform (design issue), Propose (option 1), Illustrate (option 1 on demonstration), Propose (option 2), Explain (option 2   narrative/storyboard), Justify (option 1 with criterion 1), etc.

Acquire fact from user
Question (semi-closed), User answer, Check (clarify user answer)

The analysis part of the walkthrough method is thus composed of three layers: general operational guidance, question strategies and discourse techniques for handling conversation.

2.3 Session analysis

As explained previously, different levels of analysis can be applied to data captured during the session depending on resources available and the fidelity of the required results.

If the video and audio sound track have been transcribed, text searching software can be used to create word lists with simple filters to sort facts into categories, e.g.

Domain facts: nouns are good markers for domain facts and useful indicators for constructing entity relationship models. In the case study these included "ship, compartment, cargo, hazard, crew", etc.

Functional requirements: verbs serve as reasonable markers, e.g. locate (hazard), evacuate (crew); however, verbs also indicate task actions and events (e.g. send warning), so lists have to be scrutinised with care and keywords examined in the context where they occurred.

Non-functional requirements: suffix searches (e.g. *ity, * ality, *ly, etc.) help locate adverbs and participles which frequently express non- functional criteria such as 'quickly', 'safely', 'security'.

In each case software can only help with pre-processing. Human expertise is required to extract reliable meaning from transcripts.

If transcription is not available the video is replayed and key points are examined, as well as user conversation turns which contain significant facts. The video should record the approximate contents of the screen , hence it is useful to replay the concept demonstrator at the same time so the context of the users' comment can be assessed. Video also captures gesture which can be traced to screen objects, allowing requirements for user interface requirements to be analysed.

3. Validation of the technique combination approach

The application was a shipboard emergency management system. This formed part of the research on the Esprit INTUITIVE project which aimed to develop advanced information retrieval software for multiple databases. The application was intended to help a ship's captain contain and deal with hazards such as fire, collision, spillage of dangerous chemicals, etc. An initial requirements capture exercise was carried out with ships' masters using scenario based techniques [10]. This established that decision support was required with information about the ship's cargo, identification of dangerous cargo, and procedures for dealing with fire and chemical hazards originating from cargo. One fact-capture session lasting 1/2 day was required in which 8 ships captains were present.

3.1. Concept demonstrator design

The concept demonstrator was developed in 4 days using Macromedia Director on an Apple Macintosh. This used a fictitious database of cargo, but real hazard descriptions of dangerous cargo and hazard management procedures. The system used a ship's diagram as the main user interface.   Dialogue boxes displayed an issue checklist for fire management, and pre-set questions.

The demonstrator had limited functionality. A small range of questions could be submitted to the system in three styles:

(a) Menus of pre-formed queries; these related to fire hazards and questions to find the location of dangerous cargo. Question menus were linked to the task list. This was an aide-memoire of fire management procedures designed to help the captain keep track of actions during an emergency.

(b) Concept maps of information items arranged in three hierarchical layers (see Figure 4). The user interacted with the map displays by pointing at a category/item. On higher level maps this had the effect of opening the next level, while pointing to nodes on the lowest level map submitted a query of the type 'find all in this category'.

(c) Query by menu picking; this interface allowed formation of restricted natural language questions. The user was prompted by the start of questions such as 'where is..' and a list of words to expand the question 'slot'. As the user picked a word another list appeared to help complete the question (see Figure 5). The completed query had natural language (English) syntax and semantics. Only a limited number of question paths were scripted so the range of possible questions was restricted.

The prototype was demonstrated by following a script running through the different interfaces. Six key points were planned to test design options for three task stages (analyse hazard, decide emergency procedure, and instruct crews) and three user interface functionalities (preformed queries, query by picking from map displays or menu lists, and quantity and precision of presented information).

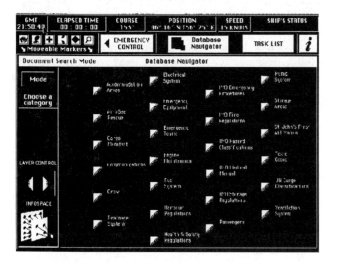

Figure 4. Concept demonstrator showing the Query by Pointing Map design option.

3.2 Session design and analysis

Eight ships captains took part in four requirements sessions, each having two captains as users, one developer acting as the requirements analyst (explainer-rapporteur) and the other running the demonstration. The captains were briefed with the grounding scenario describing a fire emergency and informed that the system would be demonstrated to illustrate two phases:

(i) Emergency phase in which decision support for dealing with the fire was the main concern

(ii) Stabilisation phase after the emergency had been contained, when information about cargo damage and status was required.

Two different developers performed the explainer-rapporteur role, (each took 2 sessions, referred to in the results as group 1 and 2). The demonstrator operator was

the same individual throughout. At key points in the scenario probe questions attempted to elicit the captains' opinions on design alternatives, e.g. requirements for organising emergency teams, selecting the appropriate emergency procedure, monitoring progress of fire fighting, the types of query interface, appropriateness of the retrieved data, effects of intelligent planning, etc. The captains were encouraged to verbalise any opinions and requirements throughout the session. In addition to the practice described in the method section, design rationale diagrams were presented to the captains after the session as a questionnaire to gather their preferences for different design options.

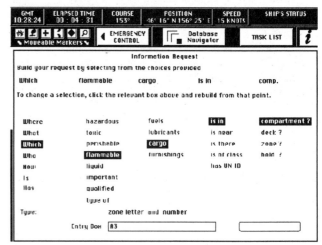

Figure 5 Restricted Natural Language Query interface

The video recordings were transcribed and analysed for different categories of facts captured, as well as looking for correspondences between the types of questions used by the requirements engineers and the quality of users' answers. Transcription of the recordings (total time 130 minutes) took 2 days, while subsequent analysis and categorisation required 3 days excluding report writing. The whole requirements analysis exercise, including session preparation and recording consumed 7.5 person days.

3.3 Results

The results are summarised in this section; for a more complete description of the results and experimental method the reader is referred to Sutcliffe [18].

User reaction. The captains responded favourably to the prototype-scenario exercise. The prototype stimulated considerable discussion between the two captains in each session which proved beneficial in uncovering domain facts and requirements. Exposure of design rationale within the demonstration was also favourably received and promoted good user participation in discussing design options. A few problems were encountered although these were largely solved during the sessions. At first the users had difficulty in thinking about information requirements, and the analyst-rapporteur had to explain the scenario in more detail. After some discussion of requirements the captains lost the thread of where they had got to in the scenario and need to be re-situated. Few non-functional requirements were volunteered in spite of prompts to consider problems in the operating environment. The design rationale questions presented after the session confused the users who did not understand the option/criteria structure even though it was explained. The different analyst styles had little perceptible effect on the users, and both groups had co-operative discussions.

Effect of analyst style. The two requirements engineers had different questioning styles in spite of the common procedure encouraged by the method. The analyst in group one explained the scenario in great depth and asked fewer questions. This individual asked proportionately more closed questions relating to the prototype as she progressed through the scenario, whereas the second analyst gave more verbal description of the design and its rationale. This was followed up by more open questions.

The group two analyst asked more questions overall, consisting of semi-closed questions to validate user information linked to the prototype display. The large number of clarification and open questions were prompted by the captains' observations about the demonstrator. Semi- closed questions also provoked considerable user reaction about procedures and information needs, e.g. 'we don't do it this way'. This led to acquisition of a large amount of useful domain knowledge, followed by many open questions to elicit system requirements in view of these new facts. This suggests that detailed questioning in the presence of a designed artefact is effective in eliciting requirements for a counter-intuitive reason; when users react strongly to what they see as being inadequate, they then say what they do want. The demonstrator acted as a stimulus to explore many new aspects of the domain rather than simply provoking reaction to the design.

User responses and facts gathered. The main differences in answer categories between the two groups was that more critiques of the concept demonstrator were elicited in group 2, whereas in group 1 requirements were stated more generally (see Table 1). More functional requirements were elicited in group one, however the number of questions in this group was small, so users appeared to have been prompted by the prototype to say what they wanted. In group 2, more domain details and design critique responses were elicited. Non-functional

requirements were infrequent in both groups, possibly because the presence of the demonstrator focused conversation towards functional issues. In group 1, 35% of the requirements were either general or ambiguous while only 16% were so in group 2. Overall the percentage of specific statements was similar for both requirements and design critiques (70-72%). In terms of obtaining specific requirements the more active questioning style of analyst 2 seemed to have paid off (significant difference group 2>1 $p=0.05$ chi square test), although this gives no guarantee that the right requirements were captured.

	Group 1	Group 2	Total
Clarification + Check	12	22	34
Design Critique	5	68	73
Function Requirement	23	13	36
Non-functional Reqs	4	0	4
Domain Detail	42	51	93
Total	86	154	240

Table 1 Summary of User Responses by Category

Topic	Group 1	Group 2
First Aid	1	12
Safety Kit	0	34
Haz Cargo	4	35
Training	0	24
Communication	1	5
Ship's plan	9	23
Interface	28	11
Checklist	0	6
Navigation	0	4
Procedures	33	0
Emergency	4	0
Cargo	4	0
Totals	82	154

Table 2 Analysis of user requirements topics

When requirements were categorised according to the topic, it transpired that most topics related to either user interface design (47%) or information requirements

(39%). A small number of topics concerned support for the users' task in general terms (9%), and non-functional requirements were rare. The main non-functional requirements concerned the system environment e.g. noise, access to the equipment, etc. The preponderance of information requirements is not surprising given the decision support type of application, e.g. information on cargo hazards, location of the fire, crew muster points, fire appliances, fire proof doors, etc. The large number of user interface requirements can be attributed to the presence of the concept demonstrator. These varied from general comments about the need for simple displays, to more detailed suggestions for format of menus, types of preformed queries, and layout of information displays.

Reaction to Design rationale and probe questions. Overall the users' reactions to the design rationale questions at key points agreed with the illustrated design options, with the exception of the query reformulation advice, where the wording and clarity of directions may have affected the users' reaction, and the users commented that browsing strategy selection depended on assumptions about retrieval volumes being correct. The rationale prompted much discussion which led to the acquisition of further facts and requirements.

The walkthrough technique for scenario validation, i.e. investigation of the various permutations of input/output event combinations and their implications, was only partially successful. The demonstrator acted as a powerful focus. Questions relating to input events or output displays that were visible at the demonstration key point were successful in validating and elaborating requirements, but attempts to elicit reaction to non-illustrated events was not so well received. The users wanted to return to the script. This suggests that when scenarios are to be used to image new usage contexts, they should be separated from scenario-demonstrations of typical user tasks.

To counteract the bias in presenting one design solution we also used design rationale diagrams after the session as a type of questionnaire on user preferences. This exercise was only a qualified success. The format of QOC style questionnaires was explained by the developers but this way of acquiring design preferences appeared un-natural to the captains. They also interpreted the scoring scales differently, as some allowed ties (e.g. 3=) on the scale, while others didn't. The simple lesson was to make instructions clearer.

Another problem was the dual phase nature of the scenario. In the first part of the scenario, (emergency management) time was important while in the second part it was not. The captains were encouraged to give their preferences in the context of the second scenario phase but they still confused the two phases. The double

scenario created problems throughout the sessions when the analysts had to explain different parts of the script and why questions were pertinent to different contexts.

5. Conclusions and discussion

This investigation has shown that technique combination for requirements analysis can be surprisingly effective, especially when combined with a detailed questioning technique. However, the inter-analyst differences were worrying. This suggests that large performance differences may be present in spite of methodical guidance. Further training might counteract these differences in performance. When the study was carried out time constraints allowed only minimal training in the walkthrough method. In particular, no training was given in discourse level techniques. In spite of this a large number of requirements and facts were captured in both groups, and user participation was successful throughout.

Some tentative guidelines may be suggested from our experience although these need testing with further examples and different applications:

- A combination of techniques is helpful and use of an artefact is advisable especially if creating a concept demonstrator is relatively inexpensive.
- Running through a demonstration scenario helps focus users' attention. A flexible approach is advised so when users react to the inadequate design, use can be made of this opportunity for capturing further domain facts and requirements. Don't be negative and defend the design.
- Set clear objectives for the requirements capture session and plan scenarios accordingly. Do not try to do too much in any one scenario otherwise users will become confused. Use one scenario per session; we encountered confusion with the two phase scenario.
- Explain the design reasoning behind the features being illustrated in the demonstration. Explaining design alternatives and rationale involves users in decision making. Keep explanations simple and present them in terms of system help for the user's task.
- Do not overload sessions with too many techniques. In particular, when conducting scenario-based validation to check system output event permutations which are not illustrated are better handed separately.
- Use a combination of open ended questions for eliciting users' design suggestions with semi-closed questions to get feedback on design features.
- Train requirements engineers in the method and questioning techniques to achieve consistent results.

Even though we expected that scenario-driven requirements acquisition would be useful for refining designs we had not anticipated eliciting extensive domain detail and further requirements. One explanation may be that the artefact acts as a focus, so if the design is wrong, it provokes users into correcting the designer's perceptions. This agrees with the increasing interest in anchoring elicitation techniques in scenario based examples [5,16]. One problem is the poor record on non-functional requirements. This may be due to the lack of a real context for the concept demonstration i.e. the bridge of a ship at sea. Taking the concept demonstrator and this approach into the field may improve discussion of performance and environmental issues.

Overall the technique combination approach proved to be an effective means for requirements capture. Furthermore, we believe, it increased user involvement in decision making and thereby became a tool for participative design. However, prior training in design rationale notation is advisable. Some adverse reaction to the complexity of rationales in diagram form has been found by other researchers [3].

Early exposure of users to design alternatives is important, as we believe that requirements definition needs to evolve from problem analysis and possible design solutions. Design rationales in combination with scenario based artefact evaluation are one way towards exposing the links between problem and solutions in a comprehensible manner. Objective testing that the technique combination approach is superior to current JAD/RAD techniques (e.g. DSDM [7]) will have to wait for comparative studies, which are costly and difficult to carry out in RE. However, the approach does offer more detailed guidance and this study demonstrates some evidence of its effectiveness. Effectiveness of novel RE techniques is best demonstrated by successful transfer of the ideas to industry, and this is the direction of our future research.

Acknowledgements

This work was carried out within the Esprit NATURE basic research and INTUITIVE RTD projects which were partially funded by the European Union's Esprit programme. It is a pleasure to acknowledge the help of Lloyds Register, UK; and especially John Hobday and Clive Bright who made access to real users possible and Mike Gilbert who developed the concept demonstrator.

References

[1] Attwood M.E., Burns B., Girgensohn A., Lee, A., Turner, T. and Zimmerman B., 1995, Prototyping considered dangerous. In Proceedings of Human Computer Interaction-INTERACT '95, Eds Nordby K., Helmersen P.H., Gilmore D.J. and Arnesen S.A., pp 179-184, IFIP/Chapman and Hall, London.

[2] Boehm B., Bose P., Horowitz E. & Lee M-J., 1994, 'Software Requirements as Negotiated Win Conditions', Proceedings of IEEE Conference on Requirements Engineering, IEEE Computer Society Press, 74-83.

[3] Buckingham-Shum S., 1995, Analysing the usability of a design rationale. In Design rationale: Concepts, Techniques and Use, Eds Moran T.P , Carroll J.M., Lawrence Erlbaum assoc, Hillsdale, NJ.

[4] Carroll, J. M., Kellogg, W. A. and Rosson, M. B.,1991, The task-artefact cycle, in J. M. Carroll (ed.), Designing interaction: psychology at the human-computer interface. Cambridge University Press. 74-102.

[5] Carroll J.M., Alpert S.R., Karat J., Van Deusen M. and Rosson M.B., 1994, Raison d-etre: capturing Design History and Rationale in Multimedia Narratives. In Proceedings of CHI-94, Human factors in Computing Systems, Eds Adelson B., Dumais S. and Olson J, pp 192-197, ACM press.

[6] DSDM-Consortium, 1995 Dynamic systems development method. Tesseract Publishers, Farnham, Surrey

[7] G. Fischer, K. Nakakoji, J. Otswald, G. Stahl & T. Sumner, 1993, 'Embedding Based Critics in the Contexts of Design', Proceedings of INTERCHI'93, ed. S. Ashlund, Computer-K. Mullet, A. Henderson, E. Hollnagel & T. White, ACM Press, 157-163.

[8] Gause D, Weinberg G., 1989, Exploring requirements. Dorset House, New York.

[9] Harker S.D.P., Eason K.D. & Dobson J.E., 1993, 'The Change and Evolution of Requirements as a Challenge to the Practice of Software Engineering', IEEE Symposium on Requirements Engineering, RE'93, San Diego, CA, Jan. 4-6, 1993, 1993, 266-272.

[10] Hobday J.S., Rhoden D., Bright C.K. and Earthy J.V., 1994, Aiding the control of emergencies on ships. IN Proceedings of IMAS-94, "Fire Safety on Ships: Development into the 21st century, pp 23-27

[11] Hughes J, O'Brien J, Rhodden T, Rouncefield M, Sommerville I., 1995, Presenting ethnography in the requirements process. In: Zave P, Harrison MD, (ed.) Proceedings of RE'95, Second International Symposium on Requirements Engineering. IEEE Computer Society Press, . pp 27-34

[12] Macaulay L., 1993, 'Requirements Capture as a Cooperative Activity', Proceedings of IEEE Symposium on Requirements Engineering, IEEE Computer Society Press 174-181.

[13] Maclean A., Young R.M., Bellotti V. & Moran T., 1991, 'Questions, Options, and Criteria: Elements of Design Space Analysis', Human-Computer Interaction 6(3&4), 201-250.

[14] Maiden N.A.M. & Rugg G., 1994, 'Knowledge Acquisition Techniques for Requirements Engineering', Proceedings

Workshop on Requirements Elicitation for System Specification, Keele UK, 12-14 July.

[15] Maiden N.A.M. & Sutcliffe A.G., 1994, 'Requirements Critiquing Using Domain Abstractions', Proceedings of IEEE Conference on Requirements Engineering, IEEE Computer Society Press, 184-193.

[16] Potts C, Takahashi K, Anton A., 1994, Inquiry based requirements analysis. IEEE Software 1994; (March): 21-32

[17] Potts C, Takahashi K, Smith J, Ora K., 1995, An evaluation of inquiry based requirements analysis for an Internet service. In: Zave P, Harrison MD, (ed.) Proceedings of RE '95: Second International Symposium on Requirements Engineering. IEEE Computer Society Press, pp 27-34

[18] Sutcliffe AG. , 1995, Requirements rationales: integrating approaches to requirements analysis. In: Olson GM, Schuon S, (ed.) Proceedings of Designing Interactive Systems, DIS '95. ACM Press pp 33-42

Session 3B

Minitutorial

Speaker

Daniel Jackson
Carnegie Mellon University, USA

"Model Checking and Requirements"

Session 4A

Inconsistencies and Exceptions

Analysing Inconsistent Specifications
Anthony Hunter and Bashar Nuseibeh

As more and more of an enterprise's information processing becomes computerised, contemporary information systems must be built to reflect the needs of multiple users and groups with different skills, motives, values, beliefs, world-views, and so on. Hence, requirements engineering methods and languages should be provided which capture and reason about these different world-views explicitly. The emerging areaof "viewpoints" in requirements engineering addresses exactly this problem. However, an important problem with viewpoints is the possibility of inconsistencies. The paper is important because it is part of a promising line of research which addresses inconsistent specifications. It is also general enough to be potentially useful for several future lines of viewpoints' research.
— *Andreas Opdahl*

A Systematic Tradeoff Analysis for Conflicting Imprecise Requirements
John Yen and Amos W. Tiao

This work falls in an area I like — the study of requirements as the dirty little things they are. In particular, non-toy projects turn up requirements that share few of the virtues we have established: consistency, completeness, non-ambiguity. This paper overturns yet another one of our cherished virtues: requirements are precise. Of course, they often are anything but. No where is this brought out more clearly as with conflicting requirements. In real projects, the fun begins when conflicts are detected. It is then that the precisness and value of the conflicting requirements is argued over heatedly by the respective camps of the project stakeholders. However, without support of a reasoning model, the conflict resolution process canbe ad hoc and prone to intuition and error. The authors are working to remedy this: they take established work from Decision Science and from Fuzzy Logic, and combine the two in a formal model. In the end, they convinced me that requirements preciseness is not a virtue but a vice — precise requirements are typically a pipedream and furthermore, they allow no room for maneuver. What we need is a representation of imprecise requirements and a model to reason about them. As I said, I like this — turning vices into virtues is a way of life when dealing with real world requirements.
— *Stephen Fickas*

Analysing Inconsistent Specifications

Anthony Hunter   Bashar Nuseibeh

Department of Computing, Imperial College
180 Queen's Gate, London SW7 2BZ, UK
Email: {abh, ban}@doc.ic.ac.uk

Abstract

In previous work we advocated continued development of specifications in the presence of inconsistency. To support this we presented quasi-classical (QC) logic for reasoning with inconsistent specifications. The logic allows the derivation of non-trivial classical inferences from inconsistent information. In this paper we present a development called labelled QC logic, and some associated analysis tools, that allows the tracking and diagnosis of inconsistent information. The results of analysis are then used to guide further development in the presence of inconsistency. We illustrate the logic and our tools by specifying and analysing parts of the London Ambulance Service. We argue that the scalability of our approach is made possible by deploying the ViewPoints framework for multi-perspective development, such that our analysis tools are only used on partial specifications of a manageable size.

1. Motivation and Background

Inconsistent specifications are an inevitable intermediate product of a requirements engineering process. Inconsistencies may arise because of the preliminary nature of elicited requirements, or indeed because of inherently conflicting customer needs (for example, because of multiple, conflicting views that these customers hold on a problem or solution domain).

A desirable product of a requirements engineering process is a formal specification which captures customer requirements. The formality of such a specification is desirable because it is amenable to formal reasoning and analysis which, in turn, also facilitate the validation of customer requirements. The process of translating informal (often vague and inconsistent) requirements statements into a precise formal (consistent) specification is a difficult one. We believe that tools which enable reasoning and analysis of inconsistent, but formal, specifications can help improve such a translation process.

In this paper, we present some formal, light-weight, logic-based tools that can provide a requirements engineer with a handle on inconsistencies in specifications. The aim of these tools is to provide additional "non-intrusive" reasoning in the presence of inconsistency and simple analysis of inconsistent information. Such reasoning and analysis can in turn provide *guidance* to the requirements engineer on what course of *action* to take in the presence of certain inconsistencies (we still believe, however, that

such action is ultimately a human-driven process). The "non-intrusive" operation of such tools is necessary because in many instances, the logic-based tools are computationally intensive and would otherwise render automated tool support unusable.

The work presented here complements our previous work on eliciting requirements from multiple perspectives using "ViewPoints" [11, 12, 25], and develops our approach of inconsistency handling in this setting [16]. Our motivation is that inconsistencies are inevitable in software development (and requirements engineering) processes and products. They provide a focus for further development (e.g., requirements elicitation), and can be regarded as "desirable" in that they highlight issues that need further attention. As such, they should be tolerated, analysed and acted upon - in other words, systematically *managed* [23].

The focus of this paper is a logic-based approach to managing inconsistent specifications. In particular, we focus on an adaptation of classical logic (termed quasi-classical, or QC, logic) that allows limited reasoning in the presence of inconsistency (§3), and extend it in simple ways that facilitate the analysis of inconsistent specifications (§4). We then develop an example to illustrate our logical tools (§5) based on the IWSSD-8 case study of the "London Ambulance Service" [15], and discuss the impact of the kind of analysis we advocate on our goal of inconsistency handling and management. We conclude with a short discussion on the role of automated tool support, and related and future work (§6 and §7). A more detailed discussion is available in [20].

2. Requirements: From fuzzy to formal

The requirements of many large software systems are characterised by imprecision. Customers often under- or over-specify their requirements, and requirements statements are often contradictory. However, in order to elicit customer requirements effectively it is essential that the needs of *all* stakeholders are captured. To this end we have used the ViewPoints framework for multi-perspective development to explicitly represent different stakeholder requirements [25].

In moving towards a precise specification that we can validate and then satisfy, there also is the need for some formal reasoning and analysis. We have found classical logic to be an appealing form of formal representation because it allows the capture of a wide range of development information, and has an existing body of tools and technology for analysis and reasoning; e.g., [1], [6]. Unfortunately, a large body of requirements information, elicited during the early part of the requirements engineering cycle, is inconsistent, and therefore there is a need to reason with inconsistent information. Classical logic, however, is *trivialised* in the presence of inconsistency; that is, by the definition of the logic, any inference follows from inconsistent information (*ex falso quodlibet*). Formally,

$$\{ \alpha, \neg \alpha \} \vdash \beta$$

To address this problem, we developed a quasi-classical (QC) logic that allows non-trivial reasoning in the presence of inconsistency.

3. QC Logic: Reasoning in the presence of inconsistency

The full formal definition of QC Logic may be found in [4]. Here we provide an informal presentation of the logic and illustrate its reasoning capabilities with a simple example[1].

The proof theory of QC logic is based on reasoning with formulae that are in conjunctive normal form (CNF). These are formulae of the following form:

$$\alpha_1 \wedge ... \wedge \alpha_n$$

where each α_i is of the form:

$$\beta_1 \vee ... \vee \beta_m$$

and each β_j is a literal.

The proof theory of QC logic (see appendix) provides the power to derive a CNF of any formula, together with the power of resolution:

$$\frac{\neg \alpha \vee \beta \quad \alpha \vee \gamma}{\beta \vee \gamma}$$

Only as a last step in any derivation is *disjunction introduction* allowed. This means that any resolvant of a set of formulae can be derived, but no trivial formulae can be derived. The proof theory is presented as a set of natural deduction rules such as:

$$\frac{\alpha}{\alpha \vee \gamma} \quad \textit{[disjunction introduction]}$$

All the QC natural deduction rules hold in classical logic, but the logic is *weaker* than classical logic in the way it is *used*. QC logic is used by providing any set of classical formulae as assumptions, and any classical formula as a query. The query follows from the assumptions if and only if there is a derivation of a CNF

of the query from the assumptions using the QC natural deduction rules.

To illustrate the use of QC in the context of the ViewPoints framework, consider the following simple example. Suppose we have two partial specifications VP1 and VP2 (representing two stakeholder ViewPoints). In VP1, there is the association:

has-exactly-one(Ambulance, Operator)

and in VP2 there is the association:

has-exactly-two(Ambulance, Operator)

If we also have the constraint:

$\forall X,Y,$ has-exactly-one(X,Y) $\leftrightarrow \neg$has-exactly-two(X,Y)

then this constraint together with VP1 and VP2 are inconsistent. However, if we also use the following additional "domain knowledge", discovered perhaps during development, or the result of agreements between developers:

$\forall X, Y,$ has-exactly-one(X,Y) $\rightarrow$ has-one-or-more(X,Y)

$\forall X, Y,$ has-exactly-two(X,Y) $\rightarrow$ has-one-or-more(X,Y)

then for both VP1 and VP2, and despite the inconsistency between them, we can still derive the potentially useful non-trivial inference:

has-one-or-more(Ambulance, Operator).

Reasoning such as in the above example is potentially useful for a range of activities in the management of inconsistencies. It allows us to go beyond *having* to remove the inconsistencies from our specifications, and then it allows us to perform the kind of analysis of inconsistent information described in the next section.

4. Logical Analysis of Inconsistent Specifications

Our analysis aims at facilitating the *tracking* and *diagnosis* of inconsistencies in order to handle the consequences of tolerating (and possibly propagating) such inconsistencies in our specifications. We present two kinds of analysis both of which are independent of the QC language itself, but which obviously make more sense in the context of inconsistent specifications. *Qualification* of inferences provides us with some intuition about our confidence in particular pieces of information in our specification, while *labelling* of specifications provides us with the infrastructure for tracking inconsistencies and identifying their likely sources.

4.1. Identifying likely sources of inconsistency

After identifying an inconsistency in a specification, our analysis attempts to indicate the likely source of that inconsistency before we decide on a further course of action. Using *labelled* QC reasoning, we obtain the labels

---

1   At this point in our work we assume a first order language without function symbols and existential quantifiers. This gives us certain computational advantages such as rendering consistency checking decidable (because, effectively, we are working with a propositional language).

of the assumptions used to derive an inconsistency. We use the term 'source' to denote the subset of the assumptions that we believe to be incorrect.

We use a labelled language to allow us to uniquely identify each item of our specification. We propagate the labels by labelling consequences with the union of the labels of the premises. This means we can identify the ramifications of each item in the reasoning, since each inference will be labelled. Labels can be used to differentiate different types of development information (e.g., constraints, domain knowledge, etc.) and in particular they can indicate the sources of information. For this paper, we adopt the following labelling strategy[2].

Definition 1. Let S be some set of atomic symbols such as an alphabet, and L a logic such as QC. If $i \subseteq S$ and $\alpha \in L$, then $i: \alpha$ is a labelled formula.

So for example, the labelled form of the resolution proof rule is:

$$\frac{i: \neg \alpha \vee \beta \qquad j: \alpha \vee \gamma}{i \cup j: \beta \vee \gamma}$$

To illustrate the use of labels, suppose we have the specification:

{a}: patient-waiting(London-Road)

{b}: ¬patient-waiting(London-Road) ∨
    ambulance-available(Hospital)

{c}: ambulance-available(Hospital)

and suppose {a}: patient-waiting(London-Road) and {b}: ¬patient-waiting(London-Road) ∨ ambulance-available(Hospital) have been a stable and well-accepted part of the specification for some time, and by contrast {c}: ambulance-available(Hospital) is just a new and tentative piece of specification. Then for the inconsistency {a, b, c}: ⊥, we could regard {c}: ambulance-available(Hospital) as the 'source' of the inconsistency.[3]

Informally then, a 'possible source' of an inconsistency is a subset of the assumptions used to obtain the inconsistency, and the remainder of the assumptions are consistent. Such a subset may be obtained by formally defining possible sources of inconsistency as follows.

Definition 2. Let Δ be a set of labelled formulae representing specification information, and let i be some label of some inference from Δ. The set of assumptions from Δ corresponding to the label is defined as follows:

$$\text{Formulae}(\Delta, i) = \{ j: \alpha \in \Delta \mid j \subseteq i \}$$

For an inconsistency $i: \bot$, $\text{Formulae}(\Delta, i)$ is a possible source of the inconsistency if $j \subseteq i$ and $\text{Formulae}(\Delta, i - j) \in CON(\Delta)$, where $CON(\Delta)$ is the set of consistent subsets of Δ (formally defined in §4.2).

There maybe a large number of possible sources of a single inconsistency, and a number of options for addressing it, such as working only with the smallest sources, or working with only the sources that have the least effect on the number of inferences from the specification. However, if we are to act on inconsistency effectively, then we really need to identify the 'likely' sources of inconsistency. To do this, we assume that for any development information, there is some ordering over that information, where the ordering captures the likelihood of the information being erroneous. So, if i is higher in the ordering than j, then i: α is less likely to be erroneous than j: β. We assume this ordering is transitive, though not necessarily linear.

Moreover, if we assume there is an ordering over assumptions, then a more likely source is the smallest possible source that contains less preferred assumptions. Assuming such an ordering over development information is reasonable in software engineering. First, different kinds of information have different likelihoods of being incorrect. For example, "method rules" are unlikely to be incorrect, whereas some tentative specification information is quite possibly incorrect. Second, if a specification method is used interactively, a user can be asked to order pieces of specification according to likelihood of correctness.

There are a number of ways that this approach can be developed. First, there are further intuitive ways of deriving orderings over formulae and sets of formulae. These include ordering sets of formulae according to their relative degree of contradiction [17]. Second, there are a number of analyses of ways of handling ordered formulae and sets of ordered formulae. These include the use of specificity [26], ordered theory presentations [27], and prioritised syntax-based entailment [3].

4.2. Qualifying inferences from inconsistent information

When considering inconsistent information, we have more confidence in some inferences over others. For example, we may have more confidence in an inference α from a consistent subset of the specification if we cannot also derive $\neg \alpha$ from another consistent subset of the specification. Therefore, we now provide formal definitions of some tools for qualifying inconsistent information. We follow these with an informal summary and an example. We begin with the definitions of some useful subsets of our specification.

Definition 3. Let Δ be a set of labelled formulae representing specification information. We form the following sets of sets of formulae:

---

2   There are many strategies that we could adopt for labelling software development information. Options include combinations of the source of the item, and time the item was inserted. For this, some mapping from labels to their associated meaning needs to be recorded. For instance, different developers could use different disjoint subsets of the labels.

3   ⊥ denotes an inconsistency.

$$CON(\Delta) = \{ \ \Gamma \subseteq \Delta \mid \Gamma \nvdash_Q i : \perp \ \}$$

$$INC(\Delta) = \{ \ \Gamma \subseteq \Delta \mid \Gamma \vdash_Q i : \perp \ \}$$

Essentially, $CON(\Delta)$ is the set of consistent subsets of Δ, and $INC(\Delta)$ is the set of inconsistent subsets of Δ.[4]

We can now define $MI(\Delta)$ as a set of sets of labels, where each set of labels corresponds to a set of minimally inconsistent formulae. A set of formulae is *minimally inconsistent* if every proper subset is consistent. Similarly, we define $MC(\Delta)$ as a set of sets of labels, where each set of labels corresponds to a set of maximally consistent formulae. A set of formulae is *maximally consistent*, if the set is consistent and adding any further formulae to the set from Δ causes the set to be inconsistent.[5]

We can consider a maximally consistent subset of a specification as capturing a "plausible" or "coherent" view on the specification. Furthermore, we consider $FREE(\Delta)$, which is equal to $\bigcap MC(\Delta)$, as capturing all the "uncontroversial" information in Δ. In contrast, we consider the set $MI(\Delta)$ as capturing all the "problematic" data Δ. Note that $MC(\Delta)$ is equal to $Labels(\Delta) - MI(\Delta)$. Thus, reasoning with $FREE(\Delta)$ is equivalent to revising the specification by removing all the "problematic" data. This means we have a choice. We can either reason with the data directly using $FREE(\Delta)$ or we can revise the data by removing the formulae corresponding to $MI(\Delta)$.

We can now use these concepts to define three qualifications for an inference from inconsistent information[6].

Definition 4. Let Δ be a set of labelled formulae representing specification information, and let $\Delta \vdash_Q i : \alpha$ hold. We form the following qualifications for inferences:

α is an **existential inference** if $\exists k \in MC(\Delta)$ such that $i \subseteq k$.

α is a **universal inference** if $\forall k \in MC(\Delta)$ $\exists j$ such that $j \subseteq k$ and $\Delta \vdash_Q j : \alpha$ holds.

α is a **free inference** if $i \subseteq FREE(\Delta)$.

Informally, a formula is an existential inference if it is an inference from a consistent subset of the specification. A formula is a universal inference if it is an inference from each maximally consistent subset of the

---

4   $\vdash_Q$ is the consequence relation for QC logic.

5   Let Δ be a set of labelled formulae representing specification information. If we define the function *Labels* as follows:

     $Labels(\Delta) = \{ \ i \mid i : \alpha \in \Delta \ \}$

     then the following set of labels can be formed:

     $MI(\Delta) = \{ Labels(\Gamma) \mid \Gamma \in INC(\Delta) \ and \ \forall \Phi \in INC(\Delta) \ \Phi \not\subset \Gamma \}$

     $MC(\Delta) = \{ Labels(\Gamma) \mid \Gamma \in CON(\Delta) \ and \ \forall \Phi \in CON(\Delta) \ \Gamma \not\subset \Phi \}$

     $FREE(\Delta) = \bigcap MC(\Delta)$

6   The approach is a derivative of argumentative logics [13].

---

specification. Finally, a formula is a free inference if it is an inference from the intersection of the maximally consistent subsets of the specification.

If α is a free inference, it is also a universal inference. Similarly, if α is a universal inference, it is also an existential inference. Clearly, if α is only an existential inference, then we are far less confident in it than if it was a universal inference. If it is a free inference, then it is not associated with any inconsistent information.

Example. Consider the following assumptions:

     {a}: accident-occurred(London-Road) ∧
           accident-reported(London-Road)

     {b}: accident-occurred(London-Road) ∧
           ¬accident-reported(London-Road)

     {c}: ambulance-available(London-Road)

This gives two maximally consistent subsets:

Set 1

     {a}: accident-occurred(London-Road) ∧
           accident-reported(London-Road)

     {c}: ambulance-available(London-Road)

Set 2

     {b}: accident-occurred(London-Road) ∧
           ¬accident-reported(London-Road)

     {c}: ambulance-available(London-Road)

From this, accident-reported(London-Road) and ¬accident-reported(London-Road) are only existential inferences, whereas ambulance-available(London-Road) is a free inference, and accident-occurred(London-Road) is a universal inference.

These kinds of qualification are useful when reasoning with inconsistent information because they provide a clear and unambiguous relationship between the inferences and problematic data. This could be useful in facilitating further development in the presence of inconsistency, since we would feel happier about relying on the less qualified inferences. Furthermore, they provide a useful vocabulary for participants in the development process to discuss the inconsistent information.

Whilst there is an overlap for existential inferencing with the approach of truth maintenance systems (for example [10], [8]), we go beyond this by adopting universal and free inferencing. Furthermore, by adopting labelling, we integrate our inconsistency management with QC reasoning and with identifying likely sources of inconsistency.

4.3.   Remarks on utility of analyses

Recall that a primary objective of our analysis of inconsistent information is to provide the developer with guidance on how to act in the presence of inconsistency. Reasoning in the presence of inconsistency that QC logic facilitates was the first step in achieving this objective. The simple analyses we have described above provide the next step. They point to likely sources of inconsistency - to which more attention could be devoted, and they give us some indication of the "quality" of our information - which again guides our actions.

The simplicity of the tools we have presented is essential for usability and construction of tool support. Nevertheless, there remains an issue of scalability which we regard as a major concern. While we are exploring the *limits* on scalability by engaging in a number of large case studies, we are not necessarily trying to prove that our techniques scale up to industrial size specifications. In our ViewPoints framework, we already have an approach for partitioning specifications into more "manageable units", which are more amenable to the kinds of analysis we propose. ViewPoints encapsulate partial specifications that can be deliberately chosen to be of a size that can be handled by our techniques. Moreover, the ViewPoints framework itself and its support tools have been constructed with the intention of tolerating inconsistency [12, 24]. Thus, the tools we have presented can be added to our framework without hindering the ViewPoint-oriented development process. So for example, while a requirements engineer is developing his/her ViewPoint specification, our tools can be happily churning out (potentially) useful inferences and analysis results.

5. Application Example

To validate and illustrate the work presented in this paper, we present excerpts of an example application: eliciting and specifying the requirements of the London Ambulance Service (LAS). This case study was the focus of, and common example used by, delegates at the Eighth International Workshop on Software Specification and Design (IWSSD-8) [15]. As mentioned earlier (§4.3), by examining this case study we are not attempting to demonstrate the scalability of our approach, but rather to demonstrate its potential usefulness when used in conjunction with a host of other tools from the requirements engineer's toolbox. Thus, we do not expect the wholesale translation of large, monolithic specifications into QC logic on which we can then perform the kind of reasoning we have described. Rather, our aim is to reduce the complexity of our reasoning and analysis by restricting them to smaller partial specifications (ViewPoints) to which we then apply our tools. The exposition below is somewhat artificial in order to illustrate the issues and contributions presented in the paper.

5.1. Requirements document

Consider the requirements for a computer-aided ambulance despatching system. A reasonable requirements engineering method involves interviews with the staff involved in order to elicit the system's requirements. In our example, stakeholder analysis yielded, among others, the following "client authorities" who laid down the procedures deployed by LAS. The requirements document for the LAS that contained information such as the

following.

Stakeholder 1: LAS Incident Room Controller

- A medical emergency is either the result of an illness or accident.

- On receipt of a phone call reporting a medical emergency, an ambulance should be despatched to the scene.

- On receipt of a phone call, if the incident is judged not to be a medical emergency, then the call should be transferred to another emergency service (e.g., police or fire brigade).

Stakeholder 2: Operations Manager

- On receipt of a phone call reporting an incident, if an ambulance is available then it should be despatched to the scene.

- On receipt of a phone call reporting an incident, if an ambulance is not available then it should not be despatched to the scene.

Stakeholder 3: Logistics Manager

- If no ambulance operators (drivers/medics) are available, then no ambulance is available.

- If no ambulances are available, then initiate a search for a free ambulance.

- If one year has passed since the maintenance work was last done on an ambulance, then perform a safety check on that ambulance.

5.2. Preliminary specification

From the above requirements document, one could generate, for example, agent hierarchies, data flow diagrams, action tables, object diagrams, and so on. Below we use QC logic (as presented in this paper), directly, to represent this specification information.

Stakeholder 1: LAS Incident Room Controller

{a}: $\forall X, Y,$ accident(X, Y) $\vee$ illness(X, Y) $\leftrightarrow$ medical-emergency(X, Y)

{b}: $\forall X, Y,$ call(X, Y) $\wedge$ medical-emergency(X, Y) $\rightarrow$ despatch-ambulance(X, Y)

{c}: $\forall X, Y,$ call(X, Y) $\wedge \neg$ medical-emergency(X, Y) $\rightarrow$ transfer-service(X, Y)

Stakeholder 2: Operations Manager

{d}: $\forall X, Y,$ call(X, Y) $\wedge$ ambulance-available(X) $\rightarrow$ despatch-ambulance(X, Y)

{e}: $\forall X, Y,$ call(X, Y) $\wedge \neg$ ambulance-available(X) $\rightarrow \neg$ despatch-ambulance(X, Y)

Stakeholder 3: Logistics Manager

{f}: $\forall X, Y,$ ¬has-one-or-more(X, Y) →
              ¬ambulance-available(X)

{g}: $\forall X, Y,$ ¬ambulance-available(X) →
              initiate-search-for-free-ambulance

{m}: $\forall X,$ over-one-year-since-last-maintenance(X)
              → initiate-ambulance-safety-check(X)

5.3. Inconsistency handling

From the above preliminary formal specification, we can now demonstrate the reasoning and analysis described in sections 3 and 4.

5.3.1. Reasoning

To check certain scenarios with respect to the preliminary specification, we must add further relevant facts (e.g., domain knowledge) to model each scenario. For example, consider the following facts:

{h}: accident(Anthony, London-Road)

{i}: call(Anthony, London-Road)

{j}: ¬has-one-or-more(Ambulance1, Operator)

{k}: ¬illness(Anthony, London-Road)

{n}: over-one-year-since-last-maintenance(Ambulance2)

From this scenario, we can generate the following (inconsistent) inferences:

{a, b, h, i}:
        despatch-ambulance(Ambulance1, London-Road)

{e, f, i, j}:
        ¬despatch-ambulance(Ambulance1, London-Road)

Using QC logic, we can still continue reasoning with the above facts, together with the preliminary specification, to generate additional inferences such as the following:

{f, g, j}: initiate-search-for-free-ambulance

{m, n}: initiate-ambulance-safety-check(Ambulance2)

So even though the assumptions are inconsistent, we can generate a useful inference. It is possible then, for example, to develop a definition for the initiate-search-for-free-ambulance procedure, without necessarily having to resolve the inconsistencies in the preliminary specification (although one may want to perform the analysis below before developing a definition for a procedure that may later have to be retracted).

5.3.2. Analysis: qualifying inferences

The following inferences are only existential inferences, and hence need to be treated with caution when the specification is revised or analysed further.

{a, b, h, i}:
        despatch-ambulance(Ambulance1, London-Road)

{e, f, i, j}:
        ¬despatch-ambulance(Ambulance1, London-Road)

In contrast, the following inference is a universal inference, and hence is less likely to be retracted when the specification is revised.

{f, g, j}:   initiate-search-for-free-ambulance

The following is a free inference from the specification, which is reassuring given the preliminary nature of the specification.

{m, n}: initiate-ambulance-safety-check(Ambulance2)

5.3.3. Analysis: identifying sources of inconsistency

For the inconsistency identified above, there are two sets of labels, in particular, that refer to problematical data. These are the labels attached to the conflicting inferences generated above, {a, b, h, i} and {e, f, i, j}. There are many possible sources of the inconsistency. However, if we assume the facts we added for the scenario, labelled from the set {h, i, j, k, n} are not causing the problem, we order these *above* the set of labels, {a, b, c, d, e, f, m}, referring to the preliminary specification. Using this ordering, we obtain a smaller subset containing the likely sources of the inconsistency, namely {a, b, e, f}. These pieces of procedural information were elicited from all three stakeholders, who need to be consulted again in order to rectify this problem (although we may also have some ordering of information according to the particular participant from which it was elicited; e.g. "the boss is always right"!).

6. Related Work

The overwhelming majority of work on consistency management has dealt with tools and techniques for maintaining consistency and avoiding inconsistency. Increasingly however, researchers have begun to study the notion of consistency in software systems, have recognised the need to formalise this notion, and have proposed techniques for tolerating or even living with inconsistencies; e.g., [2, 5, 14, 18, 22, 28, 29]. A review of this work can be found in [19, 20, 23].

Other related approaches address inconsistencies that arise in software development processes themselves. For example, an inconsistency may occur between a software development process definition and the actual (enacted) process instance [9]. Such an inconsistency between "enactment state" and "performance state" is often avoided by blocking further development activities until some precondition is made to hold. Since this policy is overly restrictive, many developers attempt to fake conformance to the process definition (for example, by fooling a tool into thinking that a certain task has been performed in order to continue development). Cugola et al. [7] have addressed exactly this problem in their temporal logic-based approach which is used to capture and tolerate some deviations from a process description during execution. Deviations are tolerated as long as they do not affect the correctness of the system (if they do, the incorrect data must be fixed, or the process model - or its active instance - must be changed). Otherwise, deviations are tolerated, recorded and propagated - and "pollution analysis" (based on logical reasoning) is performed to identify possible sources of inconsistency.

From the AI and logics communities there have been a number of other related contributions that are relevant, including fuzzy sets and non-monotonic logics (for a review, see [19, 21]). Whilst they constitute important developments that could be incorporated in our framework, they are not directly oriented to the inconsistency management issues that we consider with in this paper. In the main they are focused on resolving inconsistency by finding the best possible inferences for any given set of information, whereas we really need to be able to analyse inconsistent information, consider options, and track information to find likely sources of inconsistency.

7. Discussion, Conclusions and Future Work

Our earlier work began by providing a framework for multi-perspective software development in which multiple development participants, and the partial specifications they maintained, were represented by ViewPoints. The inconsistencies that inevitably arose between multiple overlapping ViewPoints led us to adopt an inconsistency handling approach that was tolerant of such inconsistencies. This approach relied on identifying inconsistencies, the context in which they arose, and the actions that could be performed in their presence. We further recognised that such actions did not need to remove inconsistencies immediately, but rather allowed continued reasoning and development in their presence. Keeping track of deductions made during reasoning, and deciding on what actions to perform in the presence of inconsistencies, identified the need to analyse inconsistencies in this context. This paper addressed the formal analysis of such inconsistencies..

We summarised the use of a quasi-classical logic to reasoning in the presence of inconsistency. We then examined the use of labelled QC logic to "audit" reasoning results and to "diagnose" inconsistencies. The labels facilitated the identification of likely sources of inconsistencies. We further proposed some tools for qualifying the different kinds of deductions we made during reasoning. These tools provided us with a measure of confidence in our specification information.

We believe that such logical analysis provides developers with heuristics and guidance about what actions we can perform in the presence of particular inconsistencies (for example, actions to resolve a conflict, delay resolution, ameliorate an inconsistent specification,

etc.). Our immediate research agenda is to examine these inconsistency handling actions further within our framework. In particular, we would like to examine the correlation, if any, between certain kinds of analysis results and the consequent inconsistency handling actions that could be taken.

We believe that our work provides the *foundations* for supporting a software specification process in which inconsistencies are analysed to determine the course of action needed for further development. This recognises the evolutionary nature of software development and provides a formal, yet flexible, mechanism for managing inconsistencies.

Acknowledgements
We would like to acknowledge the contributions and feedback of Alex Borgida, Gianpaolo Cugola, Elisabetta Di Nitto, Martin Feather, Anthony Finkelstein, Dov Gabbay, Carlo Ghezzi, Michael Goedicke, Jeff Kramer, Jonathan Moffett and Axel van Lamsweerde. This work was partially funded by the UK EPSRC as part of the VOILA project (GR/J15483), the British Council, and the European Union as part of the Basic Research Actions DRUMS II , PROMOTER II and the ISI project (ECAUS003).

7. References

[1]   W. Atkinson and R. J. Cunningham (1991); "Proving Properties of Safety-Critical Systems"; *Software Engineering Journal*, 6(2): 41-50, March 1991; IEE/BCS.

[2]   R. Balzer (1991); "Tolerating Inconsistency"; *Proceedings of 13th International Conference on Software Engineering (ICSE-13)*, Austin, Texas, USA, 13-17th May 1991, 158-165; IEEE CS Press.

[3]   S. Benferhat, D. Dubois and H. Prade (1993); "Argumentative Inference in Uncertain and Inconsistent Knowledge Bases"; *Proceedings of Uncertainty in Artificial Intelligence*, Morgan Kaufmann.

[4]   P. Besnard and A. Hunter (1995); "Quasi-classical Logic: Non-trivializable classical reasoning from inconsistent information"; *(In) Symbolic and Quantitative Approaches to Uncertainty (ECSQARU '95);* C. Froidevaux and J. Kohlas (Ed.); 44-51; LNCS, 946, Springer-Verlag.

[5]   A. Borgida (1985); "Language Features for Flexible Handling of Exceptions in Information Systems"; *Transactions on Database Systems*, 10(4): 565-603, December 1985; ACM Press.

[6]   M. Costa, R. J. Cunningham and J. Booth (1990); "Logical Animation"; *Proceedings of 12th International Conference of Software Engineering*, Nice, France, 144-149; IEEE CS Press.

[7]   G. Cugola, E. Di Nitto, C. Ghezzi and M. Mantione (1995); "How To Deal With Deviations During Process Model Enactment"; *Proceedings of 17th International Conference on Software Engineering (ICSE-17)*, Seattle, USA, 23-30th April 1995, 265-273; ACM Press.

[8]   J. De Kleer (1986); "An Assumption-based TMS"; *Artificial Intelligence*, 28: 127-162.

[9] M. Dowson (1993); "Consistency Maintenance in Process Sensitive Environments"; *Proceedings of Workshop on Process Sensitive Environments Architectures*, Boulder, Colorado, USA, Rocky Mountain Institute of Software Engineering (RMISE).

[10] J. Doyle (1979); "A Truth Maintenance System"; *Artificial Intelligence*, 12: 231-272.

[11] S. Easterbrook and B. Nuseibeh (1995); "Inconsistency Management in an Evolving Specification"; *Proceedings of 2nd International Symposium on Requirements Engineering (RE 95)*, York, UK, 48-55; IEEE CS Press.

[12] S. Easterbrook and B. Nuseibeh (1996); "Using ViewPoints for Inconsistency Management"; *Software Engineering Journal*, 11(1): 31-43, January 1996; IEE/BCS.

[13] M. Elvang-Goransson and A. Hunter (1995); "Argumentative Logics"; *Data and Knowledge Engineering*, 16: 125-145.

[14] M. Feather (1995); "Modularized Exception Handling"; *Draft technical report*, 9th February 1995; USC/Information Sciences Institute, Marina del Rey, California, USA.

[15] A. Finkelstein and J. Dowell (1996); "A Comedy of Errors: the London Ambulance Service case study"; *Proceedings of 8th International Workshop on Software Specification and Design (IWSSD-8)*, Schloss Velen, Germany, 22-23rd March 1996, 2-4; IEEE CS Press.

[16] A. Finkelstein, D. Gabbay, A. Hunter, J. Kramer and B. Nuseibeh (1994); "Inconsistency Handling in Multi-Perspective Specifications"; *Transactions on Software Engineering*, 20(8): 569-578, August 1994; IEEE CS Press.

[17] D. Gabbay and A. Hunter (1995); "Negation and Contradiction"; *(In) What is Negation;* Oxford University Press.

[18] T. M. Hagensen and B. B. Kristensen (1992); "Consistency in Software System Development: Framework, Model, Techniques & Tools"; *Software Engineering Notes (Proceedings of ACM SIGSOFT Symposium on Software Development Environments)*, 17(5): 58-67, 9-11th December 1992; SIGSOFT & ACM Press.

[19] A. Hunter (1996); *Uncertainty in Information Systems*; McGraw-Hill.

[20] A. Hunter and B. Nuseibeh (1995); "Managing Inconsistent Specifications: Reasoning, Analysis and Action"; *Technical report,* June 1995; Department of Computing, Imperial College, London, UK.

[21] P. Krause and D. Clark (1993); "Representing Uncertain Knowledge"; *Intellect.*

[22] K. Narayanaswamy and N. Goldman (1992); ""Lazy" Consistency: A Basis for Cooperative Software Development"; *Proceedings of International Conference on Computer-Supported Cooperative Work (CSCW '92)*, Toronto, Ontario, Canada, 31st October - 4th November, 257-264; ACM SIGCHI & SIGOIS.

[23] B. Nuseibeh (1996); "To Be And Not To Be: On Managing Inconsistency in Software Development"; *Proceedings of 8th International Workshop on Software Specification and Design (IWSSD-8)*, Schloss Velen, Germany, 22-23rd March 1996, 164-169; IEEE CS Press.

[24] B. Nuseibeh and A. Finkelstein (1992); "ViewPoints: A Vehicle for Method and Tool Integration"; *Proceedings of 5th International Workshop on Computer-Aided Software Engineering (CASE '92)*, Montreal, Canada, 6-10th July 1992, 50-60; IEEE CS Press.

[25] B. Nuseibeh, J. Kramer and A. Finkelstein (1994); "A Framework for Expressing the Relationships Between Multiple Views in Requirements Specification"; *Transactions on Software Engineering*, 20(10): 760-773, October 1994; IEEE CS Press.

[26] D. Poole (1985); "An Assumption-based TMS"; *Artificial Intelligence*, 36: 24-47.

[27] M. Ryan (1992); "Representing Defaults as Sentences with Reduced Priority"; *Proceedings of 3rd International Conference on Principles of Knowledge Representation and Reasoning*, Morgan Kaufmann.

[28] R. W. Schwanke and G. E. Kaiser (1988); "Living With Inconsistency in Large Systems"; *Proceedings of the International Workshop on Software Version and Configuration Control*, Grassau, Germany, 27-29 January 1988, 98-118; B. G. Teubner, Stuttgart.

[29] P. Zave and M. Jackson (1993); "Conjunction as Composition"; *Transactions on Software Engineering and Methodology*, 2(4): 379-411, October 1993; ACM Press.

APPENDIX A: Formal definition and proof theory of labelled QC logic

For a more complete description of QC and labelled QC logic see [4] and [20] respectively.

Language of labelled QC logic

At this point in our work we assume a first order language without function symbols and existential quantifiers. This gives us certain computational advantages such as rendering consistency checking decidable (because, effectively, we are working with a propositional language).

Definition A1. Let P be a set of predicate symbols, V be a set of variable symbols, and C a set of constant symbols. Let A be a set of atoms, where $A = \{p(q_1,...,q_n) \mid p \in P \text{ and } q_1,...,q_n \in V \cup C\}$. We call $p(q_1,...,q_n)$ a ground atom iff $q_1,...,q_n$ are all constant symbols, otherwise we call it unground.

Example. Let has-exactly-one be a predicate symbol; X,Y be variable symbols; and Cashier, Terminal be constant symbols. Then has-exactly-one(X,Y) is an unground atom, has-exactly-one(X,Terminal) is an unground atom, and has-exactly-one(Cashier,Terminal) is a ground atom.

Definition A2. Let F be the set of classical propositional formulae formed from a set of atoms A, and the $\wedge$, $\vee$, $\rightarrow$ and $\neg$ connectives. We abbreviate the formula $\alpha \wedge \neg\alpha$ by the formula $\bot$, which we read as "inconsistency". We call a formula grounded iff it is made from only ground atoms, otherwise we call it ungrounded.

Definition A3. Let L be the set of formulae formed from F, where if $\alpha \in F$, and $x_1,..,x_n$ are the free variables of α, then $\forall x_1,...,\forall x_n \alpha \in L$.

Hence the set L contains only universally quantified formulae, where the quantifiers are outermost, and ground formulae.

Definition A4. Let $\vdash_X$ be some consequence relation for some logic X, defined by some proof rules. Then, the logic X is trivialisable if and only if for all α, β in the language of X, $\{\alpha, \neg\alpha\} \vdash_X \beta$.

Note that classical logic is trivialisable according to this definition. The following two definitions are used to explain the proof theory concisely.

Definition A5. For each atom $\alpha \in L$, α is a literal and $\neg\alpha$ is a literal. For $\alpha_1 \vee .. \vee \alpha_n \in L$, $\alpha_1 \vee .. \vee \alpha_n$ is a clause iff each of $\alpha_1, .., \alpha_n$ is a literal. For $\alpha_1 \wedge .. \wedge \alpha_n \in L$, $\alpha_1 \vee .. \vee \alpha_n$ is in conjunctive normal form (CNF) iff each of $\alpha_1, .., \alpha_n$ is a clause.

Definition A6. For $\alpha_1 \wedge .. \wedge \alpha_n \in L$, $\beta \in L$, $\alpha_1 \wedge .. \wedge \alpha_n$ is a CNF of β iff $\alpha_1 \wedge .. \wedge \alpha_n \vdash \beta$ and $\beta \vdash \alpha_1 \wedge .. \wedge \alpha_n$ and $\alpha_1 \wedge .. \wedge \alpha_n$ is in CNF.

Definition A7. Let S be some set of atomic symbols such as an alphabet. If $i \subseteq S$ and $\alpha \in L$, then $i: \alpha$ is a labelled formula. Let M be the set of formulae.

Proof theory for labelled QC logic

The proof theory of QC logic provides the power to derive a CNF of any formula, together with the power of resolution. As a "last step" in any derivation, disjunction introduction is also allowed. This means that any resolvant of a set of formulae can be derived, but no trivial formulae can be derived. This proof theory is presented as a set of natural deduction rules. All the QC natural deduction rules hold in classical logic, but some classical deduction rules, such as *(ex falso quodlibet)* do not hold in QC logic. We obtain *labelled QC logic* by using only labelled formulae as assumptions, and by amending the natural deduction rules to propagate the labels. The label of the consequent of a rule is the union of the labels of the premises of the rule.

Definition A8. Assume that $\wedge$ is a commutative and associative operator, and $\vee$ is a commutative and associative operator.

$$\frac{i: \alpha \wedge \beta}{i: \alpha} \qquad \text{[Conjunct elimination]}$$

$$\frac{i: \alpha \vee \alpha \vee \beta}{i: \alpha \vee \beta} \qquad \text{[Disjunct contraction]}$$

$$\frac{i: \alpha \vee \beta}{\neg\neg\alpha \vee \beta} \qquad \text{[Negation introduction]}$$

$$\frac{\neg\neg\alpha \vee \beta}{\alpha \vee \beta} \qquad \text{[Negation elimination]}$$

$$\frac{i: \forall x\alpha}{i: \beta} \quad \text{[Universal instantiation, where } \beta \text{ is obtained from } \alpha \text{ by replacing every occurrence of x by the same constant]}$$

$$\frac{i: \alpha \vee \beta \qquad j: \neg\alpha \vee \gamma}{i \cup j: \beta \vee \gamma} \qquad \text{[Resolution]}$$

$$\frac{i: \alpha \qquad j: \neg\alpha \vee \beta}{i \cup j: \beta} \qquad\qquad \frac{i: \alpha \vee \beta \qquad j: \neg\alpha}{i \cup j: \beta}$$

[Arrow Elimination]

$$\frac{i: \alpha \vee (\beta \rightarrow \gamma)}{i: \alpha \vee \neg\beta \vee \gamma} \qquad \frac{i: \alpha \vee \neg(\beta \rightarrow \gamma)}{\alpha \vee (\beta \wedge \neg\gamma)}$$

$$\frac{i: \beta \rightarrow \gamma}{i: \neg\beta \vee \gamma} \qquad \frac{i: \neg(\beta \rightarrow \gamma)}{i: \beta \wedge \neg\gamma)}$$

[Distribution]

$$\frac{i: \alpha \vee (\beta \wedge \gamma)}{i: (\alpha \vee \beta) \wedge (\alpha \vee \gamma)} \qquad \frac{i: (\alpha \wedge \beta) \vee (\alpha \wedge \gamma)}{i: \alpha \wedge (\beta \vee \gamma)}$$

[de Morgan's laws]

$$\frac{i: \neg(\alpha \wedge \beta) \vee \gamma}{i: \neg\alpha \vee \neg\beta \vee \gamma} \qquad \frac{i: \neg(\alpha \wedge \beta)}{i: \neg\alpha \vee \neg\beta}$$

$$\frac{i: \neg(\alpha \vee \beta) \vee \gamma}{i: \neg\alpha \wedge \neg\beta \vee \gamma} \qquad \frac{i: \neg(\alpha \vee \beta)}{i: \neg\alpha \wedge \neg\beta}$$

$$\frac{i: \alpha}{i: \alpha \vee \beta} \qquad \text{[Disjunct introduction]}$$

Labelled QC logic is used by providing any set of labelled formulae (i.e., any $\Delta \subseteq M$ as assumptions, and any classical formula (i.e., any $\alpha \in L$) as a query. For a query that is a ground formula, it follows from the assumptions with some label i (denoted $\Delta \vdash_Q i: \alpha$) if and only if there is a derivation of a CNF of the query, labelled i, from the assumptions using the labelled QC natural deduction rules, remembering that disjunction introduction is only allowed as a last step in a derivation. For a query that is universally quantified, form a ground formula by instantiating it with constants that do not appear in the assumptions, and then treat the query as above.

A Systematic Tradeoff Analysis for Conflicting Imprecise Requirements

John Yen and W. Amos Tiao

Department of Computer Science
Texas A&M University
College Station, Texas 77843-3112

Abstract

The need to deal with conflicting system requirements has become increasingly important over the past several years. Often, these requirements are elastic in that they can be satisfied to a degree. The overall goal of this research is to develop a formal framework that facilitates the identification and the tradeoff analysis of conflicting requirements by explicitly capturing their elasticity. Based on a fuzzy set theoretic foundation for representing imprecise requirements, we describe a systematic approach for analyzing the tradeoffs between conflicting requirements using the techniques in decision science. The systematic tradeoff analyses are used for three important tasks in the requirement engineering process: (1) for validating the structure used in aggregating prioritized requirements, (2) for identifying the structures and the parameters of the underlying representation of imprecise requirements, and (3) for assessing the priorities of conflicting requirements. We illustrate these techniques using the requirements of a conference room scheduling system.

1 Introduction

There are at least two challenges with requirement engineering [1, 2]. First, requirements are usually *vague* and *imprecise* in nature. Computer-based analysis, however, requires an explicit formal semantics [3]. Therefore, there is a need to bridge the gap between imprecise requirements and formal specification methods. Actually, as Balzer *et. al.* pointed out, informality is an inevitable and ultimately desirable feature of the specification process [4]. Second, requirements often conflict with each other, which many conflicts are implicit and difficult to identify [5, 6]. Assessing the tradeoffs among conflicting requirements is a very challenging issue.

The imprecise nature of requirements leads to a mismatch to the existing formal specification methods. Most existing formal specification methodologies require the requirements to be stated precisely [6] or convert imprecise requirements into precise ones [4, 7]. If requirements are specified to be crisp, that is, requirements are either satisfied or not satisfied at all, their capabilities of capturing the semantics of imprecise requirements are limited.

The need to deal with conflicting system requirements has become increasingly important over the past several years. Conventional techniques tend to suppress conflicts, making any resolution untraceable and adding to the communication problem. Conflicts are inevitable part of requirement elicitation process. Viewing conflicts as positive can improve productivity, satisfaction and solution quality. Moreover, handling conflicts in more direct ways actually improves understanding of the requirements and assists with requirement validation [8, 9].

Several works have focused on conflict detection and resolution in requirement engineering [5, 9, 10]. They are, however, limited in detecting implicit conflicts and in facilitating a systematic tradeoff analysis. If requirements are crisp, one of the conflicting requirements has to be dropped or modified to resolve the conflict. However, each conflicting requirement can be satisfied to some degree if the elasticity of a requirement is captured. Hence, it is possible to explore an effective tradeoff among imprecise conflicting requirements.

In our research, we have developed a formal framework that facilitates the identification and the tradeoff analysis of conflicting requirements by explicitly capturing their elasticity. In this framework, imprecise requirements are represented by fuzzy logic so that the elasticity of constraints imposed by imprecise requirements can be captured. The conflicting relationships between two imprecise requirements can be inferred based on a reasoning scheme. We have demonstrated elsewhere the application of this inference scheme to the house of quality in concurrent engineering [2]. Once the conflicts between requirements are identified, tradeoffs should be made between conflicting requirements. In this paper, we described a systematic approach for analyzing the tradeoffs between conflicting requirements based on the techniques in decision science. The systematic tradeoff analyses are used for three important tasks in the requirement engineering process: (1) for validating the structure used in aggregating prioritized requirements, (2) for identifying the structures and the parameters of the underlying representation of imprecise requirements, and (3) for assessing the priorities of conflicting requirements.

2 Background: A Formal Foundation for Requirement Analysis

2.1 Representing Requirements Using Fuzzy Logic

Requirements represent the criteria against which the acceptability of a realization of a target system is judged. The foremost effort in requirement analysis is to represent the system requirements. The universe being constrained by a requirement is called its *domain*. Typical domains for requirements include (1) the domain containing all possible system development processes under consideration, (2) the domain containing all possible system realizations under consideration, and (3) the domain containing all possible input-output state transitions.

A requirement is *imprecise* if it can be satisfied to a degree. More formally, an imprecise requirement is a mapping from a requirement's domain to the range of satisfaction degrees (*i.e.*, the interval $[0,1]$). We call such a mapping *satisfaction function*, denoted as Sat_R:

$$Sat_R : \quad D \to \quad [0,1]. \qquad (1)$$

In essence, the satisfaction function characterizes a fuzzy subset of the domain D that satisfies the imprecise requirement R.

In our framework, the *canonical form* in Zedah's test score semantics is used as a basis for expressing imprecise requirements [11, 12]. The representation of imprecise requirements on a system development process in canonical form is established by the following definition [2, 1].

Definition 1 *Let R be an imprecise requirement on system development process in canonical form R : $A_i(p)$ is B, where p is a system development process, A_i is a property of the process, such as cost, B is a fuzzy set characterized by the membership function μ_B. Then the satisfaction function of the requirement R is defined as*

$$Sat_R(p) = \mu_B(A_i(p)).$$

An imprecise requirement about overall system behavior can be expressed using a summarization operation, such as *AVERAGE, MIN*, and *MAX* as follows.

Definition 2 *Assuming that R is an imprecise requirement about overall system functional behavior and is expressed in canonical form R : $\Psi_{R_i}(r)$ is B, where Ψ is a summarization operator, such as AVERAGE, R_i is a specific functional requirement, and r is a realization. The satisfaction degree of R is defined as*

$$Sat_R(r) = \Psi_{<s_1,s_2>\in\Gamma_2(r)}(Sat_{R_i}(<s_1,s_2>)),$$

where $<s_1,s_2>$ is a state transition and $\Gamma_2(r)$ is the set of state transitions performed by a realization r.

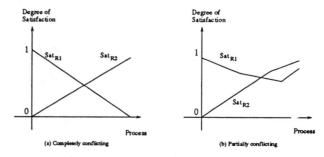

Figure 1: Conflicting imprecise requirements

2.2 Conflicts between Requirements

Once we have represented the imprecise requirements, we would like to identify conflicting relationships between requirements. To do this, we have defined four types of significant relationships between requirements: conflicting, cooperative, mutually exclusive, and irrelevant [2]. Among these four types of relationships, we are particularly interested in the conflicting relationship. Two imprecise requirements, R_1 and R_2, are said to be *conflicting* with each other if an increase in the degree of satisfaction of R_1 (R_2) *often* causes a decrease in the degree of satisfaction of R_2 (R_1). If an increase in the satisfaction degree of one requirement *always* decreases the satisfaction degree of the other, they are said to be *completely conflicting*.

In order to formally define conflicting relationships between imprecise requirements, we first assume that the two requirements being analyzed have an identical domain. Two conflicting requirements may be either *partially conflicting* or *completely conflicting* as shown in Figure 1. In order to characterize *partially conflicting* requirements, we introduce the conflicting degree between two requirements. In the definition, we consider not only the number of conflicting cases but also the conflicting extent in each conflicting case. The definition is described below [13, 2].

Definition 3 *Let R_1 and R_2 be two imprecise requirements of a target system in the domain SP of system development processes. Let U denote the set of process pairs, in which an increase in the satisfaction degree of one requirement decreases the satisfaction degree of the other, that is,*
$U = \{< p_i, p_j > |p_i, p_j \in SP, (Sat_{R_1}(p_i) - Sat_{R_1}(p_j)) \times ((Sat_{R_2}(p_i) - Sat_{R_2}(p_j)) < 0\}.$
The degree R_1 and R_2 are conflicting, denoted as $conf(R_1, R_2)$, is defined as the ratio of $\sum_{<p_i,p_j>\in U} |(Sat_{R_1}(p_i) - Sat_{R_1}(p_j)| \times |Sat_{R_2}(p_i) - Sat_{R_2}(p_j)|$ to $\sum_{(p_h \in SP)\wedge(p_k \in SP)\wedge(p_h \neq p_k)} |Sat_{R_1}(p_h) - Sat_{R_1}(p_k)| \times |Sat_{R_2}(p_h) - Sat_{R_2}(p_k)|.$

Two requirements are completely conflicting whenever their conflicting degree is one. Based on the definitions above, we can define qualitative relationships as follows.

Definition 4 *Two imprecise requirements R_1 and R_2*

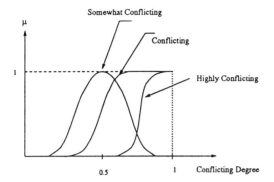

Figure 2: An illustration of "Somewhat Conflicting" and "Highly Conflicting" requirements

are said to be conflicting if and only if

$$conf(R_1, R_2) \geq 0.5.$$

By relaxing the conditions in the above two definitions, one can also define fuzzy qualitative relationship such as "highly conflicting", "somewhat conflicting" using membership functions in Figure 2.

2.3 Concepts from Decision Science

Once conflicting relationship between imprecise requirements has been identified, an effective tradeoff between them should be made. That is, we should explore a feasible requirement that maximize the overall degree of satisfaction.

Keeney and Raiffa have developed the concepts of *utility function*, *indifference curve* and *Marginal Rate of Substitution* (MRS) for the tradeoff analysis between multiple criteria in decision science [14]. A utility function maps each possible decision to the overall satisfaction degree (*i.e.*, the utility) of the decision. An indifference curve represents all the alternative combinations of two decision attributes X and Y for which a customer is equally well off. That is, the alternatives on an indifference curve all provide the same level of satisfaction.

The MRS indicates the maximal amount of a decision attribute that a customer is willing to sacrifice for a unit increase in another decision attribute. For example, if, at the point (x, y), the customer is willing to give up $\lambda\Delta$ units of X for Δ units of Y, then the MRS of X for Y at point (x, y) is λ. This is formally established by the following definition.

Definition 5 *Suppose X and Y are two decision attributes, x_1 and y_1 are values for X and Y, respectively. If the indifference curve through (x_1, y_1) is given by $v(x, y) = c$, then the marginal rate of substitution (MRS) λ at (x_1, y_1) is*

$$\lambda = -\frac{dx}{dy}\big|_{x_1, y_1} = \frac{v'_y(x_1, y_1)}{v'_x(x_1, y_1)},$$

where v'_x and v'_y are the partial derivatives of v with respect to the first and second arguments, respectively.

Generally speaking, MRSs of two decision attributes at two different decision points (*e.g.*, λ at (x_1, y_1) and λ at (x_2, y_2)) can be different. If MRSs of two decision attributes are the same for all decision points, we call it a constant MRS.

2.4 Why Introducing Fuzzy Sets into Decision Science for Requirement Engineering?

To facilitate developing the tradeoff analysis of imprecise requirements, we have employed concepts and techniques from Keeney and Raiffa's work. One natural question would be "What is the significance/contribution of adding fuzzy sets to Keeney and Raiffa's work in decision science?". In the context of requirement engineering, introducing fuzzy sets to Keeney and Raiffa's work offer three major benefits: (1) It enables requirements to be described using linguistic terms. This facilitates the transformation of informal requirements expressed in words to a formal representation (*i.e.*, the linguistic term and the associated satisfaction function). It also makes it easier to communicate and understand the requirements. This is one of the main benefits of all fuzzy logic applications. (2) It enables requirement analysts to reason about requirement relationships (*e.g.*, conflicts) using fuzzy logic (more specifically, fuzzy rule-based reasoning). A more detailed discussion on this topic can be found in [2]. (3) Finally, it enables complex requirements to be expressed/decomposed using conjunction and disjunction (*i.e.*, fuzzy AND, fuzzy OR) as well as compromise operators in fuzzy decision making.

In summary, introducing fuzzy sets into Keeney and Raiffa's decision science theory provides us a framework for requirement analysis that is easier to comprehend, able to infer partial conflicts, and more flexible in expressing and decomposing a composite requirement.

3 A Conference Room Scheduling System Example

In this section, we describe an application to requirement engineering of a conference room scheduling system that the systematic tradeoff analysis techniques introduced in the subsequent sections will be applied to. A Conference Room Scheduling System (CRSS) is a software system that generates a conference schedule used by conference room manager [2]. The requirements include:

- R_1: the cost of the development should be *low*,

- R_2: the resource used for the development should be *small*,

- R_3: the average annual profit of the schedules generated by the system should be *high*.

- R_4: the average utilization of conference rooms of the schedules generated by the system should be *high*.

- R_5: the average room type compatibility of the schedules generated by the system should be *high*.

- R_6: the average time compatibility of the schedules generated by the system should be *high*.

- R_7: the average cost of conferences should be *low*.

Because we use fuzzy logic to represent requirements, requirements can be described using linguistic terms. As we mentioned in Section 2.4, this makes it easier to communicate and understand the requirements even when the requirements are represented formally. Let p denote a development process of the conference scheduling system. *LOW*, *SMALL*, and *HIGH* are fuzzy sets, which serve as elastic constraints on the desired system. Requirements above can be represented in the canonical form in fuzzy logic. For example,

- R_1: *Development_Cost(p)* should be *LOW*.

- R_2: *Resource(p)* should be *SMALL*.

- R_3: AVG_i (*Annual_Profit(Schedule(i, Product(p))))* should be *HIGH*, where i denotes an input.

Our approach also enables complex requirements to be expressed using conjunction and disjunction operators in fuzzy logic. Assuming that the manager prefers the development cost should be low and the resource used for the development should be small, or the average annual profit generated by the system should be high. To represent the preference, we notice that the preference is composed of requirements R_1, R_2, and R_3. Thus, We can represent the preference as a requirement as follows:

$$R = (R_1 \wedge R_2) \vee R_3,$$

where $\wedge$ denotes a fuzzy AND and $\vee$ denotes a fuzzy OR. The satisfaction function of this requirement can be described as

$$Sat_R(x) = (Sat_{R_1}(x) \otimes Sat_{R_2}(x)) \oplus Sat_{R_3}(x),$$

where $\otimes$ is a fuzzy AND operator (*i.e.*, t-norm) and $\oplus$ is a fuzzy OR operator (*i.e.*, t-conorm) [15].

4 Assessing the Validity of Operator

Compromise operators are often used to achieve a tradeoff among conflicting requirements. One of the most widely used compromise operators to aggregate requirements is the weighted summation. The overall satisfaction degree of a set of requirements denoted as $(R_1, R_2, ..., R_n)$ combined using weighted summation can be expressed as follows:

$$\sum_{i=1}^{n} w_i \times Sat_{R_i}(A_i(p)),$$

where w_i is a normalized weight of requirement R_i, p is a process, A_i is an attribute of p on which requirement R_i imposes a constraint, and $Sat_{R_i}(A_i(p))$ is the degree p satisfies the requirement R_i.

In general, however, many other operators can be used to aggregate a set of requirements. Thus, it is desirable to develop techniques for validating the correctness of the chosen aggregation operator. For convenience, $A_i(p)$ is abbreviated as x_i in the following discussion.

In general, the marginal rate of substitution depends on the level of x and on the level of y. A function is *additive* if it has the form $v(x, y) = v_{R_1}(x) + v_{R_2}(y)$, where $v_{R_1}(x)$ and $v_{R_2}(y)$ are single variable functions depending only on attribute X and Y, respectively. For an overall satisfaction function combined by the weighted summation, we have the structure $\sum_{i=1}^{n} w_i \times Sat_{R_i}(x_i)$. Assuming that each weight w_i is fixed, it is easy to show that if the *corresponding tradeoff condition* below holds for all x and y, the satisfaction function is additive.

Theorem 1 *Consider four points* $A : (x_1, y_1)$, $B : (x_1, y_2)$, $C : (x_2, y_1)$, *and* $D : (x_2, y_2)$ *in an evaluation space. The corresponding tradeoff condition is said to be satisfied if given*

$$\lambda(x_1, y_1) = \lambda_A = \frac{b}{a},$$
$$\lambda(x_1, y_2) = \lambda_B = \frac{c}{a}, \text{ and}$$
$$\lambda(x_2, y_1) = \lambda_C = \frac{b}{d},$$

we have

$$\lambda(x_2, y_2) = \lambda_D = \frac{\lambda_B \cdot \lambda_C}{\lambda_A} = \frac{c}{d}. \qquad (2)$$

Proof: For simplicity, we assume the satisfaction function is additive. Thus, we can represent the satisfaction function as $Sat(x, y) = w_1 \times Sat_{R_1}(x) + w_2 \times Sat_{R_2}(y)$. From the definition of MRS, we obtain

$$\lambda_A = \lambda(x_1, y_1) = \frac{Sat\prime_y(x_1, y_1)}{Sat\prime_x(x_1, y_1)} = \frac{w_2}{w_1} \times \frac{Sat\prime_{R_2}(y_1)}{Sat\prime_{R_1}(x_1)} = \frac{b}{a};$$

$$\lambda_B = \lambda(x_1, y_2) = \frac{Sat\prime_y(x_1, y_2)}{Sat\prime_x(x_1, y_2)} = \frac{w_2}{w_1} \times \frac{Sat\prime_{R_2}(y_2)}{Sat\prime_{R_1}(x_1)} = \frac{c}{a};$$

$$\lambda_C = \lambda(x_2, y_1) = \frac{Sat\prime_y(x_2, y_1)}{Sat\prime_x(x_2, y_1)} = \frac{w_2}{w_1} \times \frac{Sat\prime_{R_2}(y_1)}{Sat\prime_{R_1}(x_2)} = \frac{b}{d}.$$

Therefore,

$$\begin{aligned}
\lambda_D &= \lambda(x_2, y_2) = \frac{Sat\prime_y(x_2, y_2)}{Sat\prime_x(x_2, y_2)} = \frac{w_2}{w_1} \times \frac{Sat\prime_{R_2}(y_2)}{Sat\prime_{R_1}(x_2)} \\
&= \frac{w_2}{w_1} \times \frac{Sat\prime_{R_2}(y_2)}{Sat\prime_{R_1}(x_1)} \frac{Sat\prime_{R_1}(x_1)}{Sat\prime_{R_2}(y_1)} \frac{Sat\prime_{R_2}(y_1)}{Sat\prime_{R_1}(x_2)} \\
&= \lambda_B \cdot \frac{1}{\lambda_A} \cdot \lambda_C = \frac{\lambda_B \cdot \lambda_C}{\lambda_A} = \frac{c}{d}. \qquad (3)
\end{aligned}$$

Thus we have proven the theorem. $\square$

Based on the works in decision science, the corresponding tradeoff condition provides us with necessary and sufficient conditions for an important result, which is described in the following theorem [14].

Theorem 2 *A satisfaction function is additive if and only if the corresponding tradeoff condition is satisfied.*

From Theorem 1 and Theorem 2, we have developed a procedure to assess the validity of the weighted summation. Once Theorem 1 holds, the overall satisfaction function can be aggregated by the weighted summation.

To illustrate how to use the above theorem in CRSS example, we first identify the relationships between requirements, which were discussed in [2]. Among them, we initially identify that R_1 and R_2 are cooperative, and R_2 and R_3 are conflicting. Using the inference mechanism developed in [2], we can infer that R_1 and R_3 have a conflicting relationship. This illustrates the second benefits of our approach discussed in Section 2.4. Having detected this conflict, we test the corresponding tradeoff condition to see if these two requirements can be aggregated by the weighted summation. Suppose a conference room manager plans to spend between $\$10K$ and $\$50K$ for building such a conference room scheduling system. Assuming further that the manager has the following tradeoff preference.

- If the development cost is low, the manager is willing to spend more money for an addtional unit gain of the average annual profit.

- If the development cost is already high, the manager hestitates to spend more money for an addtional unit gain of the average annual profit.

That is, the manager determines how much additional money he would like to invest in developing this conference room scheduling system based only on how much money he has already committed on it. By interviewing the conference room manager, the requirement engineer may identify that the marginal rates of substitution (development cost for the average annual profit) for four points in the evaluation space are $\lambda_A = 2$, $\lambda_B = 2$, $\lambda_C = 0.5$ and $\lambda_D = 0.5$. The marginal rates of substitution are illustrated in Figure 3. In the figure, it shows when the development cost is low, (*e.g.*, $\$15K$), the manager is willing to spend additional $\$2K$ for an increase of $\$1K$ of the average annual profit. However, when the development cost, say, $\$40K$, is reaching his budget limit and is conceived as high cost, he hestitates to spend more money and only willing to invest additional $\$0.5K$ for an increase of $\$1K$ of the average annual profit. From Theorem 1, we know the corresponding tradeoff condition holds. Thus, from Theorem 2, R_1 and R_3 could be aggregated by the weighted summation and their composite satisfaction function is additive.

5 Elicitation of Satisfaction Functions

One of the most important tasks in specifying imprecise requirements is to elicit the satisfaction functions from the customer. Based on techniques for assessing utility function in decision science [14], we have developed several techniques that assist requirement engineers to elicit the structures and the parameters of the satisfaction functions from the customer.

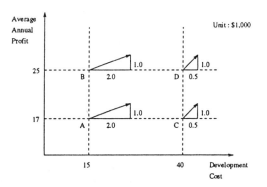

Figure 3: An example of the marginal rates of substitution for the corresponding tradeoff condition

5.1 Assessing Satisfaction Function Using MRS

Even though the customers usually know the qualitative nature of imprecise requirements, they typically find it difficult to directly specify the satisfaction function quantitatively. Hence it is desirable to develop a systematic approach to assist the customers and the requirement engineers constructing the satisfaction functions. We will first deal with the case of fixed MRS. Then we discuss assessment techniques for dynamic (*i.e.*, varying) MRS.

Suppose that customers think that the substitution rate between two requirements is constant. From the definition of marginal rate of substitution, it is easy to derive that the satisfaction function is in the following form:

$$Sat(x, y) = \beta(x + \lambda y + c),$$

where λ is the constant MRS and β and c are constants. The constant β could be viewed as the normalization factor that restricts the degree of the satisfaction to $[0, 1]$. Assuming that customers also feel that it is appropriate to use a linear satisfaction function to specify each imprecise requirement, we can obtain the parameters of the linear function using the following theorem.

Theorem 3 *We assume overall satisfaction function is aggregated using the weighted summation, $\lambda_{i,j}$ is a constant MRS for requirements R_i and R_j, and requirement R_i and R_j have a linear satisfaction function: $Sat_{R_k}(x_k) = a_k \times x_k + b_k, k = i, j$. In addition, let the relative priority between the two requirements be a constant α, i.e. $\frac{w_i}{w_j} = \alpha$, where w_i and w_j are normalized weights of requirements R_i and R_j, respectively. Then, we have*

$$a_i = \frac{\alpha + 1}{\alpha}\beta, \tag{4}$$

$$a_j = (\alpha + 1)\beta\lambda_{i,j}, \text{ and} \tag{5}$$

$$\alpha b_i + b_j = \beta c(\alpha + 1), \tag{6}$$

where β and c are constants.

Proof: Since $\lambda_{i,j}$ is a constant, we get

$$Sat(x_i, x_j) = \beta(x_i + \lambda_{i,j}x_j + c). \qquad (7)$$

Using weighted summation, we get

$$Sat(x_i, x_j) = w_i \times Sat_{R_i}(x_i) + w_j \times Sat_{R_j}(x_j).$$

Because of fixed relative priority and the linear satisfaction function assumption, we have

$$Sat(x_i, x_j) = \frac{\alpha}{\alpha+1}(a_i x_i + b_i) + \frac{1}{\alpha+1}(a_j x_j + b_j).$$

After simplification, we obtain

$$Sat(x_i, x_j) = \frac{\alpha a_i}{\alpha+1}x_i + \frac{a_j}{\alpha+1}x_j + \frac{\alpha b_i + b_j}{\alpha+1}. \qquad (8)$$

The right hand sides of Eq. (7) and Eq. (8) should be equal for all x_i and x_j, thus we have

$$\begin{aligned} \beta &= \frac{\alpha}{\alpha+1}a_i, \\ \beta\lambda_{i,j} &= \frac{1}{\alpha+1}a_j, \text{ and} \\ \beta c &= \frac{\alpha b_i + b_j}{\alpha+1}. \end{aligned}$$

After simplification, we obtain

$$\begin{aligned} a_i &= \frac{\alpha+1}{\alpha}\beta, \\ a_j &= (\alpha+1)\beta\lambda_{i,j}, \text{ and} \\ \alpha b_i + b_j &= \beta c(\alpha+1). \end{aligned}$$

Thus, we have proven the theorem. □

Requirement engineers may use the theorem to obtain the parameters of the linear satisfaction functions by solving Eq. (4), (5), and (6). Alternatively, they can validate satisfaction functions previously formulated using the theorem.

5.2 Eliciting the Structure of Satisfaction Functions Based on Dynamic MRS

In section 5.1, we have shown that parameters of satisfaction function can be constructed systematically if the satisfaction function is linear. In general, however, we need to determine the structure (*e.g.*, linear, nonlinear) of the satisfaction function before we can actually identify its parameters. Hence, we need to develop a technique for identifying structures of the satisfaction functions.

Suppose that the customer feels that the substitution rates depend on one requirement but not on the other, how can this qualitative statement help in eliciting the structures of the satisfaction function? To answer this question, we first determine the overall form of the aggregated satisfaction function. If the MRS depends on y but not on x, it is easy to derive the following form for the satisfaction function

$$Sat(x, y) = \beta(x + Sat_{R_j}(y) + c),$$

where Sat_{R_j} is a satisfaction function over attribute Y and β and c constants. That is, the amount a customer is willing to pay in x units for additional y units depends on the level of y but not on the level of x. Under such a circumstance, we can determine that the structures of the satisfaction function of attribute X is linear using the following theorem.

Theorem 4 *We assume the overall satisfaction function is aggregated using the weighted summation, and the relative priority of requirements R_i and R_j is fixed, i.e. $\frac{w_i}{w_j}$ is a constant, where w_i and w_j are normalized weight of imprecise requirements R_i and R_j, respectively. Let attributes constrained by R_i and R_j be denoted by x and y, respectively. If MRS of the two attributes satisfies the following condition*

$$\lambda(x_1, y_1) = \lambda(x_2, y_1), \forall x_1, x_2, y_1,$$

that is, λ depends on the level of y but not on the level of x, then we have

$$Sat_{R_i}(x) = a_i x + b_i.$$

where a_i and b_i are constants.

Proof: Since the MRS depends on the level of y but not on the level of x, we have

$$Sat(x, y) = \beta(x + Sat_{R_j}(y) + c), \qquad (9)$$

where $Sat_{R_j}(y)$ is a satisfaction function over attribute Y and c is a constant. Using the weighted summation to aggregate requirements, we get

$$Sat(x, y) = w_i \times Sat_{R_i}(x) + w_j \times Sat_{R_j}(y). \qquad (10)$$

Because we assumed that the relative priority is fixed, $\frac{w_i}{w_j}$ is a constant. Since the right hand sides of Eq. (9) and Eq. (10) should be equal for all x and y, $Sat_{R_i}(x)$ should be a linear structure. That is,

$$Sat_{R_i}(x_i) = a_i x_i + b_i.$$

where a_i and b_i are constants. Therefore, we have proven the theorem. □

In addition to using the theorem to elicit the linear structure of a customer's satisfaction function, we can also identify concave satisfaction function. Suppose a customer is willing to pay less and less for an additional unit in Y as the value of y increases. In other words, he might feel that

$$Sat_{R_j}(y+1) - Sat_{R_j}(y) < Sat_{R_j}(y) - Sat_{R_j}(y-1), \forall y; \qquad (11)$$

it worths less to go from y to $y+1$ than from $y-1$ to y, regardless of the value of y. It can be shown that Sat_{R_j} is strictly *concave* if λ does not depend on x and it decreases as y increases. One example of concave functions is $\sqrt{y}$. This is formally stated in the theorem below.

Theorem 5 *We assume the overall satisfaction function is aggregated using the weighted summation, and relative priority between requirements R_i and R_j is fixed, i.e. $\frac{w_i}{w_j}$ is a constant, where w_i and w_j are normalized weights of requirements R_i and R_j, respectively. In addition, we assume marginal rate of substitution satisfies the following conditions;*

(1) $\lambda(x_1, y_1) = \lambda(x_2, y_1), \forall x_1, x_2, y_1;$

(2) $\lambda(x_1, y_1) < \lambda(x_1, y_2), \forall y_1 > y_2.$

Then we have

$$Sat_{R_i}(x) = a_i x + b_i,$$

and

$$Sat_{R_j}(y) \text{ is a concave function,}$$

where a_i and b_i are constants.

Proof: The proof is similar to that of Theorem 4. □

Returning to our CRSS example, in order to find individual satisfaction functions for R_1 and R_3, we first need to determine the structures of their satisfaction functions. From the discussion in Section 4, we know the marginal rate of substitution is affected by the development cost, but not the average annual profit. Therefore, we could infer from Theorem 4 the structure of R_3's satisfaction function is linear. Suppose requirement engineer finds out that the satisfaction of the manager on the additional unit increase of the development cost drops more sharply if the development cost is already high than that if the development cost is low. Using Theorem 5, we could conclude the structure of the satisfaction function of R_1 is concave.

5.3 Elicitation of Individual Satisfaction Functions Based on Midvalue

One of the most important tasks in our methodology is to determine individual satisfaction functions. In this section, we describe a systematic procedure for eliciting individual satisfaction functions. Our approach is based on the technique for assessing utility function in decision science [14]. Let attributes X and Y be two attributes constrained by imprecise requirements R_i and R_j, respectively, and they satisfy the corresponding tradeoff condition introduced in section 4 (*i.e.*, their satisfaction function is additive). We first define the concept of *midvalue* as follows.

Definition 6 *Assume $x_a < x_m < x_b$. A value x_m is called a midvalue of an interval (x_a, x_b) if the following conditions are satisfied:*

$$Sat(x_a, y + \Delta y) = Sat(x_m, y)$$

and

$$Sat(x_m, y) = Sat(x_b, y - \Delta y).$$

The definition states that x_m is a midvalue between x_a and x_b if at any point y of attribute Y, the customer is willing to pay the same amount of Y for the increase of X from x_a to x_m as for the increase from x_m to x_b. Now we describe the elicitation procedure as follows. Let the range of X be $x_0 \leq x \leq x_1$, $Sat_{R_i}(x_0) = 0$ and $Sat_{R_i}(x_1) = 1$.

Step 1: Ask customer the midvalue, call it $x_{0.5}$, of (x_0, x_1). Define $Sat_{R_i}(x_{0.5}) = 0.5$.

Step 2: Ask customer the midvalue, call it $x_{0.25}$, of $(x_0, x_{0.5})$. Define $Sat_{R_i}(x_{0.25}) = 0.25$.

Step 3: In the same manner, we can derive $x_{0.75}$, $x_{0.125}$, etc. Based on the assessment of structure of satisfaction function and these few points, requirement engineers can fill in a smooth curve for the satisfaction function.

The above procedure could be viewed as an approximation process for an individual satisfaction function. Moreover, it could also be viewed as a consistency checking process for the structure assessment techniques developed in the section 5.2.

In the CRSS example, we have already identified the structures of the satisfaction functions for requirements R_1 and R_3. We now use the midvalue techniques mentioned above to derive the satisfaction function for each requirement. To derive the satisfaction function of R_3, requirement engineers could ask the manager about how much the average annual profit he expects. The manager may answer that the average annual profit of using this system should have at least $\$15K$ or have $\$30K$ maximal. This answer implies any average annual profit below $\$15K$ is not satisfied at all and above $\$30K$ would be cosidered fully satisfied. Since we have determined the structure of the satisfaction function is linear when the profit falls within $\$15K$ and $\$30K$, we have the satisfaction function of R_3 as follows.

$$Sat_{R_3}(x) = \begin{cases} 0 & \text{if } x \leq 15 \\ ax + b & \text{if } 15 \leq x \leq 30 \\ 1 & \text{if } x \leq 30 \end{cases}$$

Since the linear function passes through two points $(15, 0)$ and $(30, 1)$, we could use them to determine the parameters of a and b. By solving the linear equations, we have $a = \frac{1}{15}$ and $b = -1$. Therefore, the satisfaction function of R_3 is

$$Sat_{R_3}(x) = \begin{cases} 0 & \text{if } x \leq 15 \\ \frac{1}{15}x - 1 & \text{if } 15 \leq x \leq 30 \\ 1 & \text{if } x \leq 30 \end{cases}$$

To determine the satisfaction function of R_1, we need to ask the manager the midvalue of the development cost. The question would be like "Give me a development cost x' you would like to invest such that you would gain the same amount of satisfaction of the average annual profit to go from $\$10K$ to x' as from x' to $\$50K$". Assume the answer is $\$33K$, then we denote it as midvalue $x_{0.5} = 33$. In a similar manner, we could ask about midvalue $x_{0.75}$, $x_{0.25}$, etc. Suppose that $x_{0.25} = 42$ and $x_{0.75} = 23$. Since we know the structure of the satisfaction function of R_1 is concave, we could estimate the satisfaction function based on these midvalues. The satisfaction functions for R_1 and R_3 are shown in Figure 4 and Figure 5, respectively.

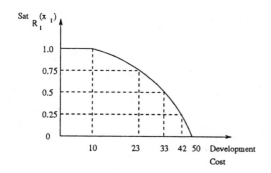

Figure 4: The satisfaction function of low development cost

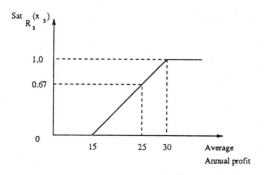

Figure 5: The satisfaction function of high average annual profit

6 Priority Assessment

In the preceding sections, we have described the techniques to acquire individual satisfaction functions and the conditions when an additive function holds. An additional task forced by the requirement engineer is to assess the priority of requirements. This is particularly important for conflicting requirements because the priorities of conflicting requirements should reflect the tradeoff the customer is willing to make. It is, however, often difficult for customers to directly provide priorities for all requirements due to complex relationships between them. Thus it is desirable to develop a technique to assist them in identifying the priority of each requirement.

The techniques we have developed are based on pairwise comparisons. Before we describe the technique, we should first address the scalability issue of our approach. Even though the number of possible pairwise comparisons is quite large (*i.e.* $N \times (N-1)$ for N requirements), the number of conflicting requirement pairs is typically smaller. Therefore, it is important to realize that our technique is intended to be used for tradeoff analysis of conflicting requirements after they have been identified. Furthermore, the procedure we describe in Section 6.2 requires only $N-1$ pairwise analysis for N requirement.

6.1 Priority Assessment Based on Constant MRS

The ratio of importance of two requirements is crucial while tradeoffs between two conflicting requirements are considered. Relative priority is usually used to represent the ratio of importance of two requirements. Intuitively, if customers are willing to sacrifice a lot in the satisfaction degree of a requirement R_2 for a small increase in the satisfaction degree of requirement R_1, R_1 should be more important than R_2. In addition, relative priority is proportional to the ratio of their changes in the satisfaction degrees. That is, the more the customers is willing to sacrifice in the satisfaction degree of R_2 for the same amount of increase in the satisfaction degree of R_1, the higher is the value of their relative priority. For example, if customers prefer a solution that increases the satisfaction degree of R_1 by 0.3, yet decreases the satisfaction degree of R_2 by 0.6, and does not change the satisfaction degrees of other requirements, then R_1 is more important than R_2 and the relative priority of R_1 to R_2 is 2.

To formally capture the intuition, we need individual satisfaction functions and the marginal rate of substitution. Individual satisfaction function is used to acquire the satisfaction degree of each requirement. The marginal rate of substitution is used to represent the amount that customers are willing to pay. In the previous section, we have developed techniques that can assess the structures and parameters of the individual satisfaction functions. Combined with the concept of marginal rate of substitution, we have developed an analytical approach to assess the relative priority. The following theorem show how to compute relative priority based on a constant MRS.

Theorem 6 *We assume that an imprecise requirement $R_k(k = i, j)$ has a linear satisfaction function which maps an attribute value to a satisfaction degree: $Sat_{R_k}(x_k) = a_k \times x_k + b_k, k = i, j$. In addition, we assume that the marginal rate of substitution of attribute x_j for x_i is a constant $\lambda_{i,j}$. Then,*

$$\frac{w_i}{w_j} = \frac{a_j}{a_i}\lambda_{i,j}.$$

Proof: Since $\lambda_{i,j}$ is the MRS of attribute x_j for x_i, the level of overall satisfaction degree should be the same before and after the substitution. Hence, we have

$$\sum_{k=1}^{n}(w_k \times Sat_{R_k}(x_k)) = \sum_{k=1, k \neq i,j}^{n}(w_k \times Sat_{R_k}(x_k)) + w_i \times Sat_{R_i}(x_i + 1) + w_j \times Sat_{R_j}(x_j - \lambda_{i,j}).$$

That is, $w_i \times Sat_{R_i}(x_i) + w_j \times Sat_{R_j}(x_j) = w_i \times Sat_{R_i}(x_i + 1) + w_j \times Sat_{R_j}(x_j - \lambda_{i,j})$. Because of the linear membership function assumption, we have

$$w_i \times (a_i \times x_i + b_i) + w_j \times (a_j \times x_j + b_j) = w_i \times (a_i \times (x_i + 1) + b_i) + w_j \times (a_j \times (x_j - \lambda_{i,j}) + b_j).$$

After simplification, we obtain $w_i \times a_i = w_j \times a_j \times \lambda_{i,j}$. That is,

$$\frac{w_i}{w_j} = \frac{a_j}{a_i} \lambda_{i,j} .$$

Therefore, we have proven the theorem. □

To illustrate how to use the theorem, suppose that

$$Sat_{R_1}(x_1) = \frac{1}{4}x_1 \text{ where } 0 \leq x_1 \leq 4,$$

and

$$Sat_{R_2}(x_2) = \frac{1}{20}x_2 - 5 \text{ where } 100 \leq x_2 \leq 120.$$

Let us assume that the marginal rate of substitution $\lambda_{1,2}$ is 2. That is, customers are willing to sacrifice 2 units of X_1 for an additional unit of X_2. Based on the above theorem, the relative priority of R_1 to R_2 is

$$\frac{w_1}{w_2} = \frac{a_2}{a_1} \lambda_{1,2} = 0.4.$$

6.2 A General Procedure for Assessing the Priorities of Requirements

We have described a technique for identifying the relative priorities of additive linear satisfaction functions. In this section, we describe a more general technique for identifying the priorities of nonlinear satisfaction functions.

Let l_i and u_i denote two extremes such that $Sat_{R_i}(l_i) = 0$ and $Sat_{R_i}(u_i) = 1$. Then we can describe the general procedure as follows.

Step 1: Rank attributes, that is, obtain the relative importance of the attributes.

Step 2: Acquire the relations between the most important attribute and another attributes using the concept of indifference. Without loss of generality, let X_1 be the most important requirement. To assess the relative priority of X_1 and other attribute X_i, we first consider the point $(l_1, l_2, ..., u_i, l_{i+1}, ...)$ (i.e., the case of satisfying only ith requirement completely, but not any other requirements). We then construct a point $(x_1, l_2, ..., l_i, l_{i+1}, ...)$ that is indifferent to the previous point by adjusting x_1. We can then obtain

$$w_i \times 1 = w_1 \times Sat(x_1),$$

that is,

$$\frac{w_i}{w_1} = Sat(x_1).$$

We can repeat the same procedure for all other requirements to obtain their relative priority to w_1.

Step 3: Normalize the priorities such that they sum to one.

To demonstrate how to utilize the procedure, assume we have 4 requirements and we have, from step 1, the relative importance of the attributes, $X_2 \succ X_1 \succ X_4 \succ X_3$, where $\succ$ means *more important than*. The order of the relative importance can be derived using technique described in section 6.1 or simply by pairwise comparisons. Hence, based on the relative importance of the attributes, we have $w_2 > w_1 > w_4 > w_3$.

To obtain the relation between w_2 and w_1, we compare point $A(l_1, x_2, l_3, l_4)$ to point $B(u_1, l_2, l_3, l_4)$ and manipulate the value of x_2 of point A until indifference is reached. There are two reasons for choosing these two particular points. First, most of the values in point A and point B is fixed and, in particular, those values are extreme values. Thus, it alleviates burden for customers to make decision about indifference of these two points. Second, we can simplify the analysis knowing that $Sat_{R_i}(l_i) = 0$ and $Sat_{R_i}(u_i) = 1$. Assume the indifference of point A and point B occurs at $x_2 = x'_2$, that is, point $A(l_1, x'_2, l_3, l_4)$ is indifferent to point $B(u_1, l_2, l_3, l_4)$. Then we have

$$Sat_R(l_1, x'_2, l_3, l_4) = Sat_R(u_1, l_2, l_3, l_4),$$

where Sat_R is the satisfaction degree of an overall system requirement. Assume the overall system satisfaction function is additive, we have

$$w_2 \cdot Sat_{R_2}(x'_2) = w_1 \cdot Sat_{R_1}(u_1) = w_1.$$

Based on the techniques of section 5, we can obtain the individual satisfaction functions. Assume $Sat_{R_2}(x'_2) = 0.7$, then we have

$$0.7 \times w_2 = w_1. \tag{12}$$

By the same tokens, we can derive the relations between w_2 and w_4 and between w_2 and w_3, say,

$$0.4 \times w_2 = w_4, \tag{13}$$
$$0.2 \times w_2 = w_3.$$

Hence, we have derived the relative priorities of attributes which represent the relative importance of requirements. Coefficients we derived here can serve another purpose. We can use these coefficients to check the consistency of ranking of attributes in step 1. For example, coefficient in Eq. (12) should be larger than that in Eq. (13) since, from step 1, w_1 is more important than w_4. We can derive the absolute values of priorities by normalizing all priorities so that they sum to 1. For the above example, we obtain the following absolute priorities: $w_1 = 0.304$, $w_2 = 0.435$, $w_3 = 0.087$, and $w_4 = 0.174$.

To determine the relative priority of R_3 to R_1 in the CRSS example, we use the procedure described above since the satisfaction function of R_1 is nonlinear. Suppose the manager considered requirement R_1 to be more important than R_3 since he had more concerns on the development cost. Following the second step of the procedure, we ask the manager about two indifference points. Assuming that two points $(42, 25)$ and $(23, 15)$ are indifferent to the manager. That is, if the average annual profit is currently \$25 and the development cost is \$42, he feels all right to decrease the average annual profit to \$15 in order to cut the development cost to \$23. Therefore, we have

$$w_{R_1} \times Sat_{R_1}(42) + w_{R_3} \times Sat_{R_3}(25) =$$
$$w_{R_1} \times Sat_{R_1}(23) + w_{R_3} \times Sat_{R_3}(15)$$

Because we have already known the satisfaction functions of both requirements, we could get the value for each point, e.g., $Sat_{R_1}(42) = 0.25$, $Sat_{R_3}(25) = 0.67$, etc. Thus, we have derived the relative priority of R_3 to R_1, i.e., $\frac{w_{R_3}}{w_{R_1}}$, is 0.75.

7 Conclusion

In this paper, we have described a formal framework that facilitates the identification and the systematic tradeoff analysis of conflicting requirements by explicitly capturing their elasticity. Based on a fuzzy set theoretic representation of imprecise requirements, this paper describes a systematic approach for analyzing the tradeoffs between conflicting requirements using techniques in acquiring utility functions in decision science. We have derived a procedure to validate whether requirements are aggregated through an additive form. We have also developed techniques to determine the structures and the parameters of satisfaction functions that characterizes the elasticity of requirements. Finally, we have a systematic approach for assessing the priorities of imprecise requirements. We have illustrated these techniques using the requirements of a conference room scheduling software project. These techniques can not only assist the requirement engineer in better understanding the tradeoffs among conflicting requirements that the customer is willing to accept, but also facilitate the customer in better understanding the implications of a requirement formulation so that they can be validated more effectively early in the software development process.

Acknowledgements

We thank Stephen Fickas and reviewers for their comments on an earlier draft of the paper. We wish to thank Xiaoqing Frank Liu and Jonathan Lee for their early work in this research. Our research has benefited from discussion with Lotfi A Zadeh, Ronald Yager, Steve Easterbrook, and Jack Callahan. This research is supported by the NSF Young Investigator Award IRI-9257293 and by the Texas Higher Education Coordination Board through the Advanced Research Program Award 999903-253.

References

[1] J. Yen, X. Liu, and S. H. Teh, "A fuzzy logic-based methodology for the acquisition and analysis of imprecise requirements," *Concurrent Engineering: Research and Applications*, no. 2, pp. 265–277, 1994.

[2] X. F. Liu and J. Yen, "An analytic framework for specifying and analyzing imprecise requirements," In *Proc. of the 18th International Conference on Software Engineering*, pp. 60–69, March 1996.

[3] C. Heitmeyer, B. Labaw, and D. Kiskis, "Consistency checking of SCR-style requirements specifications," In *Proc. of the 2nd IEEE International Symposium on Requirements Engineering*, pp. 56–63, 1995.

[4] R. Balzer, N. Goldman, and D. Wile, "Informality in program specifications," *IEEE Transactions on Software Engineering*, vol. 4, no. 2, pp. 94–103, 1978.

[5] A. Finkelstein, D. Gabbay, A. Hunter, J. Kramer, and B. Nuseibeh, "Inconsistency handling in multiperspective specifications," *IEEE Transactions on Software Engineering*, vol. 20, no. 8, pp. 569–578, Aug. 1994.

[6] J. Mylopoulos, L. Chung, and B. Nixon, "Representing and using nonfunctional requirements: a process-oriented approach," *IEEE Transactions on Software Engineering*, vol. 18, no. 6, pp. 483–497, June 1992.

[7] W. L. Johnson, M. S. Feather, and D. R. Harris, "The KBSA requirements/specification facet: Aries," In *Proc. of the 6th Annual Knowledge-Based Software Engineering Conference*, pp. 156–162, 1991.

[8] S. Easterbrook, "Resolving requirements conflicts with computer-supported negotiation," In *Requirements engineering: social and technical issues*, M. Jirotka and J. Goguen, editors, pp. 41–65, Academic press, 1994.

[9] W. Robinson and S. Fickas, "Supporting multiperspective requirements engineering," In *Proc. of the 1st International Conference on Requirements Engineering*, pp. 206–215, 1994.

[10] B. Boehm and H. In, "Identifying quality requirement conflicts," *IEEE software*, pp. 25–35, March 1996.

[11] L. Zadeh, "Test-score semantics as a basis for a computational approach to the representation of meaning," *Literacy Linguistic Computing*, vol. 5, no. 1, pp. 24–35, 1986.

[12] L. A. Zadeh, "Soft computing and fuzzy logic," *IEEE software*, pp. 48–56, November 1994.

[13] J. Lee, J. Kuo, and W. Huang, "Classifying, analyzing and representing informal requirements," In *Proc. of the sixth international fuzzy systems association world congress (IFSA '95)*, pp. 645–648, 1995.

[14] R. L. Keeney and H. Raiffa, *Decisions with multiple objectives: preferences and value tradeoffs*, Wiley, New York, 1976.

[15] H. Zimmermann, *Fuzzy set theory and its applications*, Kluwer Academic Publishers, 1991.

Session 4B

Panel

The Impact of Environment Evolution on Requirements Changes

Panel Chair
Naxim H. Madhavji, McGill University, Canada

Panelists
Ted Thompson, LTS Aviation
Periklis Loucopoulos, UMIST
Bill Agresti, MitreTek
Karel Vredenbur, IBM

Panel:
Impact of Environmental Evolution on Requirements Changes

Nazim H. Madhavji

McGill University

madhavji@opus.cs.mcgill.ca

1 Introduction

One reason why systems fail to deliver expected services is that, given certain requirements, the rest of the implementation is not faithful to the stated objectives. There are a number of reasons for this, such as weaknesses in the technologies used, lack of expertise among technical and managerial staff, perceived time pressures, market competitiveness, etc. Much is being done in research and practice to attack this problem, even though progress is slow.

Unfortunately, even if this problem can be solved – unlikely by a long shot – we still confront another problem of major proportion. This is due to the changing environment in which the system is being developed or is meant to be used. Specifically, when a system is being developed, even for a short release cycle, the environmental changes affect the survivability of the system. For example, the requirements implemented may no longer be as adequate at system completion as at the start of development. This inadequacy of the requirements manifests itself in a number of symptoms, such as functional deficiency, performance problems, inter-operability problems, usability problems, higher maintenance costs, etc. Such problems can have minor to severe consequences during system usage.

Thus, one must not only ensure building a "correct" system given a specification but must also work at identifying at the earliest point in time, perhaps continuously, how the environment has changed and how this change calls for changes in the current requirements and how to incorporate these changes into the system. One must understand this issue in a scientific manner and develop or use principles, methods, techniques and tools to address this issue.

This panel will focus on the management of requirements changes due to environmental evolution. In essence, we "are" aware of requirements changes during system development and that of system decay over time, and so this is not a new revelation. In fact, Lehman's program classification and laws of software evolution [1], for example, are an important contributor to this awareness.

Still, on a day-to-day basis, much development is "inward" focused. When it is time to plan a new release, we identify or select new requirements deemed most appropriate for the next release, and these are then implemented. These "reactive" cycles are repeated until the system is eventually decommissioned. Little thought is given to looking '"outwards" on to the environment of the system, how it is changing, and how it is affecting the current requirements. Little methodological [2] and technological work has been done to help us cope with such environmental changes in a "proactive" manner so that we can incorporate the implications of such changes in the existing system design for its longevity. The work in the area of user-centred design is a start but it needs to be generalised to all aspects of system requirements and need not be limited only to human-computer interaction.

2 Purpose of the Panel

There are numerous stages a community must go through in building a viable subfield within requirements engineering to deal with requirements changes due to environmental evolution. One of the first ones is to take stock of the known problems and current practices in this subfield. This panel brings together selected experts, who have had to deal with requirements changes, to share their experiences in diverse application domains and to exchange views with the audience.

Although domain-specific issues can dominate discussions of requirements change (e.g., system safety vs. system availability), the panel will focus first on "requirements change due to environmental evolution" and then on any differences due to domain-specific issues. Some relevant questions are

1. What lessons have been learned from practice in attempting to manage requirements change in the midst of environmental evolution?

2. Is there empirical data linking concrete environmental changes to concrete requirements changes over a period of time?

3. How should one document requirements and environment changes and their interrelationships?

4. What are the principles, methods, techniques and tools associated with requirements and environment change?

5. How do you know what has changed, or is likely to change, in the environment of the system in question (over a period of time)?

6. What effects do past (or anticipated) environmental changes have on system requirements (and hence on system behaviour)?

7. How does deregulation affect system requirements?

8. Can past environmental changes help predict future changes?

9. How often should one monitor environment changes? It costs money!

10. How can current processes (or life-cycle models) be extended to incorporate this aspect of requirements engineering?

11. How do you ensure that you have a feedback mechanism that effectively and efficiently monitors change?

3 Panelists

William Agresti (MITRETEK)
Peri Lucopoulos (University of Manchester Inst. of Science and Technology)
Ted Thompson (LPS Aviation)
Karel Vredenburg (IBM Canada)

References
[1] M.M. Lehman and L.A. Belady. *Program Evolution: processes of software change.* Academic Press, London, 1985.

[2] Vivek Nanda. *Impact of Environmental Evolution on Requirements Change.* Masters Thesis, McGill University, Montreal, Canada, May 1996.

Session 5

Keynote Address

Speaker

Colin Potts
Georgia Institute of Technology, USA

"Requirements Models in Context"

Requirements Models in Context

Colin Potts

College of Computing, Georgia Institute of Technology

potts@cc.gatech.edu

Abstract

The field of requirements engineering emerges out of tradition of research and engineering practice that stresses the importance of generalizations and abstractions. Although abstraction is essential to design it also has its dark side. By abstracting away from the context of an investigation, the designer too easily lapses into modeling only those things that are easy to model. The subtleties, special cases, interpretations and concrete features of the context of use are smoothed over in the rush to capture the essence of the requirements. Often, however, what is left out is essential to understanding stakeholders' needs. In contrast, approaches that stress context at the expense of abstraction may lead to floundering or to short-term customer satisfaction at the expense of long-term fragility of the system. What is needed is a synthesis of these two approaches: a synthesis that recognizes the complementary values of abstraction and context in requirements engineering and that does not relegate either one to a background role. Such a synthesis requires us not only to adopt new methods in practice but also to rethink our underlying assumptions about what requirements models are models of and what it means to validate them.

Most requirements engineering (RE) research and practice embodies a philosophy that I call *abstractionism*, which involves the building of simplified models of domains of discourse and proposed systems. Abstractionists make much use of formal models, such as state-based or activity-based descriptions of workflow and system behavior, entity-relationship models of domain information, and goal dependency networks for businesses and systems.

All science and engineering depends, of course, on the discovery and use of appropriate abstractions. If we could not capture what was somehow essential to the functioning or structure of a system, we could not build it. Any complex system has to be described in advance at a suitably high level of abstraction, and the world into which the system will be introduced needs to be described similarly.

Discovering the appropriate abstractions is a two-level activity. First, we must decide on the ontology of the phenomena that we wish to describe. We always have choices in picking an ontology, although devotees of specific design methods and languages seldom decide to exercise this choice. For example, a description of a piece of the world or a system that rests heavily on the concepts of events, activities and temporal sequence is more appropriate for time-constrained systems, such as real-time controllers, than an ontology that had a complementary emphasis on objects, their properties, relationships, and constraints on these properties and relationships.

Having chosen an ontology, we can then use it to represent the domain of discourse. Suppose we have chosen an ontology based on information structure and relationships (as embedded in several database design methods, for example), we are then faced with the challenge of using this ontology to represent our system. Here, too, we are faced with choices. What do we include in the model, and what do we simplify away?

The key notion underlying abstractionism is the model, a sufficiently restricted description of a phenomenon that can stand in its stead and be used to answer questions about the phenomenon. But the restrictedness of a model is both its strength and its Achilles heel. A model is more malleable than the modeled phenomenon; it can be written down on paper and stored in a relatively small repository; it can be analyzed using automated tools. But malleability is not the same as compliance, and sometimes we simplify away many of the features of the world that we may need to know about.

It is against this backdrop that several RE research projects have been conducted in recent years to investigate the role of naturalistic inquiry [1] methods in RE [4][5]. Typical of this research program has been the use of ethnographic methods to help requirements engineers understand better the use situation prior to their construction of abstractionist models.

An alternative design philosophy to abstractionism is *contextualism*, according to which the particularities of the context of use of a system must be understood in detail before the requirements can be derived. This philosophy is especially sympathetic to naturalistic methods of inquiry, which contextualists use to uncover and help interpret these particularities. Unlike abstractionists, however, contextualists do not do this just as an initial basis for model building. In fact, in the social sciences that make most use of naturalistic methods, such

1090-705X/97 $5.00 © 1997 IEEE

as anthropology, it is unusual to develop models from naturalistic data.

The differences between abstractionism and contextualism are summarized in Table 1.

A consideration of the complementary strengths and limitations of abstractionism and contextualism, which are given in Table 2, suggests that a synthesis of the two approaches would have great advantages. We would like to avoid the worst excesses of model-building and understand the context in which a system is to be used so well that it will truly fit its context rather than some simplified conception that fits our modeling ontology. On the other hand, we generally are not able to engage in laborious and time-consuming fieldwork, and our designs cannot progress reliably if a "rich description" is our only starting point.

Despite the apparent incompatibilities implied in these descriptions, I believe that the objectives of abstractionism and contextualism can be reconciled. If we cannot do this, we will continue to develop systems that, however well they are developed according to the canons of abstractionism, are less useful than their customers deserve, because they do not fit comfortably into their contexts of use.

However, to bring such a synthesis about means more than merely adopting some methods from the social sciences for data collection before the "real" work of RE starts. This weak, "uncommitted" view can lead to minor improvements in the modeling of requirements, because it entails the use of more appropriate methods for gathering background knowledge. A strong, "committed" view, however, requires that we acknowledge that most of the

Table 1: Comparison of Abstractionism and Contextualism

	Abstractionism	Contextualism
Role of description	Abstractions are powerful and general	Particularities are as informative as generalities
Design criteria	Design integrity	Contextual fit
Origin of requirements	Prescriptive recommendations	Current practice
Role of users	Management	End-users
Community of practice	RE and software engineering	CSCW and HCI

Table 2: Comparison of strengths and limitations of abstractionism and contextualism

	Abstractionism	Contextualism
Advantages	Generalization across contexts Standard methods for constructing abstractions	Accommodation of richness of contexts
Limitations	Oversimplification Overemphasis of normative cases	Descriptive, not prescriptive Overemphasis on immediate actors and current practice

phenomena contained in models of systems and domains are socially constructed, and not objectively real [1]. "Business processes", for example, do not exist in businesses at all, but in key stakeholders' interpretations of the business. They are constitutive.

If we adopt this strong view, our models become simplified descriptions of the interpretations of reality offered by the stakeholders in the system, not representations of the real world. In some situations, this distinction is too fine to make a difference (for example, in modeling physical phenomena in real-time control systems, or some legislative requirements for an information system). Generally, however, the shift is significant. The fidelity of a model no longer is as much an issue as the degree to which it represents to the stakeholders' satisfaction and agreement the islands of stability that always arise in any sea of constitutive phenomena.

Are there examples of such models? Well, yes, there are, but they are not generally to be found in the literature of RE. An example of how a strong synthesis can be brought about would be to apply Spradley's [6] methods of taxonomic elicitation and domain analysis from the discipline of cognitive anthropology to the construction of object-oriented analysis (OOA) models. Spradley's discussion of the uses to which ethnographic interviews and participant observation can be put read uncannily like the literature of OOA with two significant differences: Spradley was writing ten years before OOA caught on, and his guidelines, unlike those of most OOA methods, are rich and strongly heuristic. What results from such an inquiry into social phenomena (Spradley's favorite examples are cocktail lounges and the street life of tramps) are not models of these settings that are in any sense "correct", but rather analyses of how the participants in such social situations see their world. This is exactly how we need to look at the activity of modeling when we create models that help us develop systems to improve business processes.

Another example of a strong synthesis of contextualist and abstractionist thinking is illustrated by some work of mine with Idris Hsi [2]. We were interested in questioning the core assumption of goal-refinement approaches to RE: that a system can be said to embody a set of goals. The big contextualist questions that arise from such a claim are obvious: whose goals are we talking about? how do we gather them? what can happen to thwart these goals in a real context, and how should our system cope with the vagaries of real life? After interviewing administrative assistants about how they organize meetings and in analyzing a corpus of e-mail messages about meeting scheduling, we compared the results with the abstractions to be found in goal-refinement analyses of meeting scheduling [7]. We then developed a set of specific heuristics for doing goal refinement in conjunction with naturalistic inquiry techniques. These naturalistic techniques are not to be used initially to help develop insight about the problem that is subsequently and by some ineffable means transformed into a formal model. Nor are they used to develop models of how meeting scheduling actually happens. Rather, they are used to develop an understanding of how stakeholders interpret the process in which they are engaged and how that interpretation should shape the formal refinement process as it actually unfolds.

The strong view of synthesis leads us to ontological and epistemological commitments (commitments about what our models are models of and what it means to validate them) that conflict with the conventional use of models in RE. To validate an "as-is" model is not to test the correspondence of a model against reality but to increase one's confidence in the trustworthiness of the model [1]. Validating a "to-be" model is not a process of checking whether the model describes what the customer really wants, as if that were a stable phenomenon, but to gain or predict the approval of stakeholders from demonstrations of functionality.

References

[1] Lincoln, Y. and E. Guba, *Naturalistic Inquiry*. Sage Publications, 1985.
[2] Potts, C. and I. Hsi, Abstraction and Context in Requirements Engineering: Toward a Synthesis, *Annals of Software Engineering*, to appear, 1997.
[3] Potts, C. and W. Newstetter, Naturalistic Inquiry and Requirements Engineering: Reconciling Their Theoretical Foundations, *Proceedings ISRE'97: Third International Symposium on Requirements Engineering*. Annapolis, MD: January 6-8, 1997. IEEE Computer Society Press.
[4] Randall, D., J. Hughes and D. Shapiro, Steps toward a Partnership: Ethnography and System Design, in M. Jirotka and J. Goguen (Eds.) *Requirements Engineering: Social and Technical Issues*, Academic Press, 1994.
[5] Sommerville, I., T. Rodden, P. Sawyer, R. Bentley and M. Twidale, Integrating Ethnography into the Requirements Engineering Process. *Proceedings RE'93: International Symposium on Requirements Engineering*. San Diego, CA: January, 1993. IEEE Computer Society Press
[6] Spradley, J. *Participant Observation*. Holt Rinehart Winston, 1980.
[7] Van Lamsweerde, A. Darimont, R. and Massonet, P. Goal-Directed Elaboration of Requirements for a Meeting Scheduler: Problems and Lessons Learnt, *Proc. RE'95: 2nd International Symposium on Requirements Engineering*, IEEE Computer Society Press, 27-29 March, 1995, York, UK.

Session 6A

Foundations

Requirements for Telecommunications Services: An Attack on Complexity
Pamela Zave and Michael Jackson

A major impediment to good RE practice is the sheer complexity of the system behaviours that need to be specified. Are there techniques that could be adopted minimize and manage this complexity? This paper answers the question affirmatively and presents several techniques that have proven useful in formally specifying behaviours in the telecomunications domain. This is an important paper not only for the complexity management techniques that it presents, but also for the questions that it asks.
— *John Mylopoulos*

Naturalistic Inquiry and Requirements Engineering: Reconciling their Theoretical Foundations
Colin Potts and Wendy C. Newstetter

Much of the research in RE views the area as an enterprise dedicated for the development of better processes, techniques, tools and notations for RE. Less is devoted to thinking of the kinds of domains and the kinds of realities we are supposed to "engineer" and the differences in commitments we can and should make in intervening into these domains. This paper focuses exactly on this "thinking" part and offers a well argued and clear statement about the nature of naturalistic inquiry and its relation to RE. Among other things, the paper sheds light on how differences in the commitments we make in RE about the realities we are intervening into and the way in which we can know about them can make a difference in thinking of what we do or should be doing. The paper also clarifies the relevance of these questions (and differences in answering them) through a number of examples and quotations from the existing literature on systems development. Though the paper's conclusions on the future of naturalistic inquiry in RE are not bright, the paper does shed light brightly on the dilemmas we face in RE.
— *Kalle Lyytinen*

On the Use of a Formal RE Language: The Generalized Railroad Crossing Problem
Philippe Du Bois, Eric Dubois, and Jean-Marc Zeippen

Moving from an informal expression of a customer's needs to a precise representation is easier to recommend than to do. This paper gives a clear, step-by-step description of the process of requirements specification using a well-known case study and the Albert II specification language.
— *Robyn Lutz*

Requirements for Telecommunications Services: An Attack on Complexity

Pamela Zave
AT&T Laboratories
pamela@research.att.com

Michael Jackson
AT&T Laboratories and MAJ Consulting Limited
jacksonma@attmail.att.com

Abstract

In engineering the requirements for a telecommunications system, the greatest obstacle to be overcome is the sheer complexity of the required behavior. We present several ways of managing and minimizing this complexity, all of proven effectiveness. Most of the specification techniques result from specific application of general requirements principles to the telecommunications domain.

1. Introduction

Telecommunications services are complex and rapidly becoming more so. If we wish to apply formal methods of requirements engineering (whether forward or reverse) in the telecommunication domain, then the primary obstacle we face is the sheer complexity of the behavior to be described.

This paper focuses on one kind of telecommunications system: the long-distance telephone network (LDTN). An LDTN is owned and operated by a service provider, and provides long-distance (rather than local) access and telephone services to its subscribers. We present five ways to minimize the complexity of formal requirements for LDTNs. This attack on complexity has been developed by studying a wide range of LDTN services, and appears to bring us much closer to the goal of feasible formal requirements for telecommunications services.

All five specification techniques are applications of the principle "a formal representation should be as simple as possible, but no simpler." In addition, the first, fourth, and fifth techniques (Sections 2, 5, and 6) are also applications of more specific principles of requirements engineering or specification. We explain the relevant principles at the beginnings of these sections, in hopes that their successful application to the telecommunications domain will suggest similar applications to other domains.

In each of the next five sections, after presenting the specific principle (if any) and specification technique, we explain how it attacks complexity and compare it to approaches used in other telecommunications specifications.

The discussion in this paper is relatively informal, but the specification techniques and the integrated example (from which the examples in the paper are drawn) are not. They are based on a framework for multiparadigm specification [16] which is now being extended and embedded in Higher-Order Logic [4].

A cautionary note before we begin: The most heavily used and most hopelessly ambiguous word in all telecommunications is "call." Because it is almost impossible to discuss telephones without it, we use it extensively in casual and informal ways. Because it is hopelessly ambiguous, it does not appear—direct or disguised—in our formal specifications.

2. No description of the network

When specifying the requirements for a system, the right subject matter is not the system, but the environment of the system [18]. The phenomena of the environment include those phenomena that are shared between the environment and the system, i.e., the phenomena of the system/environment interface. There are two distinct kinds of assertion. An assertion in the indicative mood represents domain knowledge—it describes the environment as it is without or in spite of the system. An assertion in the optative mood represents a requirement—it describes the environment as we would like it to be because of the actions of the system.

Applied to an LDTN, this principle means that the subject matter of the requirements does not include the LDTN itself, except for those parts of the LDTN that are shared with its environment. This is very good news in terms of complexity, because the network is huge and varied. Figure 1 shows that the formal representation of LDTN requirements does not include the unshared parts of the LDTN. How is it possible to do this?

Call-processing requirements are specified in terms of *usages*. A *usage* is an episode during which an off-network trunk line (henceforth simply called a *trunk*) is in continuous use for the purpose of providing LDTN access

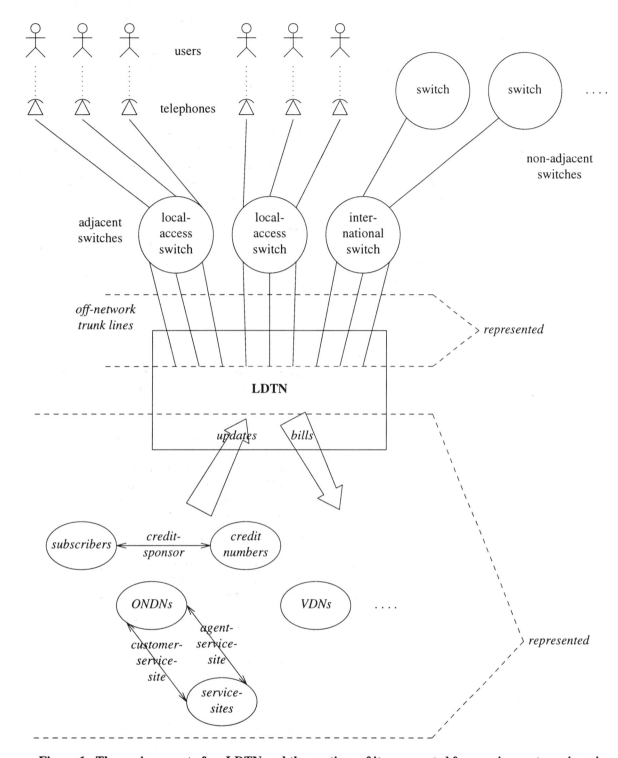

Figure 1. The environment of an LDTN and the portions of it represented for requirements engineering.

to a particular user. Each usage is either a *caller-usage* or a *callee-usage,* depending on whether or not the user initiated the episode. Voice connections are represented by the relation *connected(u_1,u_2,...,u_n,p),* where u_i is a usage and *p* is the sponsor of the connection.

To show how call processing can be specified fully in terms of usages, let us also consider a business-oriented service. Users dial a toll-free number, listen to music while they are waiting for the next available service agent, then talk to a service agent. Service agents work from home, and report for work by dialing a different toll-free number. By entering short digit sequences from their keypads, agents give commands such as "pick up next caller," "drop caller," and "initiate conference with backup agent."

In the specification of this service, there are connections among customer caller usages, agent caller usages, and backup callee usages. The toll-free numbers are associated with particular virtual *service-sites,* and these sites are used to constrain which usages get connected. For example, when a service agent executes a "pick up next caller" command, he is connected to the longest-waiting customer caller usage that is associated with the same service site he is.

The other requirements of the LDTN concern "provisioning" (which encompasses subscription, configuration, feature activation, etc.) and billing. For these functions the environment includes entities such as *subscribers, credit-numbers, service-sites,* and *directory-numbers.* Directory numbers divide into *off-network-destination-numbers (ONDNs),* which point to calling destinations outside the LDTN, and *virtual-destination-numbers (VDNs),* which point to (among other things) service sites. The environment also includes relations among these entities such as *credit-sponsor(c,s)* (meaning that the credit account with number c is billed to subscriber s), *customer-service-site(v,i)* (meaning that a customer calling VDN v will be served at service site i), and *agent-service-site(v,i)* (meaning that an agent calling VDN v will be working at service site i). The environment also includes the events through which the LDTN is notified of changes to these agreements, and the events through which the environment is notified of payments due.

As shown in Figure 1, all of these entities and subscriber agreements are represented formally. Some of the requirements state how the agreements affect call processing (by determining which usages are connected, who pays for the connection, etc.). These requirements might appear to be unimplementable, because the agreements exist in the environment and not in the LDTN. However, the requirements are augmented by domain knowledge relating the agreements to the change events, and these change events are shared with the LDTN. Implementors can use the change events to construct an image of the agreements within the LDTN, and this image can then be used to implement the requirements [18].

Figure 1 also shows that some parts of the environment—users, telephones, and external switches— are not represented in the requirements. Requirements for call processing *could* be specified in terms of telephones and users. The next step would be to specify domain knowledge about how the local-access switches behave to create a bridge between telephones and the LDTN. After reasoning about these two, we could derive refined, implementable requirements for the LDTN in terms of

behavior observable at the trunks [18].

The above approach would be reasonable for domestic call processing, but unreasonable for international calls, because the architecture of the network for local access to telephones can vary from country to country.[1] Even for domestic call processing this approach is more elaborate than necessary. Local-access switches do little processing on long-distance calls, so there is little conceptual distance between what is observable at the telephones and what is observable at the trunks. By careful description of the trunks, it is possible to get a sufficiently clear picture of what users are experiencing at their telephones.

Thus the scope of our formal representation, as shown in Figure 1, is as small as it can conveniently be. We are reducing complexity by leaving out unnecessary things. Even more important, we are avoiding the implementation bias that would follow inevitably any attempt to describe the network in the requirements: there are many possible architectures for an LDTN, and no sound basis at the requirements level for preferring any one of them. The absence of the network in the requirements is also vital to the success of two other simplifications (as will be explained in Sections 3 and 6).

Some previous specifications of telecommunications services rely on a representation of network architecture (e.g., [19]), while others incorporate events that are internal to the network and not observable from any external vantage point (e.g., [2]). Some specifications scrupulously avoid using internal events, but represent subscriber agreements as internal to the network (e.g., [3]).

3. Uniform event semantics

Not counting reference to the subscriber agreements (whose values persist much longer than a call does), the requirements for call processing are specified strictly in terms of trunk behavior. Trunk behavior is specified exclusively and uniformly in terms of events.

One obstacle to this uniformity is the distinction between in-band and out-of-band signaling on a trunk. Out-of-band signals are messages traveling on a signaling channel that is completely separate from the voice channel. In-band signals are sounds, and travel on the voice channel. For example, audible tones and announcements are in-band signals from the LDTN to its environment. DTMF tones ("touch tones") are frequently used as in-band signals from the environment to the LDTN. Automatically recognized words, and changes between sound and silence, also sometimes serve as in-

---

[1]In fact, even the access architecture for the LDTN's home country is unlikely to be as simple and uniform as we have pictured it.

band signals to the LDTN.

Obviously the implementations of in-band and out-of-band signaling are vastly different, but the roles of these signals in requirements are often identical. For example, the directory number that a caller wishes to reach usually arrives as an out-of-band signal—the local-access switch collects the dialed digits, determines that a long-distance call is being requested, and sends an *initial-access* message to the LDTN with the dialed digits in a data field. But the directory number can also arrive as an in-band signal. Suppose that a user sets up and makes a credit-card call. He then wishes to make another credit-card call without the bother of re-entering his credit information. Some LDTNs will accept the "#" DTMF tone as a signal to disconnect the current call, after which they will recognize in-band digits as the directory number of a request for a new call on the same credit account. Note that from the perspective of the local-access switch and its interface with the LDTN, a sequence of credit-card calls made in this way is a single caller usage in which a voice connection is maintained continuously between the user and the LDTN. This voice connection carries the in-band signals by which the user requests a sequence of long-distance connections.

Figure 2 illustrates how in-band and out-of-band signaling can be unified. Figure 2 is an extended finite automaton which specifies part of the interface for a callee usage. This automaton has nested states as in Statecharts [5]. A *withdraw* event ends the callee usage on the initiative of the callee, while an *EXPEL* event ends the callee usage on the initiative of the network. *EXPEL, withdraw,* and *answer* events are all out-of-band signals, which are easily represented as events.

The remaining event types in Figure 2 represent in-band signals. In-band signaling from the LDTN to the environment is put into an event format by representing the events that begin and end audible tones and announcements. If a collect call is the occasion for this callee usage, then an event of type *COLLECT-PERMISSION-NEEDED* begins the announcement that this is a collect call. An event of type *MESSAGE-ENDED* ends this announcement and begins the playing of a recorded message from the caller. Another event of type *MESSAGE-ENDED* ends this recording and begins the announcement that prompts for acceptance or refusal of the collect call.

In-band signaling from the environment to the LDTN is put into an event format by indicating when incoming sounds are significant and by indicating which incoming sounds are interpreted as which symbolic events in the specification. For example, the state labeled *collect-prompt* is also labeled *len=1*, indicating that a sequence of DTMF tones of length up to 1 may have signaling significance in this state.[2] Monitoring for DTMF tones

can end in any of these ways: (1) The state is exited because of some event having nothing to do with monitoring, for example *EXPEL*. (2) One digit is collected. (3) Since the state also has *time=4* written in it, 4 seconds elapse without the arrival of a digit.

In termination situations 2 and 3, the event that ends monitoring (a digit arrival or timeout) is an event represented in the specification. Its type in the specification depends on which of the string patterns found on transitions out of the monitoring state best matches the collected string. If "1" was collected then the final event is considered to be an *accept-collect* event. If nothing or any other digit was collected, then the final event is considered to be a *refuse-collect* event.

Another obstacle to uniform event semantics is the necessity for short-term data such as directory numbers and credit numbers. Many event-based notations are augmented with variables for just this reason (for example SDL [12] and Statecharts [5]), but we wish to avoid this complication.

Events have attributes, according to their types. For example, each event of type *initial-access* has an attribute *s* of type *keypad-string;* the value of the *s* attribute of each *initial-access* event is the string dialed by the user and collected by the local-access switch.

Event attributes are the source of all short-term data in the specification. When we refer to short-term data in the specification, we always want the value of a particular attribute of the most recent relevant event of a particular type (or a value immediately derived from it). The value of the *s* attribute of the most recent relevant[3] *initial-access* event is referred to with the syntax *initial-access.s*. The tail of this same value is referred to with the syntax *substr(initial-access.s,2)*. This notation makes variables unnecessary.

There is an attribute reference in Figure 2. The *a* attribute of a *COLLECT-PERMISSION-NEEDED* event is a recorded speech from the caller. It is used as a state label in Figure 2, indicating that the recorded speech should be played to the callee during this state (more information on this syntax will be given in Section 5).

The absence of explicit data storage in the specification of call processing is a good example of how specification techniques depend on the properties of the application domain. It works for call processing because the data is short-term, seldom manipulated, and never combined (e.g., summed or averaged). Just as the user is not expected to perform many similar transactions and

---

[2]If a state has monitoring for DTMF tones *and* an outgoing tone or announcement, then the outgoing sound ceases as soon as the first incoming digit is received.

[3]"Relevance" refers to the scope of a formal description. For example, there is a description of a callee usage from which Figure 2 is taken. When this description is being applied to callee usage *u*, only events belonging to *u* are relevant.

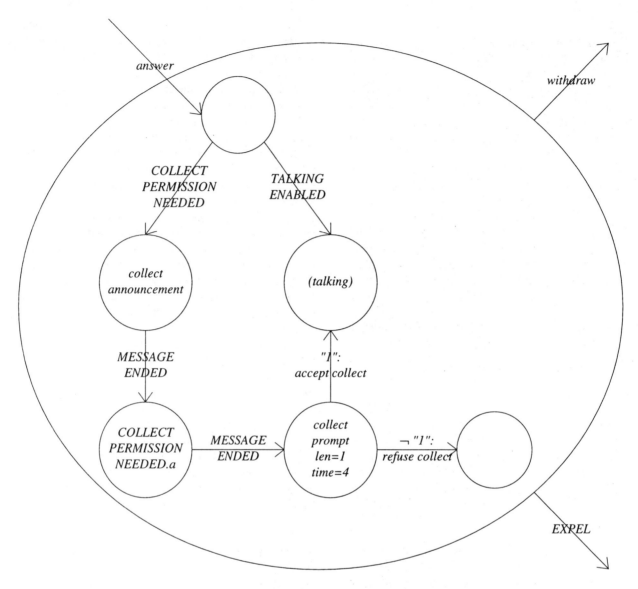

answer

withdraw

COLLECT
PERMISSION
NEEDED

TALKING
ENABLED

*collect
announcement*

(talking)

MESSAGE
ENDED

*"1":
accept collect*

COLLECT
PERMISSION
NEEDED.a

MESSAGE
ENDED

*collect
prompt
len=1
time=4*

¬ *"1":
refuse collect*

EXPEL

Figure 2. Part of the interface for a callee usage.

remember all of them, the call-processing part of the LDTN is not required to keep a history of the attributes of many similar events. (For both the user and the LDTN, remembering the most recent one is sufficient.)

A uniform event-based notation has many benefits. It is syntactically and semantically simple. An event-based notation lends itself very well to the parallel composition of modules by means of shared events, as found in process algebras. This form of composition will be used in Section 4. The absence of explicit data storage makes it unnecessary to decide which event attributes need to be saved and unnecessary to save them.

Many process algebras, for example CSP [6] and LOTOS [7], allow "hidden" or "internal" events; the behavior of a process can be nondeterministic due to hidden internal choices, which significantly complicates

the semantics of the languages. In our approach to requirements there is no need for the semantic complications of hidden events, because all events are external, as explained in Section 2. If we were using internal events, then we would be describing the required system rather than its environment.

We are not aware of any other telecommunications specification that considers in-band signaling, even though it is used extensively in LDTNs.

4. State decomposition

State decomposition is a well-known and powerful way of reducing complexity in a specification. It is an integral part of specifications in CSP [6] and Statecharts [5], it is an important aspect of multiparadigm

110

specifications [16,17], and it can also be used in languages such as Z [8].

In a specification with state decomposition, each module maintains its own state representation. The semantics of the module is a combination of constraining the current state based on the events that occur, and constraining the events that occur based on the current state. If modules share events, as they do in our specification technique, then the shared events must satisfy the constraints placed on them by all sharing modules.

Concerning the independence of the states local to the various modules, there is a continuum of possibilities from highly independent to highly dependent. If the states are highly dependent, there might be another mechanism (in addition to shared events) through which the modules may influence one another, for example state invariants [8] or "read-only" use of another module's state representation [17].

State decomposition reduces specification complexity in several ways. Each module can use a different notation, one that is well-suited to its needs for state representation and update (see Section 5). Each module has fewer relevant events, since not all events will affect its state representation or be affected by it. If the states of two modules are highly independent, and they have m and n states respectively, then a representation of $m+n$ states actually covers a state space of approximate size $m*n$.

Our specification is constructed in three modules. The *interface* module is concerned with the user interface for a user accessing the LDTN, including when access begins and ends, what the user hears, and what the user can do to affect call processing. The state of the interface module is the state of the user interfaces of all extant usages. The *connection* module is concerned with when usages are connected, and with informing usages of the status of other usages with which they are currently associated because of a connection request. The state of the connection module is the state of all extant connections and connection requests. The *agreement* module is concerned with agreements among the LDTN, its subscribers, and other service providers. It is also concerned with service, routing, and billing decisions based on these agreements. The state of the agreement module is the current state of the LDTN's agreements.

The separation between the agreement module and the other two is an obvious choice. The state of the agreement module is largely independent of the states of the other two. Its update events are completely disjoint from all call-processing events, as indicated in Figure 1. The influence of the agreement state on call processing will be discussed in Section 5.

The value of the separation between the interface and connection modules is more subtle. One benefit is that there is some independence of states and state updates, although it is limited. Consider, for example, the fragment of the connection module shown in Figure 3. This composes with the "collect" path through the callee interface in Figure 2. Note that between a *COLLECT-PERMISSION-NEEDED* event (which is shared) and an *accept-collect* event (which is also shared), Figure 3 has one state and no intervening events. Figure 2, on the other hand, has three states and two events. An interface specification often has unshared events and states that contribute to the user's comfort and elicit information from the user, without having an immediate effect on the connection.

The second benefit of the separation between the interface and connection modules is the ability it gives us to decompose even further.

Within the interface module, there are two submodules: a description of a caller interface, which applies to every caller usage, and a description of a callee interface, which applies to every callee usage and of which Figure 2 is a fragment. Within the connection module there is a submodule describing each kind of connection; a connection submodule applies to every attempt to make that kind of connection, and Figure 3 is a fragment of one of them.

Figure 4 is a call-processing example with the events depicted as small circles (time is progressing downward). The events of Figure 4 belong to four distinct usages. The four vertical boxes surrounding the events of the four usages indicate the applications of interface submodules to these usages. The boxes are labeled with the name of a submodule (*caller interface* or *callee interface*) and with a parenthetical comment indicating what the user is trying to do. The three horizontal boxes indicate applications of connection submodules to various events, so events inside two boxes are events within the scopes of two submodules.[4]

The point of Figure 4 is the limited and overlapping scopes of these submodules. This separation of concerns is made possible by the separation of the interface and connection modules. If there were no separation between interface and connection modules, then by simple transitivity, *all* of the module applications shown in Figure 4 would have to belong to a single module application. Considering the many usages, services, and overlapping cycles involved in Figure 4, this would be a complexity disaster.

Few published telecommunications specifications

---

[4]Figure 4 is inaccurate in two minor respects. When a usage is participating in a connection attempt, as has just been explained, not all the usage events are shared with the connection submodule. Also, since events are totally ordered, no two events should lie on exactly the same horizontal line. These details are simply too difficult to put into the picture.

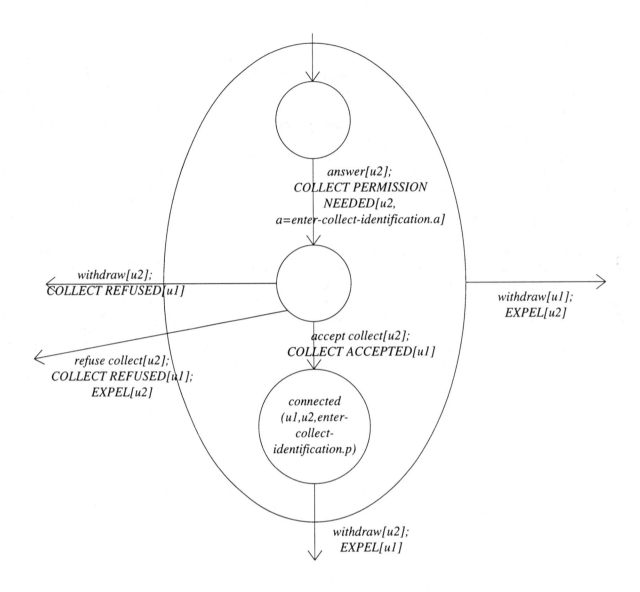

Figure 3. Part of the specification of a collect connection.

include agreements even in rudimentary form, but those that do (e.g., [3]) separate them from call processing.

It is not uncommon for previous telecommunications specifications to separate constraints on individual telephones from pairwise (connection) constraints (e.g., [3,9]), but their decompositions are crude compared to the one we are presenting here. In both referenced specifications, only the plainest of pairwise connections are considered. The cyclic patterns that make Figure 4 so interesting (a service agent talks to a sequence of service customers, a caller makes a sequence of credit-card calls) cannot occur in these specifications, so the separation between usage and connection modules is much less interesting and much less necessary. Since their usage descriptions also contain no realistic detail about the user interface, they are little more than projections of the connection description.

5. Specialized languages

There are many specification languages. Each language offers a different set of expressive capabilities, appropriate for specifying clearly and concisely a different set of properties. Each language also offers a different set of analytic capabilities, appropriate for rigorous reasoning about a different set of properties.

Thus one of the most powerful weapons we have against specification complexity is multiparadigm specification—the ability to combine languages as the problem demands [16]. We have put this weapon to use by developing specialized languages for each of the three major specification modules.

The notations of both the interface and connection modules are based on finite automata. Both have nested states as in Statecharts. In both notations the type of a

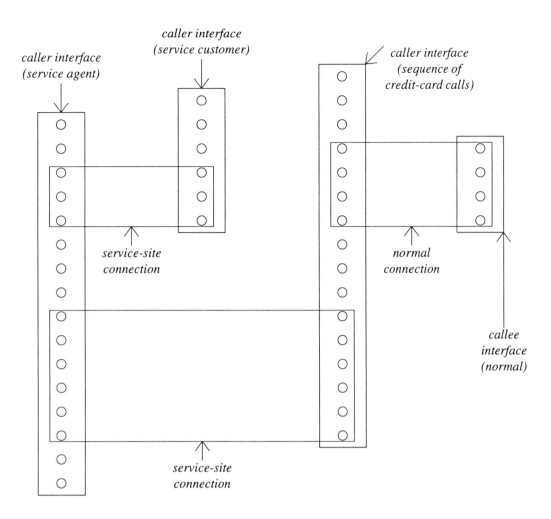

caller interface
(service agent)

caller interface
(service customer)

caller interface
(sequence of
credit-card calls)

service-site
connection

normal
connection

callee
interface
(normal)

service-site
connection

Figure 4. A call-processing example, showing the scopes of applications of submodules.

user-initiated event is spelled in lower-case characters, while the type of a network-initiated event is spelled in upper-case. In both notations attribute values for required network events can be specified. For example, Figure 3 expresses a requirement for a *COLLECT-PERMISSION-NEEDED* event to occur. An event of this type must have an attribute *a* of type *speech* (a recorded speech or announcement). The expression *a=enter-collect-identification.a* in Figure 3 indicates that the value of this attribute is the same as the *a* attribute of a previous event of type *enter-collect-identification*.

The automata of the interface module have some additional notations. As illustrated by Figure 2, a state label is the name of a tone or speech, and denotes a requirement that the sound be played on the outgoing voice channel during this state. As illustrated by Figures 2 and 5, there is a suite of notations for monitoring, recording, and parsing the incoming voice channel. In addition to the features shown here, there are also notations for parsing in-band commands of variable

length, recording voice messages, distinguishing interruptable from non-interruptable announcements, specifying error reprompts and retries, and specifying single-word voice recognition.

The automata of the connection module also have some additional notations. Since a connection submodule relates events from several usages, it is important to know which usage each event belongs to. Each event has a tag (*u1* or *u2* in Figure 3) indicating which role it belongs to; in a particular application of a connection submodule, each role is played by a specific, distinct usage.

State labels in connection automata denote requirements for connections among usages. In Figure 3, for example, the label on the third state means that during a visit to this state usages *u1* and *u2* are connected. The sponsor of the connection is determined by the *p* attribute of the most recent *enter-collect-identification* event.

In a connection submodule, a user-initiated event often stimulates a network-initiated event to inform another usage of the user-initiated event. This is so common that it has its own notation. In a transition label in a connection submodule, a semicolon is a (graphically small) state with one in-transition and one out-transition (not counting the transitions out of superstates, which are also applicable).

The agreement module contains declarations of relations representing agreements, and static constraints on them, in a notation based on the relational algebra in Z [13]. The active part of the agreement module consists of tables specifying how these agreements are used to make service, routing, and billing decisions. The syntax and semantics of these tables was developed for the purpose of specifying complex feature interactions in telecommunications systems [15].

Figure 5 is an agreement table specifying the network's response to an *initial-access* event. Each row of the table is a condition/response pair, where the response is the type of event that the network must emit, together with the proper attribute values of the event. The conditions are intended to be evaluated in order, so each condition implicitly includes the negation of all previous conditions.

Many of the conditions refer to agreement relations. For example, one of the first three conditions is true if and only if the dialed string (attribute *s* of the *initial-access* event) is one of three kinds of toll-free number. If any one of these conditions is true, then the sponsor attribute *b* of the response event gets its value from a relation associating VDNs with their sponsors.

If the first four conditions are all false, then the user is attempting either a collect or a credit-card call. The selection between them is not made by consulting the agreements, but by prompting the user for additional information, which is then checked using additional agreements. This is typical of service decisions in an LDTN—they are made using a combination of information from the agreements and information from the user. Our table language for service decisions is deliberately designed to abstract away from this distinction, so that the decision itself can be checked and understood clearly.

If the first four conditions in the table are all false, then the next condition in the table carries a network requirement to emit an event of type *CREDIT-PROMPT-NEEDED*. The condition *CREDIT-PROMPT-NEEDED* ≈≈> *refuse-cn* is true if and only if the user's response to the *CREDIT-PROMPT-NEEDED* event is a *refuse-cn* event. The condition *CREDIT-PROMPT-NEEDED* ≈≈> *enter-cn* is true if and only if the user's response is an *enter-cn* event. The user interface in Figure 6 specifies how the network should prompt the user and interpret his response.

Figure 6 is a fragment of the caller-interface submodule. Since Figures 5 and 6 share all of their events

(they have exactly the same vocabulary of event types), they have a great deal of influence on each other when composed. Without Figure 5, Figure 6 is highly nondeterministic. For example, it says that in the first state the network should emit one of five different event types, but cannot determine which one. The nondeterminism is resolved by composition with Figure 5, in which the values of the agreement relations lead unambiguously to one of the five outcomes.

The only other multiparadigm telecommunications specification we are aware of [10] uses somewhat different (although related) techniques and languages. The difference is explained by a difference in subject matter: it is concerned with specifying a local-access switch rather than an LDTN.

6. Temporal simplifications

In real requirements engineering, it is not possible to formalize everything. It is an important practical principle to apply formal methods only where they are cost-effective and obviously helpful.

One application of this principle to requirements for an LDTN is that we make several unrealistic temporal assumptions. We assume that agreements do not change during any usage that appeals to them. We assumes that there is no transmission delay in the network, so that an event is instantaneously visible to any part of the network that might be interested in it. We also assumes that otherwise possible user-initiated events do not occur during semicolon states. For example, in the transition of Figure 3 labeled *accept-collect[u2]; COLLECT-ACCEPTED[u1]*, we assume without justification that a *withdraw[u2]* event cannot occur at the semicolon.

We make these unrealistic assumptions because, without them, the specification would become so overloaded by the complexity of temporal nondeterminism and distribution issues that it would be difficult to understand what the services are doing. At least the set of specified behaviors is a subset of the set of real network behaviors. It is in fact the desirable subset: additional behaviors that violate the temporal assumptions might be tolerated, but they add nothing to the functions of the LDTN.

Note that this last attack on complexity relies heavily on the first one. If communication internal to the LDTN were described, it would be impossible to assume that communication from one end of the LDTN to the other can be implemented as instantaneous.

The practical consequence of writing an idealized specification is that the requirements are left undefined in situations that do not satisfy the temporal assumptions. These situations must be dealt with at a later (and probably less formal) stage of requirements engineering.

initial-access ≈≈>	unless *withdraw*
condition	**response**
advanced-800-caller-service(initial-access.s)=r'	*SERVICE-APPROVED* *[r=r',b=VDN-sponsor(initial-access.s)]*
advanced-800-agent-service(initial-access.s)=r'	*LOGIN-APPROVED* *[r=r',b=VDN-sponsor(initial-access.s)]*
retarded-800-destination(initial-access.s)=d'	*800-CONNECTION-APPROVED* *[d=d',b=VDN-sponsor(initial-access.s)]*
initial-access.g="1"	*NORMAL-CONNECTION-APPROVED* *[s=initial-access.s,b=initial-access.b]*
CREDIT-PROMPT-NEEDED ≈≈> refuse-cn ∧ *AT&T-DN-sponsor(initial-access.s)=b'*	*COLLECT-CONNECTION-APPROVED* *[s=initial-access.s,p=b']*
CREDIT-PROMPT-NEEDED ≈≈> refuse-cn	*COLLECT-CONNECTION-APPROVED* *[s=initial-access.s,p=initial-access.s]*
CREDIT-PROMPT-NEEDED ≈≈> enter-cn ∧ *CN-sponsor(enter-cn.s)=b'*	*CREDIT-CONNECTION-APPROVED* *[s=initial-access.s,b=b']*
true	*CREDIT-CONNECTION-DENIED*

Figure 5. Specification of the response to an *initial-access* **event.**

The goal of this later engineering stage is to make the behavior of the LDTN without the temporal simplifications as similar as possible to the behavior specified with the temporal simplifications, and it will need to take design decisions into account.

For example, consider the problem of transmission delay in a simple call, one that is specified exclusively by the caller interface, the callee interface, and a connection submodule. A reasonable design decision would be to implement most of the control logic for this type of call at the LDTN node connected to the caller trunk, and the remaining control logic at the LDTN node connected to the callee trunk. In this design, the code in the caller-access node would be based on a composition of the caller-interface submodule and the connection submodule. The code in the callee-access node would be based simply on the callee-interface submodule.

The temporal assumption that there is no transmission delay between the events of the caller-interface submodule and the connection submodule is actually satisfied by this design, since both are implemented at the same network node. There is real transmission delay, however, between the events of the callee-interface submodule and the connection submodule. This raises the possibility of internal race conditions not covered by the requirements. The best engineering approach would be to use a disciplined method for finding and neutralizing the extra behaviors made possible by the race conditions, where "neutralizing" them means making them appear to the environment as if no race condition had occurred.

Most previous specifications of telecommunications services simply enjoy the temporal simplifications without mentioning or justifying them (e.g., [11]). Blom et al. take the same approach that we do [1]. Kay and Reed's work is notable in that it does without temporal simplifications concerning the off-network trunks [9].

7. Conclusion

We use a few other techniques for reducing specification complexity that have not been covered in this paper. Most, for example reuse of modules, are common and not as interesting as the techniques featured here.

There are several reasons to believe that this attack on the complexity of LDTN requirements will be successful. We have used it to specify a substantial example (from which the fragments in this paper were taken) in ten pages of diagrams and tables. It is possible to define a large number of useful consistency checks, both application-dependent and application-independent, on requirements in this form. The approach has been motivated, in part, by a long study of the feature-interaction problem in telecommunications systems [14].

Nevertheless, much work remains to be done. We need further contact with the realities of LDTNs, including more examples and feedback from requirements engineers. We need to develop analysis algorithms and other support tools. We also need to consider the mapping from requirements in this form to network architectures, because LDTN requirements are implemented piecemeal, in terms of code installed on many different network nodes.

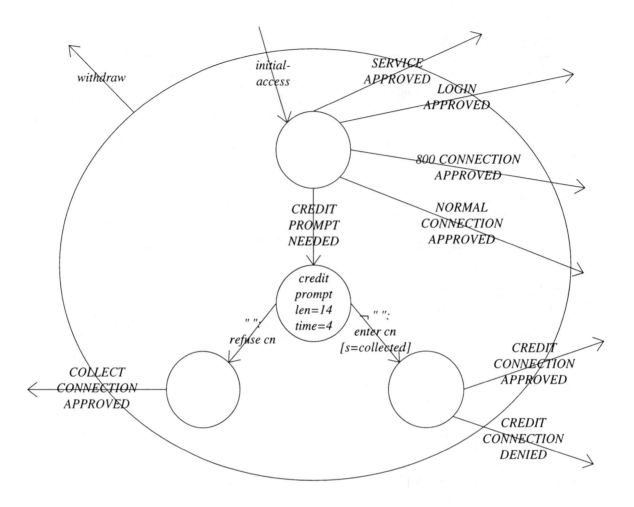

Figure 6. Part of the interface for a caller usage.

8. References

[1]   Johan Blom, Roland Bol, and Lars Kempe. Automatic detection of feature interactions in temporal logic. In K. E. Cheng and T. Ohta, eds., *Feature Interactions in Telecommunications Systems III*, pages 1-19. IOS Press, 1995.

[2]   Kong Eng Cheng. Towards a formal model for incremental service specification and interaction management support. In L. G. Bouma and H. Velthuijsen, eds., *Feature Interactions in Telecommunications Systems*, pages 152-166. IOS Press, 1994.

[3]   Mohammed Faci, Luigi Logrippo, and Bernard Stepien. Structural models for specifying telephone systems. *Computer Networks and ISDN Systems*, to appear.

[4]   M. J. C. Gordon and T. F. Melham. *Introduction to HOL.* Cambridge University Press, 1993.

[5]   David Harel. Statecharts: A visual formalism for complex systems. *Science of Computer Programming* VIII:231-274, 1987.

[6]   C. A. R. Hoare. *Communicating Sequential Processes.* Prentice-Hall International, 1985.

[7]   International Organization for Standardization. LOTOS—A formal description technique based on the temporal ordering of observational behavior. ISO 8807, 1989.

[8]   Daniel Jackson. Structuring Z specifications with views. *ACM Transactions on Software Engineering and Methodology* IV(4):365-389, October 1995.

[9]   Andrew Kay and Joy N. Reed. A rely and guarantee method for Timed CSP: A specification and design of a telephone exchange. *IEEE Transactions on Software Engineering* IX(6):625-639, June 1993.

[10] Peter Mataga and Pamela Zave. Multiparadigm specification of an AT&T switching system. In Michael G. Hinchey and Jonathan P. Bowen, eds., *Applications of Formal Methods*, pages 375-398. Prentice Hall International, 1995.

[11] Tadashi Ohta and Yoshio Harada. Classification, detection and resolution of service interactions in telecommunication services. In L. G. Bouma and H. Velthuijsen, eds., *Feature Interactions in Telecommunications Systems*, pages 60-72. IOS Press, 1994.

[12] Anders Rockström and Roberto Saracco. SDL—CCITT specification and description language. *IEEE Transactions on Communications* XXX(6):1310-1318, June 1982.

[13] J. M. Spivey. *The Z Notation: A Reference Manual*, Second Edition. Prentice-Hall International, 1992.

[14] Pamela Zave. Feature interactions and formal specifications in telecommunications. *IEEE Computer* XXVI(8):20-30, August 1993.

[15] Pamela Zave. Secrets of call forwarding: A specification case study. In *Formal Description Techniques VIII (Proceedings of the Eighth International IFIP Conference on Formal Description Techniques for Distributed Systems and Communications Protocols)*, pages 153-168. Chapman & Hall, ISBN 0-412-73270-X, 1996

[16] Pamela Zave and Michael Jackson. Conjunction as composition. *ACM Transactions on Software Engineering and Methodology* II(4):379-411, October 1993.

[17] Pamela Zave and Michael Jackson. Where do operations come from? A multiparadigm specification technique. *IEEE Transactions on Software Engineering* XXII(7):508-528, July 1996.

[18] Pamela Zave and Michael Jackson. Four dark corners of requirements engineering. *ACM Transactions on Software Engineering and Methodology*, January 1997, to appear.

[19] Israel Zibman, Carl Woolf, Peter O'Reilly, Larry Strickland, David Willis, and John Visser. Minimizing feature interactions: An architecture and processing model approach. In K. E. Cheng and T. Ohta, eds., *Feature Interactions in Telecommunications Systems III*, pages 65-83. IOS Press, 1995.

Naturalistic Inquiry and Requirements Engineering: Reconciling Their Theoretical Foundations

Colin Potts, Wendy C. Newstetter
College of Computing, Georgia Institute of Technology
{potts,wendy}@cc.gatech.edu

Abstract

A growing awareness of the need to take into account social and contextual factors requirements engineering (RE) has led to expanded use of naturalistic inquiry (NI) methods, such as ethnography, for capturing relevant data. There is little debate about the potential value of NI to the development of systems; however, most previous discussions have emphasized practical techniques and benefits. Less attention has been given to the ontological and epistemological commitments that a naturalistic research paradigm assume and the extent to which these assumptions conflict with those that pervade RE practice. In this paper we present the axioms that NI. In each case we address both the points of agreement and tension that arise when these axioms are compared with the implicit assumptions upon which RE practice and research methods are based. We illustrate the discussion with specific examples from published sources and our experience.

Keywords: *ethnography, naturalistic inquiry, qualitative methods, design*

Classification: *[1.4, 2.1,4 (research methods)] B*

In recent years, requirements engineering (RE) has witnessed a growing appreciation for and expanded use of naturalistic inquiry (NI) in requirements capture and systems design. We are using the term *naturalistic inquiry* here to refer to a suite of research methodologies which utilize the data collection techniques of participant observation, respondent interviewing, artifact collection and the generation of field notes. These techniques derive from two traditions: the anthropological tradition of ethnography and the more recent sociological tradition of ethnomethodology. Both are concerned overall with discovering the meanings and tacit understandings that participants in social contexts negotiate and derive from interactions with the artifacts and other participants there. Using these types of data during design may lead to a system to which people can adapt and be productive when the system is installed, thereby increasing the likelihood that it will be used effectively.

The cases of an air-traffic control (ATC) system and the London Underground Control Room system illustrate how using ethnographic methods can inform both original and adaptive design. In the ATC system, the design for a computer-based air-traffic control system was improved on the discovery of the controllers' practice of using paper flight strips as representational space for communication and for planning flight paths [17]. In the case of the London Underground Control Room system, ethnography was used to assess the use of a new complex computerized system complemented by an extensively modernized work space [10]. This study revealed how the new system interfered with established practices of information gathering, distribution and coordination of action among control room members. In this light, a redesign of the system was proposed to support how personnel make use of a paper timetable for communication and coordination. In both cases, NI was able to unearth the "native" uses of essential representational artifacts, the meanings and uses ascribed to them, and their function in a larger activity. Ignoring these in systems design can significantly decrease efficiency in an already stressful work environment.

Although there has been great interest in the techniques and benefits of undertaking NI, little attention has been given to the ontological and epistemological commitments that a such a research paradigm assume, and less still on the extent to which these assumptions conflict with those that pervade RE practice. The traditional RE research paradigm, in common with most engineering research and practice, is founded on the philosophical tradition of positivism, which construes knowledge as accruing through the systematic observation of stable and knowable phenomena. Such a world view is consistent with the objectives of RE, because the activity of capturing requirements assumes that stable and specifiable phenomena are out there in the customer's world available for discovery through surveys, interviews, the close reading of informal requirements documentation, participant design, modeling and even ethnographic techniques.

Requirements engineers are no more or less naive in their positivism than any other technical professionals. What happens, we claim, is that requirements engineers, being members of the culture of science, often fall back on and operate from a *tacit* positivism during RE. That modern philosophers of science and those scientists working in frontier areas of natural science repudiate naive positivism is beside the point; the point is that tacit positivism is generally a useful everyday epistemology (much more so than agonizing over the foundations of knowledge), because it is compatible with the need to capture and tie down clear-cut features that can be modeled and turned into artifacts.

As laid out by Lincoln and Guba [9], the axiomatic assumptions of NI contrast markedly with the tacit axioms of naive positivism and do not align with conventional RE practice. This means that RE practitioners who seek to adopt NI techniques cannot continue to operate within a positivistic perspective without at least making their tacit positivism explicit and therefore open to criticism and change. Lincoln and Guba's axioms can be restated as follows: (1) Identifying cause from effect is impossible because all entities simultaneously shape each other; (2) Knowledge can only be described idiographically as a working hypothesis that describes an individual case; (3) The inquirer and the phenomena under inquiry interact and influence each other, so they are inseparable; (4) Reality consists of multiple constructed realities that can be understood to some extent but cannot be predicted or controlled; and (5) All inquiry is value-laden.

In the next section of this paper, we discuss these five axioms in greater detail, contrasting them with their corresponding positivist assumptions. In doing so, we elucidate the points of agreement or tension that arise when these sets of implicit assumptions are set side by side. Our main emphasis is to show the consequences of questioning the tacit positivism of RE practice and replacing them with corresponding NI axioms. We then reexamine these correspondences as they apply to RE research. We conclude by considering the implications of the ontological and epistemological underpinnings of NI for RE practice and research and what it would mean to reconcile the apparent conflicts.

1. Axioms of NI as they apply to RE practice

1.1 Mutual influence and RE practice

Positivist view: Most actions can be explained as the result of a cause.

Naturalist view: Identifying cause from effect is impossible because all entities simultaneously shape each other

In NI, Lincoln and Guba claim, the bifurcation of events into causes and effects must be replaced by networks of mutual influence. This axiom is the most innocuous of Lincoln and Guba's challenges to customary RE practice (though not, as we shall see, to RE research). Constructing elaborate causal models is not generally important in RE practice; it is more valuable to understand relationships that exist between elements of the problem domain than to develop causal models of how things work. It is also widely understood that introducing a new system always changes the organization in which it is embedded; thus, the requirements for the system and its context of use continually shape each other as they evolve. As Dahlbom and Mathiassen [4] put it, system development is a dialectical process:

> The result of an intervention process is not a computer system representing the solution to a specified problem. A systems development project results in a changed organization. The best we can say is that a new situation has emerged. The best we can hope for is that more useful and effective procedures and tools for information processing have been introduced. The situation is different, and gradually new problems and conflicts will appear. [p. 120].

Thus RE practice enjoys the maturity of accepting that systems designed to meet requirements inevitably change those requirements.

1.2 Idiographic knowledge and RE practice

Positivistic view: Inquiry aims to develop a nomothetic body of knowledge in the form of generalizations that are "true" and will hold for other times and places.

Naturalistic view: Knowledge can only be described idiographically as a working hypothesis that describes an individual case

Most RE practice embodies *abstractionism* [], the doctrine that simplified models can capture the essence of application domains and systems. Abstractionism stems directly from the positivist approach to knowledge, in which only testable and generalizable claims are held to be worth making. Such claims are law-summarizing or *nomothetic*; they stand apart from and above the context in which they are made.

Thus, one could imagine developing a domain model of scheduling, in which abstract object classes, operations and rules that are relevant to scheduling were described generally. This domain model could then be reused, specialized, instantiated, or customized for a range of specific contexts, such as the specification of a meeting scheduler or an elevator scheduler. Or, one could write a number of specific scenarios with the intention of generalizing them into an "as-is" model of current scheduling behaviors or a "to-be" model of requirements for a scheduling support system [15]. In the abstractionist framework, descriptions of the particular are useful only to the extent to which they illuminate and help produce a description of the general.

Abstractionism does not require, of course, that every abstraction is universally applicable. The range of cases over which an abstraction is useful always depends on context. It is possible, for example, that the abstractionist might find elevator schedulers and meeting schedulers to have too little in common (perhaps because of fundamental differences among the things being scheduled or the processes involved in making scheduling decisions) to justify a common abstract model. However, more restricted nomothetic claims could still be made: for example, all meeting schedulers might have enough in common to be describable using the same basic abstractions.

An alternative design philosophy, *contextualism* [15], claims that the richness of a system's contexts of use is more important to its being understood than is an abstract model and, indeed, that it is impossible to stand apart from and above the contexts in which one operates. Context-bound knowledge is *idiographic*. According to contextualism, idiographic knowledge so dominates the inquiry situation that nomothetic abstractions tell us little and create an illusory sense of understanding, explanation, and control. Thus, a contextualist analysis of meeting scheduling might conclude that the practices of meeting scheduling in different organizations differed so much from each other that the tasks were fundamentally different and no common model of meeting scheduling could work across them all. To try to unify the scheduling practices that appear in different contexts into a family of policy abstractions of universal applicability is a quixotic quest, and the resulting abstractions will so oversimplify the domain of discourse that any system on which such a model is based would be doomed to ineffectiveness. Either the context-bound users of this general system would reject it, or they would have to develop workarounds to cope with its inflexibility, or they would have to change their behaviors to accommodate the system.

1.3 Perturbation and intervention in RE practice

Positivistic view: The inquirer and the object of inquiry are separate, discrete entities.

Naturalistic view: The inquirer and the object of inquiry interact and influence each other, so they are inseparable.

In the scientific study of most physical phenomena (other than the exotica of quantum theory) the investigator and the phenomenon investigated are treated separately. The phenomenon under study is assumed not to be perturbed by the act of studying it. If such a perturbation is possible, experimental conditions are set up in which the perturbation is eradicated or minimized. In particular, instruments are used to gather data in such a way that the investigator does not contaminate the data.

In contrast to the typical situation in the physical sciences, but in common with the well-known Hawthorne effect in social investigations, the setting for a system is very often changed by the very act of being analyzed. In RE, artificial instruments (such as surveys) are seldom used as systematically as they are in the social sciences. The requirements engineer therefore becomes a "human instrument," in that all requirements are gathered and assimilated through him or her. This has two consequences: the requirements engineer's values directly influence the way in which the system requirements are elaborated (a point to which we will return when considering Axiom 5); and the RE effort frequently becomes the focal point of power struggles within the organization, struggles which do not overlay the RE effort so much as define it. In this case, the perceived role and trustworthiness of the requirements engineer will strongly affect the types of information that he or she obtains [4].

1.4 Negotiated realities and RE practice

Positivist view: A single reality exists that is can be empirically observed, fragmented into variables and processes, predicted and controlled.

Naturalist view: Reality consists of multiple constructed realities that can be understood to some extent but cannot be predicted or controlled.

The existence and unity of physical reality is one of the most tenaciously held assumptions of tacit positivism. In everyday life, the world that we describe is the "real" world that exists equally for everyone. Thus, the fact that the world is not flat is not a matter of opinion that is true for some people but not for others. Not surprisingly, RE theoreticians and practitioners hold this tacit assumption too, and demonstrate their belief in it by their choice of methods and terminology. For example, the environment in which a system is to operate is assumed really to exist and to be unambiguously describable (albeit with considerable effort) in models and specifications. Although a model necessarily simplifies some aspects of reality while preserving others more faithfully, and although different models may capture some stakeholders' issues or perspectives better than others, RE practitioners do not habitually question the assumption that these different models are models of the same underlying phenomena and that the phenomena exist independently of the act of modeling them.

> The essence of this kind of model is twofold. First, there is some description that applies both to the machine and to the reality it models, and captures what they have in common. Second, there is a correspondence between individuals in the machine and individuals in the reality. [8]

In many RE situations, such as those involving the monitoring and control of physical processes, the positivist assumption of naive realism can and should be made without problems. For example, the relationship between the levels of the rods in a nuclear reactor and the temperature of the core may be characterized precisely in mathematical formulae, and there is no practical need to question the assumptions that the rods and core really have certain properties, that these properties exist in certain relationships to each other, and that if the control system actuates changes in the height of the rods the core will respond in predictable ways. A control system embodies (or even is) a model of these phenomena and correspondences.

However, most systems (and, for that matter, most science and everyday thinking) depend on categories that exist only because of negotiated social processes of definition and designation. To say that a meeting scheduling system responds to meeting requests by executing certain scheduling and message-sending operations sounds similar to saying that a process control system responds to temperature fluctuations by actuating control rod movements, but the two cases are very different. The designation "meeting request" is much more obviously socially constructed than "temperature fluctuation". An e-mail message or telephone call only becomes a meeting request when it is understood to be one. Unlike the temperature fluctuation, the reality of meeting scheduling is *constitutive*: it is the describing of an act as an invitation that makes it one.

A recent example of the constitutive nature of a system's environment is provided by Sachs [16], who describes the development of a problem reporting system,

TTS, in a telephone company. Management's goals were that the system would support troubleshooting and scheduling of maintenance tasks. An abstractionist perspective would hold that as far as the TTS requirements are concerned, only these activities were relevant. A contextualist analysis of the work practices within the organization showed, however, that many apparently off-task conversations between more and less experienced workers served a vital learning and mentoring role. When TTS was introduced, this function was lost because casual conversations were replaced by task-focused electronic messages.

It is usually possible with the benefit of hindsight to fix an abstractionist analysis of a problem so that insights gained from contextualist investigations seem to get taken into account. Thus, an abstractionist could propose fixing the TTS model to preserve the mentoring function of the original work practices by allowing more experienced users to send instructional messages to their less experienced colleagues. Such a hypothetical function would make explicit many of the normally tacit conventions of discourse. For example, an old hand would now have to invoke a mentoring dialog explicitly rather than weaving mentoring suggestions into an ongoing discussion. But the interactions among the workers in the pre-TTS environment, in common with interactions in most collaborative situations, were constitutive, and were inherently not specifiable or classifiable in advance. A conversation in the TTS environment can be classified as an act of mentoring only through the joint interpretation of the participants once the conversation is underway, and this is unlikely to be its only function. Workers in the original troubleshooting environment probably found such fluid instructional conversations gratifying in a way that explicit, decontextualized messages could never be.

One specific consequence of the socially constructed nature of reality for requirements and design specifications is that phenomena should be multiply describable and should be able to migrate from one description to another. In most modeling languages, a given object must be classified permanently because the reality being modeled is assumed to be fixed in structure. For example, a meeting request could not suddenly "become" an instruction. Similarly, in the TTS described by Sachs a trouble report could not suddenly become or be viewed as an opportunity for learning.

Some RE investigators are beginning to focus on these issues by emphasizing the coexistence and resolution of multiple viewpoints on business processes, entities or objectives and to propose new RE techniques that take multiple viewpoints into account [1][5]. However this line of research assumes that a common stable reality exists beneath these views, a reality that can be modeled and confirmed by an enlightened stakeholder. According to social constructionism, however, it is meaningless to talk about such things as business processes as if they exist apart from the activities of enacting and talking about them. Business processes do not exist in businesses, but in the interpretation of business actions. They are constitutive.

1.5 Value-laden inquiry and RE practice

Positivist view: Inquiry is value-free and maintained as such through use of objective methodology.

Naturalist view: All inquiry is value-laden

Values lurk behind most investigations. The positivist response is to strive to purify all inquiry to eradicate the investigator's values. According to NI, however, values cannot be left behind. Rather, the naturalistic investigator is responsible for identifying the values that affect the findings of the study and continue to be aware of them during analysis. This is why many ethnographers keep two kinds of field notes: a first set of notes that record the observations that the ethnographer made about the setting or the responses of informants during interviews, and a second set of notes that record the ethnographer's impressions about the fieldwork experience itself. This second form of notes is increasingly finding its way into print as a legitimate result of the fieldwork experience. [11].

Lincoln and Guba [9] identify several kinds of values that affect the conduct of the inquiry. The most obvious and problematic way in which values affect the process is when there is obvious value resonance or value dissonance between the investigator and the other stakeholders. In interventionist settings, such as RE, some stakeholders' interests are often served by a proposed intervention whereas others are threatened.

In standard RE the customer authority is usually taken to be the organization's management, and in business process reengineering (BPR), in which the system is part of a broader effort to redesign business activities to meet management's objectives, the organizational objectives may conflict with the requirements engineer's values. Even if there is no conflict, it is important that the stakeholders are as explicit as the political situation in the organization permits about the values to be achieved by the project. For example, most BPR projects have as their goal improvements in the efficiency, throughput or customer-perceived quality of some value-producing process, goals which could conflict with others (such as the desire of some users or management for autonomy over their work).

There are two contrasting temptations here, most dramatically illustrated by two contrasting approaches to reflecting social and contextual factors in RE. The first temptation is for the requirements engineer to adopt unthinkingly management's goals because of the contractual nature of the relationship between the requirements engineer and management. The values to be promoted by the RE activity are usually the values of those who are paying for it. Occasionally the requirements engineer will be asked to reveal sources of information that were ostensibly confidential, or one stakeholder (often management) has a hidden agenda (often reduction in staffing) that is not to be revealed to other stakeholders during the RE process.

A countervailing temptation is to adopt the value system of the group of stakeholders with whom one is most in contact. In anthropology, this is known rather disparagingly as "going native," and it is most likely to occur in RE when a requirements engineer has a long-lasting and rich involvement in one community of stakeholders. This is most likely to be the case when doing workplace ethnography or facilitating a participatory design project, because in both situations the requirements engineer is closely involved with the work as seen from the perspective of one community of stakeholders. It is inevitable that some of how these stakeholders see their work will rub off on the requirements engineer.

Designers and requirements engineers are sometimes viewed as passive conduits of requirements information rather than stakeholders in their own right. This, however, is a simplistic view that has been replaced in analogous disciplines in the social sciences. Failing to reflect on one's own values can affect a requirements effort, usually detrimentally. A resonant example of a failure of a design team to reflect on what they were learning about the user community of their system is reported by Markus and Keil

[12] in their study of an expert system that they call CONFIG. This system grew out of a research project into knowledge-based system design, and the application was selected largely to impress company management with the relevance of the technology to the company's objectives. However, the specific context in which CONFIG was used, computer sales, imposed demands on the functionality of CONFIG that its designers steadfastly rejected over a period spanning several years:

> The developers of CONFIG didn't build what the users said they wanted; its intended users repeatedly made it clear that they wanted CONFIG to be integrated with PQS [a mainframe-based pricing quotation system], but the developers rejected this request. The developers cited technical concerns, but it may be more accurate to say that they could

not see the value in the users' request because it conflicted with their own preferred design concept - of CONFIG as a stand-alone system. [p.23].

A second intrusion of values into the investigation identified by Lincoln and Guba is the paradigm adopted by the investigator. In the case of RE, there are different paradigms for conducting RE that reflect alternative views of what the process is and what its goals are. Some of these paradigms assume tacitly a commitment to a value system, whereas others do so more explicitly. The conventional, contractual model, for example, views RE as a means of clarifying the requirements of a single, privileged group of stakeholders to whom the requirements engineer has a contractual obligation. The nature and force of this obligation is usually accepted implicitly. Soft systems methodology [3] views any planning and design project as serving the needs of several categories of stakeholders who must be distinguished carefully from the outset: the customer (or ultimate beneficiary), the owner (contractual sponsor or party who can declare the project a success or failure), and the actors (end-users and direct consumers of the system's results). Participatory design, on the other hand, reflects a strong ideology that end-users should be involved in the design of the system, not just because the system will be more effective by management yardsticks but also because it is morally right for them to participate in design.

The point here is not so much that different paradigms assume different value systems, and still less that some value systems are superior to others, but that the requirements engineer adopting one of these paradigms and using it effectively necessarily adopts some of its value system too. It is impractical, for instance, to adopt participatory design practices purely as technique without a commensurate adoption of the values of workplace democracy that underpin them.

A third form of value-ladenness is the influence of the substantive theory adopted by the investigator. In the case of RE, this amounts to the ontology underpinning the requirements method being used. For example, object-oriented and structured methods assume completely different perspectives on what a system is. When adopting an object-oriented method, the requirements engineer (who is a human instrument) is more finely tuned to pick up information about the structure of the task domain than the function or purpose of what happens there. A requirements engineer adopting a structured method, on the other hand, is more finely tuned to information about processes, steps and events, and less attuned to the structure of information. (To a child with a hammer, everything is a nail.) It may seem to stretch the meaning of the word "value" to regard as values the core ontological foci of object-oriented or structured methods,

but these ontological emphases dictate how the requirements engineer interprets the context and what is worth giving attention.

2. Axioms of NI applied to RE research

RE research exhibits the same telltale signs of tacit positivism as RE practice. Next, we go through Lincoln and Guba's axioms once more, this time attending to the research implications.

2.1 Mutual influence in RE research

It is common to encounter to encounter claims in the RE research literature that an intervention (such as the adoption of an RE method) will cause improvements in productivity or quality. Thus the nature of the project and the initial result of the intervention are pretended not to affect significantly the intervention itself.

One consequence of this asymmetry in the research community's view of cause and effect is that empirical research that seeks to show the influence of a method on an outcome measure is more often valued as "research" in comparison with accounts of how experience with a method shapes subsequent applications of it, accounts that are marginalized often as "experience reports" or lessons learned during technology transfer. The first is allegedly principled scholarship, whereas the second is mere process improvement. Our view is that research goals and results shape each other much more directly than the pure cause-effect dichotomy implies and that valuable innovations can only come from tight intertwining of theory and practice, inquiry and interpretation [13].

Investigators in analogous fields see this intertwining as both inevitable and desirable. In qualitative social research, such as ethnography such intertwining goes by the name of "grounded theory" [6], theory that arises from and is shaped by a close interpretation of early findings in a study rather than by a fixed theoretical framework or set of hypotheses that the investigator adopts before the investigation starts and adheres to throughout. In the application of qualitative methods to organizational objectives, it is known as "action research" [3], a style of inquiry that deliberately mixes research and invention and involves organizational members as participants in and shapers of the research objectives.

2.2 Idiographic knowledge in RE research

Abstractionism is strongly advocated in RE research. The entire basis of methodological research and tool construction is predicated on the belief that what is good for one system is good for many (if not all). Very few proposed RE techniques, methods or tools are ever described as being restricted to a small subset of RE contexts.

The reflective case study summarizes the idiographic knowledge gained from investigating a single use context and/or intervention in detail. Unlike the controlled experiment or the supposedly general RE method, which rest on the ability to generalize to multiple cases, the case study resides mainly at the level of idiographic knowledge.

2.3 Perturbation and intervention in RE research

The breakdown of the requirements engineer's neutrality is acknowledged and embraced in the Industry-as-Laboratory [13] and action research [3] paradigms of inquiry. In contrast to positivistic research, in which a phenomenon is assumed to be stable and is controlled as far as possible during the time span of the investigation; in action research, the investigator's interventions are driven by the evolution of his or her grounded theories.

Incremental intervention of this kind can be criticized on the grounds that failing to hold a phenomenon constant while a case study unfolds makes it impossible to generalize from the case study to other situations in which the phenomenon is of interest. However, it is unrealistic to try to apply the standards of quantitative, experimental science to most studies of interventions in RE projects. For the experimental situation to be interesting, it must be realistic. But no two real projects are similar enough ever to rule out the interpretation that it is differences among the contexts themselves, not the intervention being studied, that is the possible source of any differences in outcome. Moreover, stakeholders in a real project have personal and political interests in the outcome of the project and cannot therefore be treated like guinea pigs. They may demand that interventions be made or stopped when they judge them to be effective or harmful. The inevitability of these demands makes a controlled experiment almost impossible to carry out.

2.4 Negotiated realities in RE research

The implications of this axiom to RE research are also far-ranging. The RE literature is replete with claims about the reality of system development that should better be seen as claims about continually negotiated social agreements. Consider, for example, the concept of requirements "ambiguity." It is usually considered good to specify requirements unambiguously where unambiguity is defined by the IEEE Standard (830-1984) as:

An SRS [software requirements specification] is unambiguous if - and only if - it has only one interpretation. [7]

Tacit positivism leads to the assumption that ambiguity inheres in a requirement and can therefore be engineered into a new specification language, rather than being a process of negotiated interpretation involving the writers and readers of a requirement. Designers of requirements specification languages typically take it for granted that ambiguity (in common with other desirable and undesirable features of a specification) is "out there" in a definitive semantic definition of the language. A reader or writer can only use the language correctly or incorrectly. According to the NI tradition, however, semantics of languages are created and resolved during events of language use, and a language designer's intended meaning or the formal semantics embedded in language support tools cannot claim absolute definitiveness. If all the consumers of a requirement interpret a requirement identically, it is unambiguous; if they interpret it differently, it is ambiguous. This suggests that RE research should investigate the interpretation and use of specification languages rather more, and investigate the formal properties of such languages rather less.

Similarly, requirements engineers are often encouraged to identify the "real" customers for a system, even though, in the case of packaged software, customers may be defined by the post-delivery act of buying the system, not by any pre-development agreement. For example, some recent empirical work [2] studied the process of conceptual design in a telecommunications development project. The project in question was a middleware platform on which future applications were to be built. Because it was removed from the concerns of application users, this project was hampered by not clearly identifying who the customers were, and the project team converged only very slowly on a vision of the product's requirements. It was inevitable that the definition of who the customers were should have evolved slowly during a process of discussion and negotiation. Searching for the "real" customers is futile when customers are created by the system and the system has not yet been created itself. Thus the emphasis of some RE research should move from the elicitation of requirements from customers, to the discovery and invention of features that will later be judged useful. (Such a shift would make RE research look more like design research.)

2.5 Value-laden inquiry in RE research

In RE research, different stakeholders, including the research investigators themselves, their peers, the reviewers of publications, funding agencies and potential adopters of research innovations conspire unconsciously to make RE seem to be a value-free pursuit. The usual way to judge whether a piece of research is good is to compare it with the model of the physical sciences, a comparison that relies on a set of values derived from positivism. Soundness, repeatability, formal rigor and value-independence are values, and, like all values are open to question.

It is legitimate to worry that replacing or weakening the acceptance of these values might lead to research in which anything goes, in which casual interpretations of unverifiable data are presented as findings. Lincoln and Guba [9] address the issue of research trustworthiness in great detail, both undermining the purported trustworthiness of positivism-inspired values in the social sciences and discussing in detail practices that increase the trustworthiness of naturalistic research. There is no reason why similar practices could not be adopted in RE.

3. Implications

The previous discussion sets forth the challenges posed to RE by the epistemological and ontological foundations of NI. Clearly there are some correspondences where unproblematic alignment is possible, but there are more conflicts where alignment of any kind seems impossible. In this section, we concentrate on those themes that apparently make impossible any reconciliation of tacit positivism and NI. Accepting that the two approaches are irreconcilable would mean that they could not coexist effectively and that the RE community must repudiate one or the other (almost certainly NI). We contend that to sweep away the challenges that the axioms of a naturalistic research paradigm raise is to miss an important opportunity to rethink the whole enterprise of RE.

We frame this discussion around Table 1, which articulates the *Uncommitted*, or what might called "weak" view of RE, and the *Committed* or "strong" view. We use these terms to refer to either a willingness or reluctance to embrace the epistemological and ontological groundings of naturalistic inquiry and the resulting outcomes for RE.

The Uncommitted view construes NI as just another set of techniques that provide data unobtainable through other methods. Here the axioms convenient to the RE enterprise are brought on board while those raising challenges are jettisoned. Requirements engineers who use naturalistic data collection techniques but fail to embrace the axioms of NI gloss multiple realities as either customer fickleness or user inarticulateness. Moreover those ungeneralizable constitutive processes are nothing more than elusive causal chains merely needing to be tracked down and subordinated to the RE process. Overall, in the Uncommitted view, the enterprise still must pre-

specify conditions as though such things were stable and to build systems based on those conditions.

The Committed view accepts the five axioms as fundamental to the RE enterprise. In this view, data

Table 1: Contrasts between the uncommitted (weak) and committed (strong) views of the role of NI in RE

Uncommitted (weak) View	Committed (strong) View
Axioms taken as practical warnings for RE technique	Axioms taken as fundamental to RE enterprise
Use ethnography and other qualitative methods because work processes can be richer than they look. The resulting system will be fitter for its purpose.	Human activity is inherently flexible and adaptive rather than rule-governed and pre-planned. A system based on abstract process models will be unsatisfactory.
Customers are fickle, so nail down as many requirements as feasible in advance and plan for change.	Intervening in an organization by introducing a system always changes the organization, surfaces , some contradictions between its objectives, and leads to new system requirements.
The information that customers tell you depends on how you ask for it and whose interests they think you are furthering.	Organizations' processes and requirements are constructed and modified by the act of reflecting about them.
To-be supported processes are difficult to elucidate. The RE has to manage the negotiation among multiple views.	To-be supported processes are constitutive. They only exist through an activity of negotiation.
Multiple models are necessary to describe the system to different stakeholders. It is important to be aware of organizational politics and cut through it to the customer's rational requirements.	System development is always value-driven. Conflict among stakeholders is inevitable. The RE also promotes values.

collection, contexts, requirements, and design must be understood differently. There is no reality out there to specify, only multiple constructed realities. Stable processes cannot be captured because all interactions are constitutive and negotiable. Systems can never be value-free; they are inherently saturated with someone's ideology, theory, past and present. Relevance is prized in place of the spurious sense that phenomena can be generalized and controlled. Such a paradigm shift means that the bedrock of RE must give way to shifting and unpredictable sands, a rather unappetizing prospect.

We realize that this Committed view leaves the RE community in a conundrum. If, as most would agree, systems design must be informed by both the technical and social elements, but NI eschews most of the very things that RE seeks to pin down, how can RE and NI be reconciled? If idiographic knowledge obtained during RE practice and research cannot be generalized, how can it then be useful? The key distinction to be made here is between generalizability and transferability [9]. In generalization, investigation of the controlled context gives rise to nomothetic knowledge, which can subsequently be applied to multiple application contexts. The success of this process depends on the abstraction being appropriate (that is, essential information being preserved in the abstraction while only inessential information is lost) and on the abstraction being relevant to the application (that is, the nomothetic knowledge encapsulated in the abstraction being salient to the new context). In transferability, in contrast, there is no abstraction, only a sending and a receiving context. This is much more like using analogies or cases to solve a problem. Whether idiographic knowledge gleaned from an investigation of the sending context can be transferred to the receiving context depends critically on what Lincoln and Guba [p. 124] call "fittingness," that is the degree of congruence between the sending and the receiving contexts. An appreciation of this degree of congruence can only arise from a rich knowledge of both contexts, and not from faith in spuriously general abstraction. Thus transferability sacrifices the largely illusory quality of generality in favor of the more substantive quality of relevance.

The second possible strategy is to trade the notion of isomorphism for "islands of stability". No longer would the aim be to map the "real" world onto a set of requirements devoid of subtle nonspecifiable entities. Rather it would be to search for places where fluid or constitutive activities occur, where negotiated meaning and understanding were arrived at, where the boundaries of varied activity types exist. The "islands" would be the boundaries or frames provided by the continually replayed types of events or activities identified by the respondents.

The "seas" would be the large spaces available for constitutive behavior. The Nynex TTS [16] study discussed earlier begins to get at what this would look like. In that case, ethnography identified a sequence of established interactions that the new system by-passed. A reconceived system would provide intentionally seas of meaning-making that supported an already existing set of well-practiced activity types.

Another example is an educational learning environment we are developing at Georgia Tech that scaffolds middle-school students learning to design and learning science through design. From an ethnographic study of undergraduates in mechanical engineering learning design, process elements in a design sequence were identified. Cognitive phases such as problem formulation and understanding, criteria extraction, conceptualization, criteria application and decision-making, and resource gathering, to name a few, were identified. Tools are now being built to support these processes. Case studies of similar design problems and simulation tools will also be available to the teams. However, the function of this learning environment is not to prescribe a fixed route through a design space. Instead it is to support knowledge-building and meaning-making as the design team moves from an imperfect problem statement to a built artifact. When and how students choose to use the suite of tools is not fixed but negotiated as they wrestle with the design. These tools instantiate the frames that constitute complex problem-solving, but within these frames we expect fluid constitutive activity to occur.

Undoubtedly reconciling the foundational differences between RE and NI is problematic in a way that is not true of the incorporation of isolated NI techniques into RE methods. But naturalistic techniques cannot be seen merely as a set of slick tools for better and more sensitive data collection. Naturalistic methods can be valuable in practice, but to deliver this potential, they must be taken with their philosophical baggage intact.

References

[1] Boehm, B. and In, H. "Identifying Quality-Requirement Conflicts." *IEEE Software*, March, 1996. pp. 25-36.

[2] Catledge, L. & Potts, C. "Collaboration during Conceptual Design." *Proc. 2nd Int. Conf. Requirements Engineering*, Colorado Springs, CO, April 15-18, 1996. IEEE Computer Soc. Press, pp. 182-190.

[3] Checkland, P. & Scholes, J. Soft Systems Methodology in Action, Wiley, 1990

[4] Dahlbom, B. and L. Mathiassen, *Computers in Context: The Philosophy and Practice of Systems Design*, NCC Blackwell, 1993.

[5] Easterbrook, S. and Nuseibeh, B., "Managing Inconsistencies in an Evolving Specification." *Proc. RE'95: 2nd Int. Symp. Requirements Engineering*. York, UK, March 27-29, 1995. IEEE Computer Soc. Press, pp.48-55.

[6] Glaser, B. & A. Strauss, *The Discovery of Grounded Theory*. Aldine, 1967.

[7] IEEE Guide to Software Requirements Specifications (ANSI/IEEE Std. 830-1984). IEEE Inc., 1984

[8] Jackson, M. A. *Software Requirements and Specifications*, Addison-Wesley, 1995.

[9] Lincoln, Y.S. and Guba, E.G. *Naturalistic Inquiry*, Sage Publications, 1985.

[10] Luff, P., Heath, C. & Greatbatch, D. "Work, Interaction and Technology: The Naturalistic Analysis of Human Conduct and Requirements Analysis," in M. Jirotka & J. Goguen (Eds.) *Requirements Engineering: Social and Technical Issues*, Academic Press, 1994, pp. 259-288.

[11] Malinowski, B. *A Diary in the Strict Sense of the Term*, Harcourt, Brace & World, 1967.

[12] Markus, M.L. and Keil, M. "If We Build It, They Will Come: Designing Information Systems that People Want to Use," *Sloan Management Rev.*, Summer, 1994. pp. 11-25.

[13] Potts, C. "Software Engineering Research Revisited," *IEEE Software*. September, 1993. pp. 19-28.

[14] Potts, C. "Using Schematic Scenarios to Understand User Needs," Proc. DIS'95: Designing Interactive Systems, Ann Arbor, MI, August 23-25, 1995, pp. 247-256.

[15] Potts, C. & I. Hsi, Abstraction and Context in Requirements Engineering: Toward a synthesis, *Annals Software Eng.* (to appear).

[16] Sachs, P. "Transforming Work: Collaboration, Learning, and Design." *Comm. ACM, 38(9)*, September, 1995. pp. 36-45.

[17] Sommerville, I., T. Rodden, P. Sawyer, R. Bentley and M. Twidale, Integrating Ethnography into the Requirements Engineering Process. *Proceedings RE'93: International Symposium on Requirements Engineering*. San Diego, CA: January, 1993. IEEE Computer Society Press

[18] Suchman, L. "Making Work Visible," *Comm. ACM, 38(9)*, September, 1995. pp. 56-64.

On the Use of a Formal RE Language

The Generalized Railroad Crossing Problem[*]

Philippe Du Bois, Eric Dubois, and Jean-Marc Zeippen
Computer Science Department, University of Namur
21 rue Grandgagnage, B-5000 Namur (Belgium)
pdu@cediti.be – edu@info.fundp.ac.be – jmz@cediti.be

Abstract

In this paper, we report on the use of the **AlbertⅡ** *specification language through the handling of the Generalized Railroad Crossing case study. This formal language is based on an ontology of concepts used for capturing requirements inherent to real-time, distributed systems. Its essential feature comes from its* naturalness, *i.e. the possibility of a direct mapping of customers' informal needs onto formal statements, without having to introduce artificial elements. The language relies upon formal grounds (real-time temporal logic) which support the reasoning of the analyst during the elaboration of the specification. These reasoning capabilities are illustrated in the context of a goal-oriented approach adopted for the elaboration of the case study.*

1 Introduction

AlbertⅡ is (yet another) formal specification language proposed to support RE activities. The design of the language started around 1992 within the framework of the Esprit II project *Icarus* [11]. The validation of the language constructs has been achieved through the handling of non trivial case studies (ranging from CIM to advanced telecommunication systems) performed by the members of the team who developed the language. At the time being, the language is subject of several technology transfer initiatives. In particular, it is being used by two industrial partners in the context of the development and the evolution of two large, distributed, software-intensive, heterogeneous systems (a video-on-demand application and a satellite-based telecommunication system) [1].

Like other pionneering RE languages (e.g., [4, 13, 14]), **AlbertⅡ** shows high degrees of *expressiveness* and *formality*. The language is based on an ontology of concepts that has been proven useful for capturing functional requirements inherent to real-time, distributed, composite (or agent-oriented) systems [12]. Its underlying formal framework is based on **Albert-CORE** [7], an object-oriented variant of real-time temporal logic where the concept of action has been introduced to solve the *frame problem* [3]. Besides these two properties, the most important, original characteristic of **AlbertⅡ** comes from its *naturalness*, i.e. the possibility offered by the language to straightforwardly map the informal statements provided by the customers onto formal statements expressed in the language. At the stage of RE modelling, we feel essential to avoid the introduction of artificial elements in the formal statements. One could easily measure the semantic distance existing between a customer's informal statement like *"The received message has to be resent within 3 minutes. "* and its formal counterpart expressed in terms of a timed state/transition diagram stating that: *"Received messages are time-stamped and stored in the persistent data structure WaitingMessages. At each clock tick, messages which are stored for more than 3 minutes are removed and sent again."* If the notation does not allow to keep formal statements close to informal ones, the use of verification and validation tools may result in conclusions which are not understandable by the customers because they are referring to the artificial elements (like *WaitingMessages* in our example). Moreover, some important elements of the informal description (like the mapping of received messages onto sent messages) can be hidden in the formal description because of their 'encoding'.

In Sect. 2, we illustrate the naturalness and expressiveness properties of **AlbertⅡ** through the handling of the standard *Generalized Railroad Crossing* (GRC) case study [16]. In particular, we suggest how useful is the *declarative* style provided by **AlbertⅡ** in that respect. The presentation of the formal semantics of the language is presented elsewhere [7]. We introduces to the so-called *System Requirements Document* (SRD) of the GRC. It contains the specification of the system expected by the customers. Obviously, the SRD has not been written in one shot : its results from a RE process which can be viewed as consisting in deriving the SRD from customers' global requirements. In Sect. 3, we sketch a 'faked' view of this process,

---

[*] Supported by the DGTRE Région Wallonne, CAT project #2791.

illustrating how **Albert II** can be used as the modelling language, and discussing how the formality offered by the language is helpful for tackling the reasoning activities the analyst has to face. Finally, the paper concludes with a discussion of issues related to Design Engineering (DE), a short comparison of **Albert II** with similar approaches, and an outline for further work.

2 Using **Albert II** to Write the SRD

2.1 The SRD

The SRD is the output of the RE activity. It includes the identification and the specification of different components of the system, each of them belonging either to the *environment* (or *domain knowledge*) —considered to be there—, or to the components corresponding to the *machine* [19] —having to be installed—. In the context of the GRC case study one may distinguish among:

- two components belonging to the environment, namely *Train* (designating the trains travelling on the multiple tracks that are controlled) and *Gate* (the operated crossing gate);

- the *Controller* component which corresponds to the controlling system to be installed for operating the gate.

This document is at the basis of the contractual relationship established between:

- the customers (i) who have agreed on the infrastructure and functionalities of the system to be installed and (ii) who guarantee the behaviour of the components belonging to the environment;

- and the designers who are in charge of the implementation of the system. Since the machine can be a composite one (hardware, devices, software, humans, etc.), the software designer is only in charge of the implementation of the identified software component[1].

As such, the SRD document should be readable by both parties. In practice, this means that, with little exception, only natural language can be used. However, for the purpose of veryfing and validating the SRD content, it is important for the analyst to be able to manage a formal counterpart of this informal description. As already noted, it is essential that this formal description is 'close to' the informal one. Besides the argument developed in the previous section, the use of an expressive formal language should help the analyst to manage the *pre-traceability* link to be maintained between the two descriptions.

---

[1]In some SRDs (see, e.g., the ESA PSS-05 standard), customers may want to identify the different components of the machine but, in some other cases, the machine can be seen as a single component.

2.2 The **Albert II** Language

The design of **Albert II** has been essentially influenced by the need for pre-traceability. Its expressiveness and naturalness come from its *declarative* style of specification.

The ontology on which **Albert II** is grounded considers a collection (or *society*) of *agents* interacting with each other in order to provide the expected behaviour. Each agent is characterized by *actions* that may change its *state* of knowledge of the external world. Actions are performed by agents to discharge contractual obligations expressed in terms of *local* and *cooperation* constraints.

An **Albert II** specification is thus organized into units corresponding to the agents specifications. Logical statements appearing into an agent specification define the set of admissible *behaviours* (or *lives*) the agents may experience. These statements are classified into categories for which a specific *template* (i.e. pattern of formula) is defined. This provides methodological guidance to the analyst in the elicitation and structuring of requirements[2].

Basically, a specification in **Albert II** is made up of (i) a graphical part where the vocabulary is *declared* (together with some general properties) and (ii) a textual part where the logical formulae *constraining* the admissible behaviours are stated.

The language is illustrated in the following two subsections for the GRC problem. Giving all the details about the language is outside the scope of this paper : the interested reader may refer to [7].

2.3 Graphical Declarations

Figure 1 shows the declaration of the society of agents associated with the GRC system. An agent (as well as each society which is in fact a compound agent) is represented by an oval. Shaded oval represent a *class* while plain ovals represent *individual* agents. Here, the *GRC* society is made of different agents, viz. several *Train* agents, one *Controller* agent and one *Gate* agent.

The details of Fig. 1 show the declaration of the *state* structure (i.e. the memory) of each agent as well as of the *actions* which may happen during its possible lives. State components (graphically depicted with rectangles) are typed; actions (graphically depicted with ovals) can have typed arguments. Types may vary from simple (built-in) data types to complex (user-defined) data types (recursively built using type constructors). All along the rest of this paper, we will only construct simple types, namely three enumerated types:

GATE-EXP-STATUS = ENUM[Up,Down]
STATUS = ENUM[Up,GoingDown,Down,GoingUp]
POSITION = ENUM[Elsewhere,R,I]

Note also that types associated with each class of agents (here: *TRAIN*) are automatically defined. They correspond

---

[2]The usefulness of such templates had already been identified in, e.g. RML [13] (which is built on top of first order predicate logic).

to the type of agents identifiers. Constants are also automatically defined to refer to the individual agents (here: *GATE* and *CONTROLLER*). We will also need the following constants defined as 0-ary operators (*GammaDown*, *GammaUp*, *Delta*, *Epsilon1*, and *Epsilon2*) ranging over non negative integer, and verifying (i) $Epsilon1 \leq Epsilon2$ and (ii) $GammaDown < Epsilon1$.

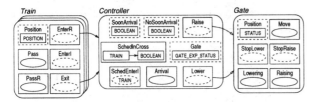

Figure 1: SRD - Declaration of the Agents of the GRC Society

Let us now comment upon the state components and the actions declared in Fig. 1 (based on the accompanying specification dictionary[3]).

- *Position* of *Train* is a single time position (dashed rectangle) that may vary with time and take the value *Elsewhere* ("not in the section of interest"), *R* ("in the region of interest"), or *I* ("in the railroad crossing"). *Position* of *Gate* can be similarly interpreted. Its value can be *Up*, *GoingDown*, *Down*, or *GoingUp*. The *Gate* component of *Controller* is similar: its values can be *Up* or *Down*.

- *SchedInCross* of *Controller* is a central component keeping track of the trains supposed to be in the crossing. It is modelled as a table, where *TRAIN* is the type of the index (domain) and *BOOLEAN* is the type of the elements (range).

- *SoonArrival* and *NoSoonArrival* are two auxiliary boolean predicates whose meaning is explained in the following subsection.

- *EnterR*, *EnterI* and *Exit* (dashed ovals) are three events, i.e. instantaneous actions. These events are associated with the entry of the train in the region, its entry in the crossing and its exit from the crossing, respectively. *PassR* and *Pass* are two actions with a duration (or processes). As we shall see later on, they are associated with action composition constraints.

- For the purpose of the gate management, we have introduced the expected events (*Lower*, *StopLower*,

---

[3]To save space, we omit this documentation here. However, when using the **Albert II** graphical editor, the analyst is encouraged to develop a lexicon attaching an informal meaning to the different components of the specification.

Raise, *StopRaise*) and three processes (*Move*, *Lowering*, *Raising*)

- To manage the scheduled trains, an event (*SchedEnterI*) and a process (*Arrival*) are declared. Their roles are defined later.

Finally, Figure 1 gives information about the responsibility that the different agents have on state components and actions. For example, one can read that the *Lower* action is under the control of the *Controller* agent and that there is a *Position* state component controlled by the *Gate* agent. Moreover, as noted in [19], a suitable RE language should also make clear whether an action (or state component) controlled by an agent is part or not of the inferface with another agent. In **Albert II** graphical declarations, arrows show how agents may influence each other through exportation/importation of *information* (i.e. actions and/or state components). Here one notices that:

- events *EnterR* and *Exit* are controlled by trains. They are part of the interface between the trains and the controller which can perceive them (like sensors);

- events *Lower* and *Raise* are controlled by the controller. They are part of the interface between the controller and the gate since they may bring some effects on it (like actuators).

Conditions under which an agent brings information to the outside or perceives external information are given as constraints.

2.4 Textual Specification of Constraints

The textual constraints in an **Albert II** specification are used for pruning the (usually) infinite set of possible lives associated with the agents of a system.

Constraints are classified into **Local** constraints on the internal behaviour of an agent and **Cooperation** constraints on the interaction of agents within a society. To guide the analyst in the elicitation and structuring of requirements specifications, the constraints are further classified into categories for which a characteristic *template* is defined. For example, **State behaviour** constraints express restrictions on the possible agent behaviours only in terms of the values that can be taken by its state components, while **Action Composition** constraints express restrictions only in terms of admissible sequences of actions. Details about the different templates offered by the language are given in [7]. In the sequel, we will see examples of a few of them.

Appendix A gives the complete textual specification of the GRC system. Due to the lack of space, the informal comment corresponding to a direct, 'natural' paraphrase of each formal statement is only provided for the *Controller* agent[4]. By having a look at the specification, we hope that the reader is able to judge the expressiveness and the naturalness of the language.

Let us now comment the specification of the central agent, namely the *Controller*. To write it, we have considered the descriptions provided by Heitmeyer and Lynch in Sect. 5 of [17] as our source of information. They are giving a formal specification of the controller in terms of an extended timed automaton[5]. This style of specification is also supported in **Albert II** and thereby it would have been straightforward to produce an analogous specification. However, we have preferred to base our specification on the informal comments given by the authors together with their formal, *operational* specifications.

The authors give the an informal statement associated with the expected behaviour of the *Controller* as follows:

"The system lowers the gate if the gate is currently up (or going up) and some train might soon arrive in I ("soon" means by the time the computer system can lower the gate)."

The formal **Albert II** counterpart is given by the two following clauses:

CAPABILITY
$\mathcal{XO}$(Lower / Gate = Up $\wedge$ SoonArrival)
```
There is an obligation to lower the gate
if and only if it is up and there is an
arrival soon.
```
STATE BEHAVIOUR
SoonArrival
$\Longleftrightarrow \exists$ t: ($\neg$ SchedInCross[t])
$\wedge \Diamond_{=GammaDown}$ SchedInCross[t])
```
SoonArrival means that there is a train
which is not scheduled to cross now but
will be scheduled within the next
GammaDown seconds.
```

The second essential property is given by:

"The system raises the gate if the gate is currently down (or going down) and no train can soon arrive in *I*. ("soon" means by the time the gate can be raised plus the time for a car to pass through the crossing plus the time for the system to lower the gate)."

which, in turn, is formally reformulated as:

CAPABILITY
$\mathcal{XO}$(Raise / Gate = Down $\wedge$ NoSoonArrival)

```
There is an obligation to raise the gate
if and only if it is down and there is
no arrival soon.
```
STATE BEHAVIOUR
NoSoonArrival
$\Longleftrightarrow \forall$ t: ($\neg$ SchedInCross[t])
$\wedge \Box_{\geq GammaUp+Delta+GammaDown}$ $\neg$ SchedInCross[t])
```
NoSoonArrival means that there is no
train scheduled to cross now, neither
during at least the next GammaUp +
Delta + GammaDown seconds.
```

Note that the informal statement given by Heitmeyer and Lynch has been slightly modified to "... *and no train in I* and no train can soon arrive in I ...". This is required since trains already in I represent also a danger!

Other formal statements presented in the specification of *Controller* deal with the management of the *SchedIn-Cross* state component associated with the table of trains which are expected to be in the crossing. Changes of this component are entailed by the *Exit* and *SchedEnterI* actions[6]. *Exit* is an external action perceived from the outside while *SchedEnterI* is an internal action whose occurrence time is computed from the time of occurrence of a perceived *EnterR* action (see formal statements under **Action Composition** and **Action Duration** headings).

As a final comment, we would like to stress the importance of the **Cooperation** constraints. Such constraints include (i) the templates (**Action Perception** and **State Perception**) describing, respectively, how an agent perceives actions and parts of the state of other agents, as well as (ii) the templates (**Action Information** and **State Information**) describing, respectively, how an agent lets other agents know about the actions it performs and how it show parts of its state to other agents. Perception and information constraints provide the analyst with a way to add a dynamic, time-varying dimension to the importation and exportation relationships between agents expressed in the declaration part. From the specification given in Appendix A, we can see that communication is highly reliable among the different agents. In a more realistic situation, possible failures should have been considered, like in the perception by the gate of *Raise* and *Lower* events issued by the controller. Such aspects can be easily expressed using **Albert II**.

3 From Requirements to the SRD Using Albert II

The elaboration of the SRD results from several stages going from initial elicitation of organizational goals up to the final system specification. At the level of the elab-

---

[4]We are currently developing a tool helping in the semi-automatic reformulation of the formal statements in a natural language.

[5]There are doing so to be able to use their formal verification framework, however they mention an auxiliary version of the specification should be provided to the customers.

---

[6]Such actions are required to deal with the frame problem. Note however that *NoSoonArrival* and *SoonArrival* are specific state components for which the frame rule does not apply.

oration of **AlbertⅡ** specifications, our experience has showned the importance of some key stages.

In the first part of this section, we illustrate these stages in the elaboration of the GRC system requirements. The different stages are presented following a 'faked' process perspective (inspired, among others, by [8, 6, 19]). In the second part of this section, we suggest how the formal semantics associated with the language can support the reasoning activity of the analyst during these different stages.

3.1 Elaboration of System Requirements

We distinguish four stages in the elaboration of system requirements. As already mentioned, the process does not appear as such in real life situations. Indeed, the four stages we will describe are interwined. They should be seen as *co-processes*; furthermore *backtracking* has to be considered.

Stage 1: Modelling the Problem Domain. The modelling of the *problem domain* (or *universe of discourse*) relies upon the identification of the real-world phenomena of interest.

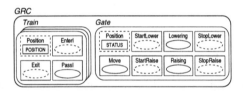

Figure 2: GRC Problem Domain

In our running example, the relevant information is presented in Fig. 2 using **AlbertⅡ** notations. It is important to note the following.

- We only consider two agents (the class of trains and the gate). Each of them is equipped with the actions that may describe its behaviour. At this stage, we are not discussing who has the responsibility for such actions : an agent has to be seen just as an *object* in formal languages for object-oriented analysis (like Troll [23] or LCM [25]), i.e. a way of structuring information.

- As far as the *Train* agent is concerned, there are only two relevant phenomenas, viz. the entering of the crossing (*EnterI*) and the exit (*Exit*). The value type associated with *Position* includes only two distinguished values (the train is in the Crossing or out); it is defined as: $POSITION = ENUM[Elsewhere, I]$

Lack of space prevents us not to present the textual specifications associated with the identified objects.

Stage 2: Identifying the Requirements. At this stage, we are specifying the *requirements* (or *optative* properties [19]). Those are the 'high-level'[7] goals expressed on top of the problem domain that we would like to be met by the future system. In terms of **AlbertⅡ** this consists in introducing constraints concerning action occurrences and state components evolution. Note that it usually consists in synchronizing the standalone phenomena identified at Stage 1.

As an example, we can consider the following basic requirement in our example (expressed using the elements declared in Fig. 2):

> *Safety Property:* When a train is in the crossing, the gate must be down.

STATE BEHAVIOUR
$t.\text{Position} = I \Longrightarrow g.\text{Status} = \text{Down}$

> *Utility Property:* If the gate is not up, this is due to three possibilities: (i) there is a recent preceding state in which there was a train in the crossing, or (ii) a train is soon entering the crossing, or (iii) the period since the last train exited the crossing and until a new train enters it is not long enough to allow a car to pass through the crossing.

STATE BEHAVIOUR
$g.\text{Status} \neq \text{Up} \Longrightarrow$
$\exists\, t: \blacklozenge_{\leq GammaUp}\, t.\text{Position} = I$
$\vee\, \exists\, t': \lozenge_{\leq GammaDown+(Epsilon2-Epsilon1)}\, t'.\text{Position} = I$
$\vee\, \exists\, t'': \blacklozenge_{=s}\, t''.\text{Position} = I \wedge \exists\, t''': \lozenge_{=s'}\, t'''.\text{Position} = I$
$\qquad \wedge s' - s \leq GammaUp + Delta + GammaDown$

Stage 3: Designing the Infrastructure. During this stage, a high-level design (or engineering) activity takes place: the analyst identifies the *actual* agents of the system, i.e. those who will be responsible for the different actions and state components identified in the problem domain, as well as for ensuring the requirements identified in the previous stage.

Analogously to [19], we make a distinction between the agents belonging to the environment and those associated with the machine (subsystem to be implemented): the behaviour of the agents belonging to the environment is specified (the so-called *indicative* properties) while the behaviour of the machine remains undefined (for now).

Figure 3 presents the graphical declaration of the GRS infrastructure, i.e.:

- the *Controller* (the system to be installed) together with its associated responsibilities related to the control of the *Lower* and *Raise* actions.

---

[7]This notion is relative depending on one's perspective, as noted in [19]. However in the context of a given problem domain, it should be fixed.

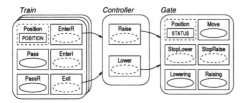

Figure 3: GRC Infrastructure

- a new *EnterR* action under the control of the *Train* agent but shared with the *Controller* agent. The identification of this action is central because it is associated with the perception that the *Controller* should have about the entrance of a train in the critical region R. (The design rationale is that some time is needed to lower the gate.)

The specifications of the indicative properties are part of the descriptions presented in the SRD (see Appendix A).

Stage 4: Specifying the Machine. This last stage consists in developing the specification of the machine. Its specification may include the introduction of new internal machine actions and/or state components. As pointed out in [18] for control intensive systems, most of these new elements are related with the internal representation that the system maintains about the environment.

The result of this stage was already presented in Sect. 2 (see the constraints of the *Controller* agent in Appendix A).

3.2 Reasoning on Albert II Specifications

Reasoning is central in the process that has been sketched above: for instance, the analyst must prove that the requirements (identified at Stage 2) are met by the SRD (developed during Stages 3 and 4).

In general, complete automation of such proofs is not possible, due to the basic properties of Albert II and its underlying logic. Furthermore, even in situations where automation is applicable (e.g. finite state space), we do not think that the produced output is sufficient. As pointed in [2], a detailed formal proof is poorly readable by the analyst. It is due to: (i) the presence of too low-level details and (ii) the lack of feedback on the 'why' does a proof does hold or not —in particular in the context of the evolution of specifications (e.g. to evaluate the impact of a requirements change in terms of consequences on the SRD).

However, due to the large amount of information to be managed during a reasoning process and the disastrous effects 'small' errors might introduce, we hardly envision it as being practical without any tool support.

This suggests the following strategy to support the analyst in his/her reasoning activity upon Albert II specifications:

1. To support the analyst in the design of the proof, yielding a *high-level proof.*

2. To automate the formal proof of localized, restricted properties (which have been identified in the high-level proof).

We are just starting the development of such framework. However, we give some preliminary insights in the next two paragraphs.

High-Level Proof. A possible strategy to design a proof is to follow the specification statements and the axioms associated with the templates of the Albert II language.

As an example, let us consider the safety property (see Stage 2 in Sect. 3.2) and show a fragment of the proof that the SRD meets this property.

A. Assume that the *Position* of a given train t is I at time s_1.

B. From the **specification** *(Train – Init Valuation)*, we known that $t.Position = Elsewhere$ at initial time. Therefore according to the **frame axiom** associated with the language, there has been a time s_2, $s_2 < s_1$, where an action affecting I to $t.Position$ happened and such that *Position* kept value I since then. Given the **specification** *(Train – Effects of Actions)*, there is only one such action: *EnterI*.

C. If action *EnterI* happens at s_2, the **specification** *(Train – Action Composition, Action Duration)* tells us that *EnterR* happened at $s_3 \leq s_2 - Epsilon1$.

D. The **specification** *(Train – Action Information; Controller – Action Perception)* implies that $t.enterR$ in *Controller* also happens at s_3.

E. From the **specification** *(Controller – Action Composition, Action Duration and Effects of Actions)*, we deduce that the event *SchedEnterI(t)* happens at $s_3 + Epsilon1$ and sets *SchedInCross[t]* to *TRUE*.

F. From **E.** and the **specification** *(Controller – State Behaviour; Train – Action Composition, Capability)*, we deduce that *SoonArrival* is verified at s_3.

G. Using a similar reasoning, we show that $g.Status = Down$ at latest at $s_3 + GammaDown$.

H. From **B., C., E., G.,** and the **specification** *(Operators, GammaDown < Epsilon1)*, we derive that $g.Status = Down$ before s_2.

I. To **conclude**, we still have to show that the gate remains closed until s_1. This is done using the **specification** *(Controller – Capability* associated with the *Raise* action).

133

Local Formal Proof. To support low-level, formal proofs, we are planning to develop tools based on theorem proving and model-checking techniques.

For the theorem proving part, we are investigating the PVS system [5] and are trying to embed Albert-CORE in the higher-order logic of PVS. Similar work has already been tackled for the duration calculus [24], for instance.

For the model-checking part we have already implemented a prototype of a decision procedure using the sharing-tree data structure. The decision procedure is defined for a decidable fragment of real-time temporal logic. Therefore, to practically use this decision procedure, we have to investigate and adapt abstraction techniques to transform part of Albert II specifications into formulas of such a decidable logic. The PVS theorem prover will also be used to prove the soundness of the abstractions used. Such a method has already been applied sucessfully for proving the correctness of concurrent programs [15].

4 Conclusion

In this paper, we have illustrated the use of the Albert II RE language to specify the requirements of a real-time composite control system. To conclude, we discuss the transition to the design phase, relate Albert II to other approaches, and sketch the work we are currently tackling.

RE vs. DE. The Albert II specifications presented in the previous sections deal with issues related to RE and not to DE. In particular, the meaning of our concept of agent should be stressed. In the GRC case study, the specification of the *Controller* agent (machine to be built) is given. Having introduced it, freedom is left for the design and implementation. For example, the controller can be implemented as a network of sensors monitored by a piece of software, or as a human being operating the gate when he/she notices a train is approaching, ... In other words, at the RE level, we define only the *role* of agents but give no detail about architectural design considerations.

Starting from the SRD (i.e. the specification of the machine and of its environment), the designer will be faced with multiple complex issues when deriving the design specification document. Besides the architectural issued mentioned above, we can mention two of them.

- Usually, design specifications are formulated in a more *operational* style. To this end, *declarative* and *non deterministic* aspects of the Albert II specification will have to be removed. In particular, transformations are required for statements including references to the future, like, "this action occurs if the value of this state component remains the same during the next three minutes."

- When designing the machine, we are shifting from a 'closed world' view (i.e. the machine and its environment) to the 'open world' view (i.e. the design specification of the machine only). To achieve this, some important information coming from the problem domain (like "trains never reverse direction") will have to be considered by the designer.

Related Work. There are several formal specification languages initially designed to support the DE phase which include facilities for modelling real-time aspects as well as powerful analysis tools. Examples include timed automata, timed statecharts or timed Petri nets. However, such languages are not always well suited to RE modelling because they rely on an *operational* style of specification which in many cases does not fit well the *declarative* style adopted by customers when expressing their wishes. Moreover, these languages rely sometimes on modelling artefacts which are not compatible with real-world modelling (e.g., the interdiction of having simultaneous events).

Among the very few formal declarative languages explicitly designed for RE purposes and including built-in facilities for dealing with real-time issues, we may cite ERAE [10], TRIO+ [21] and PARTS [20]. ERAE includes only minimal syntactical facilities for structuring specifications and the non-resolution of the frame problem makes the writing of specifications cumbersome. TRIO+ is very close to Albert II but lacks from methodological guidance. It is worth noting that different prototypes of tools supporting verification and validation have already been developed for TRIO (the ancestor of TRIO+, which does not support the OO facilities). Finally, the philosophy taken in PARTS is rather similar to the one adopted for Albert II. PARTS is grounded onto RTTL [22] but equipped with high level constructs making possible to organize requirements. However, these constructs supports a functional view of the system rather than an agent-oriented view.

Further Work. There is an on-going national project related to the development of tools supporting Albert II.

A graphical/textual editor as well as a checker guaranteeing minimal consistency and completeness have already been developed. Besides this tool, we are also developing tools supporting the verification and validation of specifications. They are based on:

- the development of a 'framework' to support high-level reasoning;

- the use of the PVS theorem prover together with model-checking and abstract interpretation techniques to perform local, formal reasoning;

134

- the development of an animation tool allowing the exploration of the different possible lives associated with a specification. A prototype of this tool has already been considered [9].

As noted before, **Albert II** deals with functional, real-time requirements only. However, we are working on the integration of our language with the i^* framework proposed for modelling organizational goals and non-functional requirements [26].

Acknowledgments

We would like to thank François Chabot and Jean-François Raskin for their input concerning the discussion of tools for formal reasoning. François helped us also in the final proofreading. We are also indebted to the other members of the team (Patrick Heymans, Bernard Jungen and Michaël Petit) for many fruitful discussions.

References

[1] 2RARE (2 real applications for requirements engineering), esprit project programme (contract number 20.424), November 1995. Information available at http://www.info.fundp.ac.be/~phe/2rare.html.

[2] James Armstrong and Leonor Barroca. Specification and verification of reactive system behaviour: the railroad crossing example. *Real-Time Systems*, 10(2):143–178, March 1996.

[3] A. Borgida, J. Mylopoulos, and R. Reiter. On the frame problem in procedure specification. *IEEE Transactions in Software Engineering*, SE-21(10), October 1995.

[4] J.A. Bubenko. Information modeling in the context of system development. In S.H. Lavington, editor, *Information Processing 80*, pages 395–411. North-Holland, 1980.

[5] Judy Crow, Sam Owre, John Rushby, Natarajan Shankar, and Mandayam Srivas. A tutorial introduction to PVS. Technical report, Computer Science Lab, SRI International, Menlo Park CA, April 1995.

[6] A. Dardenne, A. van Lamsweerde, and S. Fickas. Goal-directed requirements acquisition. *Science of Computer Programming*, 20:3–50, 1993.

[7] Philippe Du Bois. *The Albert II Language: On the Design and the Use of a Formal Specification Language for Requirements Analysis*. PhD thesis, Computer Science Department, University of Namur, Namur (Belgique), September 1995. Available at http://www.cediti.be/~pdu/thesispr-uk.html.

[8] Eric Dubois. Logical support for reasoning about the specification and the elaboration of requirements. In *The role of artificial intelligence in databases and information systems*, pages 28–48, Guangzhou (China), July 1988. IFIP WG2.6/W68.1.

[9] Eric Dubois, Philippe Du Bois, and Frédéric Dubru. Animating formal requirements specifications of cooperative information systems. In *Proc. of the Second International Conference on Cooperative Information Systems – CoopIS-94*, pages 101–112, Toronto (Canada), May 17-20, 1994. University of Toronto Press inc. Available at ftp://ftp.info.fundp.ac.be/publications/RP/RP-94-009.ps.Z.

[10] Eric Dubois, Jacques Hagelstein, and André Rifaut. A formal language for the requirements engineering of computer systems. In André Thayse, editor, *From natural language processing to logic for expert systems*, chapter 6. Wiley, 1991.

[11] Eric Dubois, Jacques Hagelstein, Axel van Lamsweerde, Fernando Orejas, Jeanine Souquières, and Pierre Wodon. A guided tour through the ICARUS project. *Software Engineering Notes*, 20(2):28–33, 1995. Available at ftp://ftp.info.fundp.ac.be/publications/RP/RP-95-017.ps.Z.

[12] Martin S. Feather. Language support for the specification and development of composite systems. *ACM Transactions on Programming Languages and Systems*, 9(2):198–234, April 1987.

[13] Sol J. Greenspan, Alexander Borgida, and John Mylopoulos. A requirements modeling language. *Information Systems*, 11(1):9–23, 1986.

[14] D. Harel, H. Lachover, A. Naamad, A. Pnueli, M. Politi, R. Sherman, and A. Shtul-Trauring. STATEMATE: a working environment for the development of complex reactive systems. In *Proc. of the 10th International Conference on Software Engineering – ICSE'88*, pages 396–406, Singapore, April 11-15, 1988. IEEE.

[15] K. Havelund and N. Shankar. Expirements in theorem proving and model checking for protocol verification. In *Proc. of the FME'96 Symposium on Industrial Benefit and Advances in Formal Methods*, Oxford (UK), March 18-22, 1996.

[16] C.L. Heitmeyer, R.D. Jeffords, and B.G. Labaw. A benchmark for comparing different approaches for specifying and verifying real-time systems. In *Proc. of the 10th International Workshop on Real-Time Operating Systems and Software*, May 1993.

[17] C.L. Heitmeyer and N. Lynch. The generalized railroad crossing: A case study in formal verification of real-time systems. In *Proc. of the IEEE Real-Time Systems Symposium*, San Juan, Puerto Rico, December 7-9, 1994.

[18] Michael Jackson. *System Development*. Prentice Hall, 1983.

[19] Michael Jackson and Pamela Zave. Deriving specifications from requirements: An example. In *Proc. of the 17th International Conference on Software Engineering – ICSE'95*, pages 15–24, Seattle WA, April 23-30, 1995. ACM Press.

[20] Kyo C. Kang and Kwang-Il Ko. PARTS: A temporal logic-based real-time software specification and verification method. In *Proc. of the 17th International Conference on Software Engineering – ICSE'95*, pages 169–176, Seattle WA, April 23-30, 1995. ACM Press.

[21] Angelo Morzenti and Pierluigi San Pietro. Object-oriented logic specifications of time critical systems. *ACM Transactions on Software Engineering and Methodology*, 3(1):56–98, January 1994.

[22] Jonathan S. Ostroff. Formal methods for the specification and design of real-time safety critical systems. *The Journal of Systems and Software*, pages 33–60, April 1992.

[23] Gunter Saake, Ralf Jungclaus, and Thorsten Hartmann. Application modelling in heterogenous environments using an object specification language. In *Proc. of the International Conference on Intelligent and Cooperative Systems – ICI-CIS'93*. IEEE CS Press, 1993.

[24] Jens Ulrik Skakkebæk and Natarajan Shankar. Towards a duration calculus proof assistant in PVS. In *Proc. of the 3rd International School and Symposium on Formal Techniques in Real Time and Fault Tolerant Systems*, pages 660–679, Lubeck (Germany), September 19-23, 1994. LNCS 863, Springer-Verlag.

[25] Roel Wieringa. LCM and MCN: Specification of a control system using dynamic logic and process algebra. In C. Lewerentz and T. Lindner, editors, *Case Study Production Cell – A Comparative Study of Formal Software Development*, Lecture Notes in Computer Science. Springer-Verlag, 1994.

[26] Eric Yu, Philippe Du Bois, Eric Dubois, and John Mylopoulos. From organization models to system requirements - a "cooperating agents" approach. In *Proc. of the Third International Conference on Cooperative Information Systems – CoopIS-95*, Vienna (Austria), May 9-12, 1995. University of Toronto Press inc.

A   SRD – Textual Constraints

Train

BASIC CONSTRAINTS
INITIAL VALUATION
Position = Elsewhere

LOCAL CONSTRAINTS
EFFECTS OF ACTIONS
EnterR: Position := R
EnterI: Position := I
Exit: Position := Elsewhere
CAPABILITY
$\mathcal{F}$(Pass / Position $\neq$ Elsewhere)
ACTION COMPOSITION
Pass $\leftrightarrow$ PassR ; Exit
PassR $\leftrightarrow$ EnterR ; EnterI
ACTION DURATION
Epsilon1 $\leq$ | PassR | $\leq$ Epsilon2

COOPERATION CONSTRAINTS
ACTION INFORMATION
$\mathcal{K}$(EnterR.c / TRUE)
$\mathcal{K}$(Exit.c / TRUE)

Gate

BASIC CONSTRAINTS
INITIAL VALUATION
Position = Up

LOCAL CONSTRAINTS
EFFECTS OF ACTIONS
c.Lower: Position:= GoingDown
StopLower: Position:= Down
c.Raise: Position:= GoingUp
StopRaise: Position:= Up
CAPABILITY
$\mathcal{F}$(StopLower / Position $\neq$ GoingDown)
$\mathcal{F}$(StopRaise / Position $\neq$ GoingUp)
ACTION COMPOSITION
Move $\leftrightarrow$ Lowering ; Raising
Lowering $\leftrightarrow$ c.Lower ; StopLower
Raising $\leftrightarrow$ c.Raise ; StopRaise
ACTION DURATION
0 < | Lowering | $\leq$ GammaDown
0 < | Raising | $\leq$ GammaUp

COOPERATION CONSTRAINTS
ACTION PERCEPTION
$\mathcal{XK}$(c.Lower / Position = Up)
$\mathcal{XK}$(c.Raise / Position = Down)

Controller

BASIC CONSTRAINTS
INITIAL VALUATION
Gate = Up
> The initial expected status of the Gate is Up

SchedInCross[t] = FALSE
> There is no scheduled train to pass the crossing at the initial time

SoonArrival = FALSE
> There is no expected SoonArrival at the initial time

NoSoonArrival = TRUE
> There is NoSoonArrival at the initial time

LOCAL CONSTRAINTS
STATE BEHAVIOUR
SoonArrival $\Longleftrightarrow$
$\exists$ t: ($\neg$ SchedInCross[t]) $\wedge$ $\Diamond_{=GammaDown}$ SchedInCross[t])
> SoonArrival means that a train which is not scheduled now but which will be scheduled within GammaDown seconds

NoSoonArrival $\Longleftrightarrow$
$\forall$ t: ($\neg$ SchedInCross[t]) $\wedge$ $\Box_{\geq GammaUp+Delta+GammaDown}$ $\neg$ SchedInCross[t])
> NoSoonArrival means that there is no train scheduled to cross now neither during the next GammaUp + Delta + GammaDown seconds

EFFECTS OF ACTIONS
SchedEnterI(t): SchedInCross[t] := TRUE
> After the occurrence of a SchedEnterI event for a train t, this train belongs to the list of trains scheduled to pass the crossing

t.Exit: SchedInCross[t] := FALSE
> After the occurrence of an Exit event issued by a train t, this train does not belong to the list of trains scheduled to pass the crossing

Lower: Gate := Down
> After the occurrence of a Lower event, the
> expected status of the Gate is Down

Raise: Gate := Up
> After the occurrence of a Raise event, the
> expected status of the Gate is Up

CAPABILITY

$\mathcal{XO}$(Raise / Gate = Down $\wedge$ NoSoonArrival)
> There is an obligation to Raise the gate if
> and only if it is Down and there is
> NoSoonArrival

$\mathcal{XO}$(Lower / Gate = Up $\wedge$ SoonArrival)
> There is an obligation to Lower the gate if
> and only if it is Up and there is a
> SoonArrival

$\mathcal{F}$(Arrival(t) / SchedInCross(t))
> It is forbidden for an Arrival of train t to
> happen if this train is already in the
> crossing

ACTION COMPOSITION

Arrival(t) $\leftrightarrow$ t.EnterR ; SchedEnterI(t)
> The Arrival of a train t is decomposed in
> its EnterR in the region (issued by the
> train) followed by an event associated
> with its SchedEnterI in the crossing

ACTION DURATION

| Arrival | = Epsilon1
> The arrival of train takes Epsilon1 seconds

COOPERATION CONSTRAINTS

ACTION PERCEPTION

$\mathcal{K}$(t.EnterR / TRUE)

$\mathcal{K}$(t.Exit / TRUE)
> The EnterR and Exit events issued by a train
> are perceived in all circumstances

ACTION INFORMATION

$\mathcal{K}$(Raise.g / TRUE)

$\mathcal{K}$(Lower.g / TRUE)
> Raise and Lower events are made visible to
> the gate in all circumstances

Session 6B

Applications and Tools 2

GRAIL/KAOS: An Environment for
Goal-Driven Requirements Analysis, Integration and Layout

R. Darimont

E. Delor

P. Massonet

A. van Lamsweerde*

The KAOS methodology provides a language, a method, and meta-level knowledge for goal-driven requirements elaboration. The language provides a rich ontology for capturing requirements in terms of goals, constraints, objects, actions, agents, etc. Links between requirements are represented as well to capture refinements, conflicts, operationalizations, responsibility assignments, etc.

The KAOS specification language is a multi-paradigm language with a two-level structure: an outer semantic net layer for declaring concepts, their attributes and links to other concepts, and an inner formal assertion layer for formally defining the concept. The latter combines a real-time temporal logic for the specification of goals, constraints, and objects, and standard pre-/postconditions for the specification of actions and their strengthening to ensure the constraints.

The method roughly consists of (i) identifying and refining goals progressively until constraints that are assignable to individual agents are obtained, (ii) identifying objects and actions progressively from goals, (iii) deriving requirements on the objects and actions to meet the constraints, and (iv) assigning the constraints, objects and actions to the agents. Meta-level knowledge is used to guide the elaboration process; it takes the form of conceptual taxonomies, well-formedness rules and tactics to select among alternatives.

GRAIL is an environment under development to support the KAOS methodology. The GRAIL kernel combines a graphical view, a textual view, an abstract syntax view, and an object base view of specifications. The current version integrates a graphical editor, a syntax-directed editor, an hypertext navigator, and a LaTex report generator. GRAIL has been used for the requirements reengineering of a large, complex telecommunication system.

The presentation of the tool will be accompanied by a brief presentation of the KAOS methodology that is necessary to understand it. Tool functionality will be illustrated with extracts from an industrial case-study. The presentation will conclude by discussing current limitations of the tool together with ongoing and future developments.

*The first three authors are from the CEDITI Tech Transfer Center and the Université Catholique de Louvain. The fourth author is from the Département d'Ingénierie Informatique, Université Catholique de Louvain. The development of GRAIL is partially supported by the "Fonds Européen de Développement Régional" and the "Ministère de la Région Wallonne" (DGTRE, Convention Objectif 1 No. 2905 - CEDITI).

Requirement Metrics—Value Added

T. Hammer (GSFC), L. Rosenberg (Unisys GSFC), L. Huffman (Unisys GSFC),
and L. Hyatt (GSFC)
thammer@pop300.gsfc.nasa.gov

Actions in the requirements phase can directly impact the success or failure of a project. It is critical that project management utilize all available tools to identify potential problems and risks as early in the development as possible, especially in the requirements phase. The Software Assurance Technology Center (GSFC) and the Quality Assurance Division at NASA Goddard Space Flight Center are working on developing and applying metrics from the onset of project development, starting in the requirements phase. This talk will discuss the results of a metrics effort on a real, large system development at GSFC and lessons learned from this experience. The development effort for this project uses an automated tool to manage requirements decomposition.

This report focuses on the metrics used to assess the requirement decomposition effort and to identify potential risks. The objective of the requirement assessment was to: determine how requirements were being distributed across planned releases, determine to which release the most requirements were allocated, and characterize the expansion of requirements from one level to the next. This report illustrates the requirement metrics development and its application with examples from a large software development effort and shows how the derived metrics were used to identify some areas of project risk.

This report discusses how metric analysis in the requirements phase can be accomplished on any government or industry project. The use of an automated tool to manage requirement development facilitation of metrics for improved insight into development and risk assessment will also be expanded.

Eliciting Requirements: Beyond the Blank Sheet of Paper

Haim Kilov and Ian Simmonds
Insurance Research Center
IBM T J Watson Research Center
{kilov,isimmond}@watson.ibm.com

This report will show how we have successfully avoided the "blank sheet of paper" problem in eliciting business requirements. We did this in a context in which we had explicitly separated business specification ("analysis") from solution specification ("design"); in producing a precise, compact, understandable, yet complete business specification for a non-trivial business problem; and in which business analysts of a non-consulting company became self-reliant and comfortable with the approach in a reasonably short period of time. More generally, "research" terminology (e.g., "invariant", "generic relationship") essential for writing such a specification was quickly understood and freely used by all industry participants.

The "blank sheet of paper" problem can appear when a requirements engineer is entering an uninvestigated business area, or is bombarded by a large, unstructured body of material. Elicitation of requirements demands the adoption of an appropriate frame of reference and "units of thought". In the absence of existing materials, the "unit of thought" is absent and candidates may be unclear. Conversely, when bombarded with inappropriately structured materials (frequently including concepts beyond-the-scope of the business problem) in the form of examples, scenarios, state machines, message definitions and so on, there may be too many candidate units of thought, all of which may be inappropriate for elicitation.

We will show how business patterns help to produce complete, explicit and rigorous business specifications understandable by both business users and system developers. These specifications require rigorous expressions of behavioral semantics - that is, assertions - rather than loose, "intuitive", descriptions. We will present examples of both elementary patterns - such as "composition" - and non-elementary patterns - such as "assessment" and "information gathering". Unlike typical programming constructs, instantiations of business patterns are inherently interactive and so must adapt to their changing environment.

Selection of an appropriate pattern for eliciting the requirements of a particular business situation provides a structured - and unambiguous - way to understand the problem and specify an appropriate information management solution: the sheet of paper is no longer blank. Equally, in using requirements specified in this way, readers familiar with the patterns (perhaps encountered elsewhere in the requirements for the same project) can reuse their own knowledge; all readers benefit from the clear separation of overall structure from details specific to a particular application of the pattern.

Session 7

Keynote Address

Speaker

John Rushby
SRI International, USA

"Calculating with Requirements"

Calculating with Requirements[*]
(Extended Abstract)

John Rushby
Computer Science Laboratory
SRI International
Menlo Park CA 94025 USA

1 Automated Formal Methods

Requirements *elicitation* is concerned with discovering what is wanted; it necessarily depends on social processes such as discussion, introspection, review of documentation, and experimentation. Requirements *engineering*, on the other hand, is concerned with turning the products of elicitation into precise, unambiguous, and complete descriptions of what the system under consideration is to do. Although they shade into and complement each other, and may be part of a larger iterative process, requirements elicitation and engineering are different activities that need to be supported by different techniques and tools.

I maintain that *formal methods* provide techniques and tools that are appropriate—and effective—for the requirements engineering activity. They are effective because they allow certain questions about requirements to be reduced to calculations, and this is valuable because it allows *reviews* to be supplemented or replaced by *analyses*. I am using these terms in the sense in which they are employed in the guidelines for software on commercial aircraft [10, Section 6.3]: reviews are processes that depend on human judgment and consensus, while analyses are objective "mechanical" processes such as testing or calculation. Of course, certain questions do require human judgment, and some decisions require consensus, but many other issues are better addressed by analyses than by reviews: analyses are systematic, can be checked by others, and can even be automated. Especially when automated, analyses can be more reliable and thorough than reviews, and cheaper.

Formal methods are often advocated for the intellectual framework that they provide, and the methodological benefits that are believed to accompany their use.

While I certainly agree that formal methods can help designers to conceptualize and formulate issues and requirements in a perspicuous and productive manner, I am skeptical that assurance based on human review is any more reliable for formal specifications than for informal descriptions. Instead, I claim that the distinctive benefit of specifically *formal* methods is that they support analysis through formal deduction—that is by theorem proving and related methods such as model checking. These are systematic processes that have the character of calculation and, like numerical calculations, they can be automated. The reasons for favoring mathematical modeling and calculation are the same in computer science as in other engineering disciplines: they allow the consequences of requirements and the properties of designs to be accurately predicted and evaluated prior to construction.

In most engineering disciplines, it is automation that releases the full potential of mathematical modeling: the highly efficient wings of a modern airplane could not be designed without the massive automation of computational fluid dynamics, finite element analysis, and several other mathematical modeling techniques. Automating the calculations underlying formal methods yields similar benefits for computer science: many questions that are currently examined by intensive but unreliable reviews, or by massive but necessarily incomplete testing, can be settled by automated calculation—thereby providing analysis that is more reliable and more comprehensive than at present, and liberating human reviewers for more creative and challenging tasks that truly require their judgment.

2 Experiments

Over the last few years, application of specialized but pragmatically effective theorem proving techniques, and of model checking and related methods, has made it possible to subject formal requirements specifications to several kinds of automated analysis. Some

[*]This work was partially supported by NASA Langley Research Center under contract NAS1-20334, by the Air Force Office of Scientific research, Air Force Materiel Command, USAF, under contract F49620-95-C0044, and by the National Science Foundation under contract CCR-9509931.

of these have been applied experimentally to requirements documented in "Change Requests" (CRs) for the flight software of the Space Shuttle by a team involving the group at Lockheed Martin (formerly IBM) that develops this software, several NASA centers (Johnson, Langley, and JPL), and SRI. These experiments, running alongside what is generally considered an exemplary process for requirements review, provide useful anecdotal evidence for the effectiveness of automated formal analyses.

2.1 Very Strong Typechecking

One of the most basic "theorems" that can be proved about a specification is type-correctness. For programming languages, this theorem is "proved" by simple typechecking algorithms, but specification languages can use general theorem proving for this purpose with a corresponding increase in the strictness in the notion of type-correctness that can be enforced [9]. By simply describing its interfaces and functional requirements in such a very strongly-typed specification language, 86 issues (including 11 considered "major") were detected in several iterations of a CR for installing Global Positioning System (GPS) navigation on the Shuttle [3].[1]

2.2 Completeness and Consistency of Tabular Specifications

Rather stronger than typechecking are completeness and consistency checks for tabular specifications of the kind advocated by Parnas [14]. The idea is that the rows and columns of the table should partition the input space into distinct regions; if the circumstance in which each column (*resp.* row) applies is specified by a logical formula, then consistency requires that the pairwise conjunctions of the formulas should be false, and completeness requires that their overall disjunction should be true. Depending on the logic and theories used in the formulas, the deductive capabilities needed to discharge these proof obligations range from propositional tautology checking, though decision procedures for ground linear arithmetic, to full interactive theorem proving [11]. The capabilities at the lower end of this spectrum can be automated very effectively and can scale up to very large tables; those at the upper end are less automated and scale less well. Rapid progress is being made on developing balanced techniques that combine efficient and scalable automation with adequate expressiveness [4]. Applying some of

these methods to a Shuttle CR for the "Heading Alignment Cylinder" (HAC), revealed that several rows in one table had overlapping conditions, and that several unstated assumptions needed to be articulated before the completeness and consistency of others could be established [2]. The Shuttle requirements analysts considered discovery of the missing assumptions to be particularly valuable. Similar successes with automated completeness and consistency checking have been reported for the RSML [6] and SCR [7] requirements methods.

2.3 State Exploration and Model Checking

Deeper analysis of tabular specifications is possible using model checking to examine whether certain desired properties are invariant or reachable. Atlee and Sreemani have demonstrated that such methods scale to quite large examples [13]. Model checking can also be used to perform analyses that are intermediate between testing individual cases (as can be performed by prototyping or animation), and proving general theorems. The idea is to "downscale" (aggressively simplify) the description of the system and its environment so that they become finite state. Finite state exploration or model checking methods can then be used to examine all the reachable states. Although the necessary simplification may be too aggressive for verification (i.e., properties of the simplified model may not be true of the full system) it is often adequate for refutation (i.e., bugs found in the model will also exist in the full system). Anecdotally, it seems that exploring *all* the states of a simplified system specification is much more effective at finding bugs than sampling those of the full system by prototyping or animation. These techniques are applied routinely (and very effectively) to hardware and protocols, but their application to software and to requirements is relatively new (see, for example [8]). Applying them to a Shuttle CR for three-engines-out abort sequencing revealed 19 errors, of which only 1 (albeit the most significant) had been discovered by human review [1].

2.4 Formal Challenges

Some of the most searching examinations of requirements specifications can be performed by "challenging" the specification with a putative theorem: "if this specification captures my intent, then the following ought to follow." Suppose, for example, that we had specified the operation of sorting a sequence; we might then challenge the specification by asking whether sorting an already sorted sequence leaves the

---

[1]At an earlier stage in this activity, formal methods had detected 30 issues, including 7 considered major, compared with 4, of which 1 was considered major, for the human review-based process [1].

sequence unchanged (i.e., we attempt to prove the theorem $\forall s : sort(sort(s)) = sort(s)$). Such a challenge would reveal a deficiency in many of the early specifications of sorting, which required the output to be ordered, but neglected to stipulate that it should also be a permutation of the input [5]. Applying this approach to the mature requirements specification for a Shuttle function known as "Jet Select" suggested the challenge "no jet that is designated as failed shall be selected for firing" [12]. The attempt to prove this was unsuccessful, revealing thereby a significant and previously undiscovered problem with the existing requirements specification.

3 Summary

Many issues in requirements engineering can be explored and analyzed using automated formal methods. At present, the tools supporting these analyses are not ideal: considerable knowledge and experience are required to select the most appropriate tool for a given task, to formulate the problem in a suitable manner, and to coax the tool into divulging a useful result. However, engineering challenges in the integration and scaling of formal methods tools are being rapidly overcome and these difficulties should soon be significantly reduced. Judicious use of automated formal calculations will allow many issues to be explored more completely and reliably than at present, and will allow the talents of human reviewers to be reserved for those problems in requirements analysis that truly require them.

References

Papers by SRI authors can generally be retrieved from http://www.csl.sri.com/fm.html.

[1] Judith Crow and Ben L. Di Vito. Formalizing Space Shuttle software requirements. In *First Workshop on Formal Methods in Software Practice (FMSP '96)*, pages 40–48, San Diego, CA, January 1996. Association for Computing Machinery.

[2] Judith Crow and Ben L. Di Vito. Formalizing space shuttle software requirements: Four case studies. Submitted for publication, 1997.

[3] Ben L. Di Vito. Formalizing new navigation requirements for NASA's space shuttle. In *Formal Methods Europe FME '96*, volume 1051 of *Lecture Notes in Computer Science*, pages 160–178, Oxford, UK, March 1996. Springer-Verlag.

[4] David L. Dill. Closely cooperating decision procedures. In M. Srivas, editor, *Formal Methods in Computer-Aided Design (FMCAD '96)*, Palo Alto, CA, November 1996. To appear.

[5] S. L. Gerhart and L. Yelowitz. Observations of fallibility in modern programming methodologies. *IEEE Transactions on Software Engineering*, SE-2(3):195–207, September 1976.

[6] Mats P. E. Heimdahl. Experiences and lessons from the analysis of TCAS II. In Steven J. Zeil, editor, *International Symposium on Software Testing and Analysis (ISSTA)*, pages 79–83, San Diego, CA, January 1996. Association for Computing Machinery.

[7] Constance Heitmeyer, Alan Bull, Carolyn Gasarch, and Bruce Labaw. SCR*: A toolset for specifying and analyzing requirements. In *COMPASS '95 (Proceedings of the Tenth Annual Conference on Computer Assurance)*, pages 109–122, Gaithersburg, MD, June 1995. IEEE Washington Section.

[8] Daniel Jackson and Craig A. Damon. Elements of style: Analyzing a software design feature with a counterexample detector. *IEEE Transactions on Software Engineering*, 22(7):484–495, July 1996.

[9] Sam Owre, John Rushby, Natarajan Shankar, and Friedrich von Henke. Formal verification for fault-tolerant architectures: Prolegomena to the design of PVS. *IEEE Transactions on Software Engineering*, 21(2):107–125, February 1995.

[10] *DO-178B: Software Considerations in Airborne Systems and Equipment Certification*. Requirements and Technical Concepts for Aviation, Washington, DC, December 1992. This document is known as EUROCAE ED-12B in Europe.

[11] John Rushby. Mechanizing formal methods: Opportunities and challenges. In Jonathan P. Bowen and Michael G. Hinchey, editors, *ZUM '95: The Z Formal Specification Notation; 9th International Conference of Z Users*, volume 967 of *Lecture Notes in Computer Science*, pages 105–113, Limerick, Ireland, September 1995. Springer-Verlag.

[12] John Rushby. Automated deduction and formal methods. In Rajeev Alur and Thomas A. Henzinger, editors, *Computer-Aided Verification, CAV '96*, volume 1102 of *Lecture Notes in Computer Science*, pages 169–183, New Brunswick, NJ, July/August 1996. Springer-Verlag.

[13] Tirumale Sreemani and Joanne M. Atlee. Feasibility of model checking software requirements. In *COMPASS '96 (Proceedings of the Eleventh Annual Conference on Computer Assurance)*, pages 77–88, Gaithersburg, MD, June 1996. IEEE Washington Section.

[14] A. John van Schouwen, David Lorge Parnas, and Jan Madey. Documentation of requirements for computer systems. In *IEEE International Symposium on Requirements Engineering*, pages 198–207, San Diego, CA, January 1993.

The views and conclusions contained herein are those of the author and should not be interpreted as necessarily representing the official policies or endorsements, either expressed or implied, of the Air Force Office of Scientific Research or the U.S. Government.

Session 8A

Case Studies

Integrated Safety Analysis of Requirements Specifications
Francesmary Modugno, Nancy G. Leveson, Jon D. Reese, Kurt Partridge, and Sean D. Sandys

The safety analyst has a cornucopia of different techniques available from which to choose. Guidelines as to the strengths and weaknesses of each choice are harder to come by, however. This paper reports the use of several safety techniques on the requirements of a large system. The comparative evaluation of techniques, the concrete examples of errors found, and the clearly stated conclusions will be useful to the practitioner.
— Robyn Lutz

Formal Methods for V&V of Partial Specifications: An Experience Report
Steve Easterbrook and John Callahan

Science is there to remove myth. The bag of myths associated with the term "formal method" is so heavy, that the large majority of practitioners is shying away, if something close to this term comes up. This paper puts away quite a number of such myths: Formal methods *can* be applied in a simple, yet formally correct manner, formal verification *is* beneficial, even if one never strives for completeness, "shadow" activities might provide a helpful skeleton for a project. These are just samples of the experiences described in the paper, and there is a lot more. This work rises above parochial boundaries of software engineering. Come, listen, and squeeze out more of the authors' rich and valuable experience.
— Roland Mittermeir

Extended Requirements Traceability: Lessons from an Industrial Case Study
Orlena Gotel and Anthony Finkelstein

Many of us complain about the lack of good, industrial strength data to back up models and methodologies proposed in the RE field. The authors, indeed, have proposed yet another RE model, so we can feel quite in our rights to ask them pointed questions about the new model's usefulness and need in a field already crowded with untested models. This paper gives a sharp reply to such questions! The paper backs up the model, called contribution structures, with a three year study of a software project undertaken by a commercial communication service provider. The heart of the paper is the recording of the *dynamic*
set of personnel who were associated with the project, either as customers, end-users, developmental or managerial staff. By following the project through multiple years, with all of its changes in personnel, one is able to see the efficacy of the authors' model in a way that would just not be convincing in a small, academic exercise. Oh yes, the paper is also very well written. I applaud the authors' efforts and perseverance.
— Stephen Fickas

Integrated Safety Analysis of Requirements Specifications*

Francesmary Modugno, Nancy G. Leveson, Jon D. Reese
Kurt Partridge, and Sean D. Sandys

Computer Science and Engineering
University of Washington
Seattle, WA 98195-2350

Abstract

This paper describes an integrated approach to safety analysis of software requirements and demonstrates the feasibility and utility of applying the individual techniques and the integrated approach on the requirements specification of a guidance system for a high-speed civil transport being developed at NASA Ames. Each analysis found different types of errors in the specification; thus together the techniques provided a more comprehensive safety analysis than any individual technique. We also discovered that the more the analyst knew about the application and the model, the more successful they were in finding errors. Our findings imply that the most effective safety-analysis tools will assist rather than replace the analyst.

Keywords: software safety, software safety analysis, software requirements specification.

Introduction

Although there are well-established techniques and procedures for analyzing electro-mechanical systems for safety, only relatively recently have researchers begun to develop similar techniques for software or for systems that contain software. The long-term goal of our research is to provide system developers with comprehensive support for software-safety analysis. To accomplish this, we must do more than simply propose techniques and demonstrate them on small examples, we also need to demonstrate the feasibility of applying them to realistic systems and to demonstrate their effectiveness. We can then apply what we have learned to improve the methods and tools, determine whether they can be effectively applied by others, and learn how to make them more usable by system developers.

Leveson and colleagues have been developing a series of techniques and tools to analyze safety-critical software [2, 8, 9, 10, 12]. We have demonstrated the techniques and shown that each can be effective indi-

vidually for the analysis of safety-critical systems. For example, Heimdahl [2] found several sources of dangerous incompleteness and nondeterminism in the specification of TCAS II (Traffic Alert and Collision Avoidance System), an airborne collision avoidance system required on all U.W. aircraft carrying more than 30 passengers. To explore the feasibility of automated analysis, we also applied some prototype tools to an automated highway system model [14]. We did not, however, study the interaction between the analysis results nor did we study the analysis process itself.

To our knowledge, there has been no effort to apply all these techniques in an integrated safety analysis of a system or to compare the results of using different hazard analysis techniques on the same requirements specification. Moreover, there is little information on the analysis process itself. In this paper, we explore the feasibility of such a comprehensive safety analysis on a specification of a guidance system for a high-speed civil transport being developed at NASA Ames [5]. The goal of this case study was to determine the feasibility of performing such analyses and to evaluate the techniques and their contribution to the safety-analysis process. For example, we were interested in seeing how the results of the individual analyses interacted—what type of information each analysis provided and how that information complemented, was different from, or was supported by the results of the other analyses. We were also interested in learning about the analysis process: how difficult the techniques were to learn and apply, what problems we would encounter when learning and applying the techniques as a group, how difficult it would be to interpret the results and integrate them, etc.

This paper reports the results of this empirical evaluation. First we provide a brief description of the guidance system. The model is much too large to include more than just a brief overview in this paper (the entire model is over a hundred pages long). More information about the specification can be found in the appendix. The main body of the paper presents an overview of software-safety analysis and the results of

*This work has been partly funded by NASA/Langley Grant NAG-1-1495, NSF Grant CCR-9396181, and the California PATH Program of the University of California.

1090-705X/97 $5.00 © 1997 IEEE

the analyses. We conclude with a discussion of our results and experiences along with planned future work.

The Guidance System

Figure 1 shows an overview of the Vehicle Management System (VMS) for a high-speed civil transport being developed at NASA Ames [5]. The VMS assists the pilot with tasks such as on-board flight planning, navigation, guidance and flight control. From the flight-path information received from the Air Traffic Controller, the Translator computes the Reference Flight Path (RFP), which completely specifies the aircraft's planned trajectory in space (i.e., its desired altitude, velocity, and flightpath angle, which is the number of degrees from the horizon in an earth reference frame) along with information such as the flap and gear extension control points, top of descent point and distance to next way point.

During automatic flight, the guidance system compares the aircraft's actual trajectory (as determined by sensor inputs) with the planned RFP trajectory and generates both lateral and vertical motion flight-control commands to null trajectory errors. These commands are sent to the Flight Control, which directly executes them, producing a change in the aircraft's velocity, altitude, or flightpath angle. During manual flight, the pilot can directly issue commands to the Flight Control in order to produce the desired effect.

The guidance system can also be operated jointly by the underlying computer and the pilot. In this mode, the pilot can define the reference flight path by entering the desired altitude, velocity and flightpath angle into the Mode Control Panel (MCP) and can select the operational mode of the guidance system by depressing the appropriate mode buttons on the MCP.

During both automatic and joint flight modes, the guidance system selects its operational mode based on (1) the pilot's selections from the Mode Control Panel, (2) the aircraft's position and velocity relative to the desired trajectory, and (3) the previous operational mode. Using this information, it computes the desired flight control commands, which include the commanded flight path angle and the commanded thrust. The guidance system specification is described briefly in the appendix.

Types of Safety Analyses

A *safe system* is one that is free from accidents or unacceptable losses [9]. The heart of analyzing a system from a safety perspective is identifying and analyzing the system for *hazards*, which are states or conditions of the system that combined with some environmental conditions can lead to an accident or loss event [9]. Once the hazards are identified, steps can be taken to eliminate them, reduce the likelihood of their occurring, or mitigate their effects on the system.

In addition, some hazard *causes* can be identified and eliminated or controlled. Although it is usually impossible to anticipate all potential causes of hazards, obtaining more information about them usually allows greater protection to be provided.

A hazard analysis requires some type of model of the system, which may be an informal model in the mind of the analyst, a written informal or formatted specification of the system, or a formal mathematical model. Different models allow for different types of analyses and for additional rigor and completeness in the analysis. For the system evaluated in this paper, we use a state-machine model of the system's required black-box behavior. We model the requirements instead of an actual software design because many accidents involving software systems can be traced to requirements flaws [13]. In addition, by modeling and analyzing the requirements, we can find problems early in the development cycle when they are more easily and effectively addressed.

A fundamental tenet of linear control theory is that every controller is or contains a model of the controlled process. In our specification language, this model describes how the controller will behave with respect to various state changes in the controlled process. It is also used to determine the current process state given the previous state and new sensor measurements (readings) of various process variables. Because a model is an abstraction, it is necessarily incomplete.

Hazards and accidents can result from mismatches between the software view of the process and the actual state of the process—that is, the model of the process used by the software gets out of synch with the real process. This mismatch may occur because the internal model is incorrect or incomplete with respect to important properties or the computer does not have accurate information about the process state.

Safety then depends on the completeness and accuracy of the software's model of the process. A state-machine specification of system or software requirements explicitly describes this model and the required behavior of the software. A goal of software-safety analysis is to ensure that the model of the controlled process (i.e., the requirements specification) is sufficiently complete that it specifies safe behavior in all circumstances in which the system will operate. Jaffe [7, 8, 9] has defined general criteria for determining whether such models satisfy this goal. The results of applying these criteria to our specification are described below.

The specification (model) can also be analyzed with respect to specific, known hazards. We accomplish this goal using what we call *State Machine Hazard Analysis* (SHMA) [9]. SMHA, like most safety analyses, involves some type of search. How that search is performed depends on the structure of the model and the goal of the search. One classification for such search techniques is forward or backward [9].

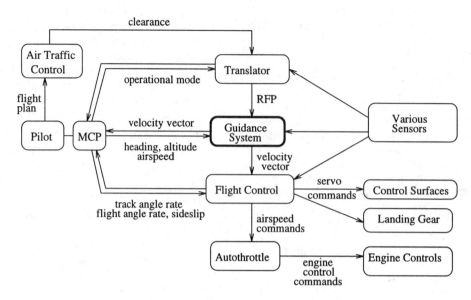

Figure 1: Overview of the Vehicle Management System for a high-speed civil transport being developed at NASA Ames. At the center is the guidance system, which automatically controls the aircraft's flight.

A *forward* (sometimes called *inductive*) search takes an initiating event (or condition) and traces it forward in time. The result is a set of states or conditions that represent the effects of the initiating event. An example of such a search is determining how the loss of a particular control surface will affect the flight of an aircraft.

Tracing an event forward can generate a large number of states, and the problem of identifying all reachable states from an initial state may be unsolvable using a reasonable set of resources. For this reason, forward analysis is often limited to only a small set of temporally ordered events.

In a *backward* (also called *deductive*) search, the analyst starts with a final event or state and identifies the preceding events or states. This type of search can be likened to Sherlock Holmes reconstructing the events that led up to a crime. Backward search approaches are useful in accident investigations and also in eliminating or controlling hazards during system development by, in essence, investigating potential accidents before they occur.

The results of forward and backward searches are not necessarily the same. Tracing an initial event forward will most likely result in multiple final states, not all of which represent hazards or accidents. Because most accidents are caused by multiple events, to be fully effective the forward analysis must include more than a single initiating event. Combinatorial explosion usually makes an exhaustive search of this type impractical and limits the number of initiating events that can be considered. The advantage of this type of search is that hazards that have not previously been identified can theoretically be found.

Tracing backward from a particular hazard or accident to its preceeding states or events may uncover multiple initiating or contributing events, but the hazards usally must be known. System engineers are quite effective in identifying system hazards, of which there are usually a limited number. Finding all the causes of such hazards is a much more difficult problem. It is easy to see that if the goal is to explore the precursors of a specific hazard or accident, the most efficient method is a backward search procedure. On the other hand, if the goal is to determine the effects of a specific event, a forward search is most efficient.

In this case study, we performed both forward analyses (simulation and deviation analysis) and a backward analysis (fault tree analysis), as described below.

Analyzing the Guidance System Model

Errors were found when constructing the model, checking the consistency and completeness criteria, and performing the forward and backward analyses. We note that the NASA guidance system specifications we used are in the research stage and still under development. The fact that we found errors or potential hazardous states reflects only on the preliminary nature of the specification.

Knowing the backgrounds of the modelers and analysts is helpful in interpreting the results. Modugno, who did the bulk of the model construction as well as overseeing the analyses and synthesizing the results, is a computer scientist with no physics training. Within computer science, her background is human–computer interaction and only recently has she begun to work in software safety. Reese is trained as a computer scientist and created one of the forward analysis methods [15]. Sandys and Partridge are Ph.D. students in computer science, studying software safety. In addition, Sandys holds a bachelor's degree in physics and helped construct the parts of the model that required physics knowledge. Finally, Leveson has been working in the area of safety for 15 years. She helped with the initial stages of model development although she did not help perform analyses in order to determine how well they could be done without her expertise.

The original guidance system specification was written by a NASA expert in aircraft control and guidance systems. He has a Ph.D. in physics, is a licensed pilot, and has been working in this area for over 30 years. We interacted with him while constructing the model, and he provided some support during the analyses, especially in determining appropriate environmental data to use in the formal simulation of the model. The next sections describe the errors found during model construction and analysis.

Errors Found While Constructing the Model

The specification was developed using three different documents provided by the NASA Ames researchers: 1) a description of the overall goals of the guidance system design along with a description of the vertical operation modes using a combination of English, Laplace diagrams and state machines; 2) a Jackson Charts [6] specification of the vertical operation modes; and 3) pseudo-code used to create a program to control a vertical motion simulator for pilot testing of the guidance system.

In the process of constructing the model, we found errors in the original guidance system specification. Discovering these errors points to the utility of going through the process of developing a model of specifications as well as to the utility of the model itself. We note again, however, that these documents are still in development so many of the errors we found might have been found in a later stage by the developers.

For example, we found a line in the pseudo-code in which the system transitioned to one state but illuminated a different button on the display. This error was probably due to quick editing on a word processor when writing the pseudo-code (i.e., copying a portion of the code and editing it incorrectly). Nonetheless, a programmer might not notice the error. Indeed, we might not have noticed it either had our modeling language not forced us to think about the relationship between the computer logic and the display information; that is, the form of our model caused us to think about the requirements specification in a particular way.

As a more serious example, we found places in the pseudo-code in which the system prompted the pilot for a particular data value when in fact another value was required by the guidance system logic. Again, we discovered this error because, when constructing the model, we had to trace the flow of the pilot's inputs through to the computer logic.

We also found several omissions in the NASA specification when constructing the model. For example, nowhere in the documents did it say how often the inputs would arrive. We obtained this information from conversations with the designer as we were constructing the model and attempting to understand the logic of the guidance system. The notations used for the original NASA specifications did not require this information to be included.

Note that these errors were not necessarily found because of any mathematical notation in our language—in fact, many in the formal methods community would not consider our language to be "formal." We believe that we found these errors in translating from the Jackson Charts and pseudo-code into our modeling language (called Requirements State Machine Language or RSML) mainly because the process required that we consider how to represent each Jackson Chart action or line of pseudo-code in RSML and then determine where it fit into our model. The type of thinking and analysis required to construct a state-machine model provides a basis for uncovering some types of requirements errors. The same type of benefit may be found using other modeling techniques, but will depend upon the modeling methodology and language and is not automatically a property of all models.

In summary, the form of our RSML model (i.e., the state decomposition we decided on), the process of constructing the model from the various informal specifications provided by the NASA Ames researchers, and the resulting structured model helped us to discover errors in the specification.

Completeness and Consistency Checks

Jaffe and colleagues [8, 9] have defined a set of formal criteria to identify missing, incorrect, and ambiguous requirements for process-control systems. Briefly and informally, the criteria ensure (1) completeness of transitions and default values during normal and nonnormal operation, including startup and shutdown; (2) complete specification of all inputs and outputs; (3) complete specification of the interaction between the computer and the operator; (4) complete description and handling of all inputs including essential value and timing assumptions about these inputs; (5) complete specification of the output conditions with respect to timing and value, including environmental capacity, data age, and latency requirements; (6) complete specification of the relationship between inputs and outputs, including feedback loops and graceful degradation; (7) and complete specification of the paths between states with respect to desirable properties such as basic reachability, recurrent or cyclic behavior, reversible behavior, reachability of safe states, preemption of transactions, path robustness, and consistency with required system-level constraints.

Analyzing a specification in terms of these criteria depends on the form of the specification (its size, language, etc.). In some cases the criteria can be enforced by the syntax of the specification language, while in other cases, the criteria can be checked by manual inspection or with the assistance of automated tools. For example, Heimdahl [2] has automated the checking of RSML specifications for two of the 47 criteria, i.e., those to ensure robustness and nondeterminism. Heitmeyer and colleagues [3] provide tools similar to Heimdahl's to check for the internal consistency of specifications in SCR [4], a state-based specification language that uses an assortment of tabular notations to define state transitions.

Without realizing it, Modugno had begun to do some of these checks informally when constructing the model and later when examining the completed model. The syntax of the modeling language is designed to make some omissions obvious. She later realized that her efforts in examining the model in this way were really a part of the planned completeness criteria checking. We also found errors relating to completeness by performing forward simulation and the other hazard analyses.

Some of the errors that were apparent immediately from examining the model involved missing transitions. For example, by reviewing the RSML graphical representation of the guidance system, we noted that several components had states with no transitions between them. Upon further examination, we discovered there was no detailed specification for these transitions in the documents. As a specific example, once both the Glideslope Lock and Altitude Lock were set, they were never unset, i.e., in examining the model we noticed that there was a transition from the state Glideslope-Lock-Off to Glideslope-Lock-On and from Altitude-Lock-Off to Altitude-Lock-On, but there were no transitions in the other direction. This type of omission is defined by one of our path completeness criteria, i.e., *reversibility.*

In addition to missing transitions, examining the model revealed input not used and missing output. When constructing the model, we had created an input (output) interface for each input (output) in the pseudo code. A simple search on each variable name helped us determine if and how each input was used or output was produced. Several inputs were never used, and two output values were never produced.

After completing the model, Modugno manually examined it for violations of each of the criteria. She believes that her familiarity with the model (from having constructed it) greatly facilitated the analysis. We have not yet had an avionics expert try to apply the criteria, but plan to do so soon.

Several omissions were detected during the manual check of the completeness criteria. For example, the specification for the Primary Flight Display did not detail how the displayed geometric figures mapped to actual data values, how long they were to be displayed and what triggered them to appear or disappear. Similarly, as we noted above, some of the mode annunciators were missing trigger events. Both these omissions were found by analyzing the specification using the human–computer interface criteria.

Several incompletenesses in the state specification were found, among them uninitialized input variables, such as measured speed and measured altitude (which are necessary to model the system during take off), and unspecified timing intervals between inputs. The NASA documents did not specify how often the sensors would report their data values nor was there any specification on how long the system should wait for the pilot's input once the prompts appeared, although a timeout was implied in the Jackson Chart and pseudo-code specifications.

Additionally, we found several states that did not specify a response to certain inputs while in that state. For example, if the pilot selects a particular vertical guidance mode, the system will prompt him or her to enter the reference speed and altitude. However, there was no specification of the system behavior if the pilot then selected another vertical guidance mode without first entering the requested data.

With respect to robustness, the current NASA specification for the guidance system contains no checks for legal ranges or timing values and thus no behavior is specified for out-of-range input values or late, early, or missing inputs. Similarly, range and timing values were missing for flight control command outputs along with data age limits (the length of time before input data or output commands are considered obsolete).

Finally, by examining the relationship between the outputs and inputs, we discovered that there were no feedback loops to check on the response of the aircraft to the generated flight control commands, and thus no behavior was specified to respond to either expected input values or unexpected, early, late or missing input values. These omissions could cause a failure in the system to go unnoticed.

The NASA guidance system is still under development and many of these omissions probably would have been caught and corrected without our analysis. However, these types of requirements specification flaws are common and often persist until late in development or actual use of the software. For example, Lutz found that the Jaffe completeness criteria covered most of the 192 requirements errors identified as safety-critical that were not detected until system integration testing of the Galileo and Voyager spacecraft [13]. We found manual checking of the criteria to be helpful in finding important specification omissions and believe that automated tools to assist the analyst in checking the criteria might detect errors not found by our manual process.

Forward Simulation

Forward simulation was performed by Modugno and Sandys. The RSML simulator was used to execute several scenarios to observe the states of the controller given specific inputs. The extensive physics background of Sandys along with help from a NASA expert were required to generate reasonable scenarios. The quality of the results of the simulation will, in general, be limited by the quality of the test scenarios.

The depth-first search property of forward simulation makes using simulation alone infeasible to detect reachable hazardous states. If, as is reasonable, hazards are assumed to be infrequent, then the probability of blindly finding a hazard using this type of search is low. However, forward simulation can be useful when used in conjunction with other analysis methods. For example, we found we could use other analysis techniques to narrow the search space and then use forward simulation to investigate the smaller region.

Despite these limitations, by simply tracing the transitions forward for individual components, we did find a hazardous state within the normal operation of the requirements specification. The state was one in which the flaps were not DOWN but the plane was in a landing approach. During landing, the flaps help slow the plane. In certain instances, the plane could get "stuck" in this hazardous state, potentially leading to an accident. The problem arises from an assumption in the specification that inputs triggering the flap motion will arrive in a particular order. In particular, the specification assumes that the information about the point in space where the pilot is supposed to initiate the flap extension (FLAP-INIT-POINTS) will arrive before the boolean input indicating that flap extension can begin. Yet there is no indication of this assumption anywhere in the specification. Specifications should either clearly state such assumptions, or they should be written to handle their violation.

Uncovering this assumption revealed another assumption within the FLAPS component—namely that the FLAP-INIT-POINT input will arrive before the plane reaches that point in space. Nothing in the specification indicates what should happen if the FLAP-INIT-POINT information arrives after passing it. In fact, a similar situation may have contributed to the recent crash of a Boeing 757 in Cali, Columbia. In that accident, the pilots entered a way point into the guidance system that the aircraft had already passed. In an attempt to reach that way point, the guidance system commanded the plane to turn around, causing the aircraft to crash into a mountain.

These two assumptions are related to some of the errors revealed by the completeness and consistency checks. Theoretically, they are covered by the completeness criteria, but we did not find them during our manual inspection of the model using the criteria as a checklist, pointing to the need for better manual inspection methods or automated assistance to the analyst.

Deviation Analysis

Deviation Analysis is a new type of forward search technique that takes its inspiration from HAZOP (HAZards and OPerability analysis), a very successful analysis procedure used in the chemical process industry. Both techniques are based on the underlying system theory concept that accidents are the result of deviations in system variables.

Deviation analysis takes a formal model of the system along with deviations in the system inputs from their normal or expected values and examines the effects of these deviations on the system's behavior. It can also help identify potential system hazards resulting from deviations in system inputs.

To assist the analyst, Reese [15] has developed a tool for performing a deviation analysis on an RSML specification. The analyst first selects one or more inputs to deviate and describes how to deviate them (high, low, very high, very low, and so on). The analyst then marks states, functions or outputs for which a deviation is considered hazardous. The tool produces a set of scenarios in which the deviated inputs would produce a deviation in a marked state, function or output. Each scenario lists assumptions on other input values that lead to the deviation and what effect the initial deviations and additional assumptions have on the affected state, function or output. For example, an input that is deviated high could cause the system to enter a hazardous state by, for example, a function computing a value that is too high or the software producing an incorrect output.

The deviation analysis was performed by Reese. By deviating different inputs, he found several potential hazards in the model. For example, he wanted to determine what effect a too-high measured speed would have on the commanded flight path angle (recall that the measured speed is the speed of the aircraft as determined by the sensors, and the commanded flight path angle is one of the guidance system outputs—see Figure 1). Using the deviation analysis tool, he deviated the measured speed "high" and marked the function that computes the commanded flight path angle. The deviation analysis tool initially produced five scenarios. Each scenario showed that deviating the measured speed alone would not cause a deviation in the commanded flight path angle. However, they also indicated that the initial deviation, coupled with one or more other deviations in the input, would cause the commanded flight path angle to be incorrect.

In performing the deviation analysis, Reese was limited by his unfamiliarity with the guidance system model. He was initially unable to determine which

```
┌─────────────────────────────────────────────┐
│   Controller prompts pilot for speed and altitude │
│                                                 │
│ Configuration:                                  │
│     Controller Aircraft-Model Gears : up        │
│     Controller Vertical-Modes                   │
│         High-Level-Mode : NOT minimum-time      │
│     MCP-Inputs Reference-Speed : waiting        │
│     MCP-Inputs Reference-Altitude : waiting     │
│     MCP-Inputs Armed-Speed : waiting            │
│ Event:                                          │
│     prompt-for-reference-speed                  │
│     prompt-for-reference-altitude               │
│     prompt-for-armed-speed                      │
└─────────────────────────────────────────────┘
                        ▲
                        │
┌─────────────────────────────────────────────┐
│      Pilot initiates minimum time approach      │
│                                                 │
│ Configuration:                                  │
│     Controller Aircraft-Model Gears : up        │
│     Controller Vertical-Modes                   │
│         High-Level-Mode : NOT minimum-time      │
│     MCP-Inputs : idle                           │
│                                                 │
│ Event: guidance-mode-received-event             │
│                                                 │
│ Condition:                                      │
│     guidance-mode = minimum-time                │
│     time > top-of-descent-time + one-minute     │
└─────────────────────────────────────────────┘
```

Figure 2: The sequence of events leading to the hazardous state in which the plane is in minimum-time mode and the landing gear are not deployed is initiated by the pilot selecting the minimum-time approach button when the plane is more than one minute away from the top-of-descent time.

states or functions to mark as important items to analyze when deviating a particular input. Only after understanding the model in detail was he able to interpret the scenarios output by his tool and use them to investigate other deviations. His experience supported our observations thus far of the importance of application knowledge in performing the analysis.

Backward Search

In backward search, the analysis starts with a hazardous state and builds trees showing the events that could lead to this state. In order to perform this analysis, we first had to come up with a list of potentially hazardous states. Using these configurations, Modugno then did a backward search, examining the results for potential hazards under expected and failure behaviors. She found the process to be labor intensive and required in-depth understanding of the model. In addition, the trees grew quickly and the presentation quickly became unwieldy. This same problem has been noted with other types of automated analysis applied to hardware models [1].

As an example, consider the hazard of the landing gear not deploying during a minimum-time approach landing. The top node of the generated tree

thus represents the case in which the system is in the minimum-time landing mode and the gears are up.

There are two changes that can lead to this configuration—a change in the gear position from DOWN to UP or a change in the vertical mode from any other mode to minimum time. By analyzing the state transitions, we determined that the only way the gears can go from DOWN to UP is if the mode changes from minimum time to take-off/go-around. Therefore, the only way the guidance system can get into the hazardous state is when the gears are already up and the plane enters minimum-time mode. Because there were a large number of potential paths to the hazardous state, we used forward simulation to help prune the number of paths that needed to be considered.

For the guidance system to transition to minimum-time mode, the pilot must initiate a minimum-time approach by selecting the minimum-time mode button on the MCP and then entering the reference speed, reference altitude, and armed speed into the MCP within a certain time period. Figure 2 is a portion of the tree showing the partial configuration of the guidance system that gives rise to this series of events: the gears are up; the Controller's high-level vertical mode is not minimum time; and the MCP Inputs from the pilot are idle (i.e., no input is expected). The figure also shows the event and conditions under which the guidance system can transition out of this configuration. The event that triggers the transition is a guidance-mode-received-event, which indicates that the pilot has selected a new vertical guidance mode. The conditions that must be satisfied for the transitions to fire are that 1) the selected guidance mode must be minimum time, and 2) the selection must be made more than one minute before the top-of-descent time. If these two conditions are met, then the guidance system will prompt the pilot to enter the reference speed, reference altitude and armed speed.

Hence, there is a normal sequence of events that can lead to the hazardous state. By analyzing the specification, we found three conditions under which the system will remain in this hazardous state under normal operation (which means that the plane will land without the gears extended). For example, for the gears to transition from UP to DOWN, the measured speed must be 250 knots. If that exact speed is never reported by the sensors, the transition will never occur. Using this analysis, we were able to change the conditions under which the gears UP to DOWN transition occurs to avoid remaining in the hazardous state. Similar analyses and changes addressed the other hazardous conditions.

Discussion and Conclusions

We have described the results of an empirical evaluation of a set of hazard analysis techniques. Several conclusions follow from our results.

First, these techniques can be applied to real systems. We were able to build a model of a complex system and to apply completeness criteria and forward and backward hazard analysis techniques to that model. We have shown that these techniques work on more than very small, research-paper examples. In the process of demonstrating this, we obtained additional insight into the analysis process, which we will use to improve the techniques in the future.

Second, the techniques were able to find real errors in the specification. Again, we note this does not imply anything about the quality of the NASA specifications. The project developing the guidance system is a research project, and we were modeling early designs for the system. The emphasis in this NASA project has not been on getting complete specifications, but rather in examining ways to reduce modes and hence mode confusion in flight management computers. However, the errors or omissions we found are typical of those found in real system specifications and designs, including errors that have been found to have contributed to accidents in the past. Our goal was to determine whether our techniques could find important errors in real systems, and they did.

We also do not want to overstate the results. The process of carefully reviewing any specification or attempting to understand a system specification well enough to model it is likely to uncover errors, no matter what techniques are used. We cannot determine how effective our techniques are, especially with respect to alternatives, without more careful comparisons and perhaps controlled experiments.

Third, we found that the various hazard analysis techniques provided us with different and complementary types of information. While checking the completeness criteria manually, we found some potentially important omissions in the system specification. In the automated forward analyses, we detected some initial states and events that could lead to hazardous system states. In the backward analyses, we started with hazardous states and conditions and identified several possible paths leading to those hazards. All these types of analysis were useful. The backward analyses forced us to think about potential hazards of the system and examine their causes. The forward analyses forced us to think about potential failures and examine their effects. Together the analyses provided us with a more complete safety analysis than any of them alone provided.

Some of the analysis techniques support other ones. For example, in the fault tree analysis, taking a step back from an initial configuration provides a set of possible previous states. By performing a forward simulation from each of these states, we were able to prune away those states that cannot really precede the initial configuration. Similarly, both forward and backward simulation assisted in the checking of some of the completeness criteria.

Fourth, the design of the modeling language can support or hinder the analysis process. We discovered a number of important errors in the NASA documents simply by constructing the state-machine specification. We credit this result to the type of thinking and analysis required when constructing this type of model.

In addition, we believe this particular representation facilitated the analysis process; the specification provided a structure that supported the different techniques. For example, the organization of the model provided an orderly way to examine the specification for completeness and consistency errors. Also, the state abstraction and organization facilitated the fault tree analysis: Having transitions between states and the conditions under which those transitions are taken located together makes tracing the events backwards through the specification simple. In both the Jackson Chart specification and the pseudo-code, this information is dispersed throughout, making the backward tracing of events and conditions difficult. For example, often the conditions for a particular state change in the pseudo-code are buried within the conditions of a group of nested `if-then-else` statements and function calls. Tracing these back could be difficult (and we note, was difficult and error prone; back tracing through function calls and nested `if-then-else` statements is how we constructed the model initially).

In addition, the fact that a subtle typographical error in one specification language translates to an obvious omission in another language suggests that analysts can be provided with specification languages that focus their attention on the places where we have found that errors are likely to be made. The same can probably be said of analysis—the form of the analysis results will most likely affect the ability of analysts to interpret them with respect to their own application expertise and knowledge.

A conclusion that might be drawn here is that the design of a modeling or specification language should reflect the type of analyses to be performed on it, i.e., an understanding of the type of analyses desired should precede language design. In addition, many if not most errors will be found during expert review rather than by automated tools, so readability and organization of the specification to enhance the ability of the expert to find errors are also important considerations in specification language design. Our modeling language (RSML) is changing and evolving as we learn more about the modeling and analysis process.

Finally, we found that using the hazard analysis techniques requires application expertise, and hence they are probably best applied by application experts. When we began this project, none of us had any experience with guidance systems. In addition, the main analyst (Modugno) was new to the area of software safety and had no prior experience with the techniques. Although we found that the techniques were easy to learn, applying them successfully depended on an understanding of the application domain. Even building the model required a detailed understanding of the physics of the plane's motion. Moreover, translating from the designer's notation into RSML required an understanding of engineering notation not commonly used in computer science.

Whether application experts can use our modeling and analysis techniques and will find them useful, however, is yet to be determined. We have ascertained the feasibility of performing an integrated safety analysis and the utility of such an analysis. Our next step is to determine whether others can use the techniques to perform an analysis. Then we can begin to determine what tools we need to develop to support analysts.

Our findings so far suggest that the most effective safety-analysis tools will assist rather than replace the analyst. We should therefore focus on building tools that augment human abilities rather than attempting to do the analysis completely with tools alone. For example, because both the backward and forward analysis techniques require searching through a large space, tools to help the analyst prune that space and navigate through it could be helpful. Conversely, expert knowledge of the application allows the analyst to select the most plausible search paths and to eliminate immediately scenarios that are physically or logically impossible in the system being analyzed. We hope to have a NASA guidance system expert apply our techniques and tools to compare the results obtained in terms of number and type of problems found. A partnership between application, safety, and tool experts will most likely turn out to be the most effective way to produce useful results.

We plan to study further the whole issue of tool design. For example, the type of support that automated tools and techniques should provide will depend on an understanding of the cognitive demands of the particular analysis techniques and an understanding of how automation can be used to lessen those demands. Studying the process of how experts learn and use the analysis techniques should help us gain some of this understanding, as well as using results of research in other fields such as cognitive psychology and cognitive engineering.

We also plan to extend the modeling and analysis techniques to include ways to model the human operator and to provide techniques for human-error analyses. As computers have become more sophisticated and their role in process control has changed from simply an interface to the process to an autonomous controller, a large number of accidents in safety-critical systems have been blamed on "human error." We need to understand why these errors occur and how we can prevent them. Modeling the operator and analyzing the entire system—automated controller *and* operator—for potential hazards can help us reach this goal.

Acknowledgments

We would like to thank Charlie Hynes at NASA Ames for all his help and for providing the NASA guidance system specifications.

References

[1] P.K. Andow, F.P. Lees, and C.P. Murphy. The Propagation of Faults in Process Plants: A State of the Art Review. *7th International Symposium on Chemical Process Hazards*, University of Manchester, 1980.

[2] M.P.E. Heimdahl and N.G. Leveson. Completeness and Consistency Checking of Software Requirements. In *IEEE Transactions on Software Engineering*, vol. 22, no. 6, June 1996.

[3] C.L. Heitmeyer, B.L. Labaw, and K. Kiskis. Consistency checking of SCR-style requirements specifications. In *Proceedings of the International Symposium on Requirements Engineering*, 1995.

[4] K.L. Heninger. Specifying software for complex systems: New techniques and their application. *IEEE Transactions on Software Engineering*, 6(1):2–13, January 1980.

[5] C. Hynes. An example guidance mode specification. Technical report, NASA Ames, 1995.

[6] M.A. Jackson. *Principles of Program Design*. Academic Press, 1975.

[7] M.S. Jaffe. *Completeness, Robustness, and Safety of Real-Time Requirements Specification*. Ph.D. Dissertation, UCI, June 1988.

[8] M.S. Jaffe, N.G. Leveson, M.P.E. Heimdahl, and B.E. Melhart. Software requirements analysis for real-time process-control systems. *IEEE Trans. on Software Engineering*, 17(3):241–258, 1991.

[9] N.G. Leveson. *Safeware: System Safety and Computers*. Addison-Wesley, 1995.

[10] N. Leveson, S. Cha, and T. Shimeall. Safety verification of ada programs using software fault trees. *IEEE Software*, 8(7):48–59, 1991.

[11] N.G. Leveson, M.P.E. Heimdahl, H. Hildreth, and J.D. Reese. Requirements Specification for Process-Control Systems. *IEEE Transactions on Software Engineering*, 20(9):684–707, 1994.

[12] N.G. Leveson and J.L. Stolzy. Safety analysis using Petri nets. *IEEE Transactions on Software Engineering*, SE-13(3):386–397, 1987.

[13] R. Lutz. Targeting safety-related errors during software requirements analysis. In *Proceedings of the First ACM SIGSOFT Symposium on the Foundations of Software Engineering*, 1993.

[14] V. Ratan, K. Partridge, J.D. Reese, and N.G. Leveson. Safety analysis tools for requirements specifications. *Compass 96*, Gaithersburg, Maryland, June 1996.

[15] J.D. Reese. *Software Deviation Analysis*. Ph.D. Dissertation, UCI, 1996.

Appendix: The Guidance System Specification

We specified the control behavior of the guidance system using a parallel state-machine model. The modeling language is essentially RSML [11], which we previously developed to specify an aircraft collision-avoidance system, although the language is evolving as we gain more experience in modeling real systems and determine what is required and desirable to build understandable and analyzable models.

In our modeling language, components are modeled by parallel state machines. Each state machine is composed of states connected by transitions. Transitions are triggered by input or by internal events and are only taken when their guarding condition is true. A triggered transition produces an output or an internal event, which can in turn trigger transitions in other state machines.

In addition to the graphical specification of each state machine, the RSML model defines the guarding conditions on a transition using a logic table (AND/OR table). Figure 3 shows an example. The far-left column of the AND/OR table lists the logical phrases. Each of the other columns is a conjunction of those phrases and contains the logical values of the expressions. The table evaluates to true if one of its columns is true. A column evaluates to true if all of its elements are true. A dot denotes "don't care."

As shown in Figure 4, the guidance system can be modeled using three parallel state machines: the MODE CONTROL PANEL, which models the pilot's selections on the Mode Control Panel; the DISPLAYS, which models the state of the displays; and the CONTROLLER, which models the software logic.

Controller. The CONTROLLER issues commands to control the lateral and vertical motion of the plane. The particular command issued is dependent on the current CONTROL MODE of the guidance system, which indicates whether the guidance system is in fully automatic mode or jointly (pilot and computer) controlled mode, the current LATERAL MODE, the VERTICAL MODE, and the controller's model of the state of the aircraft.

For space reasons, we describe only the VERTICAL MODES state machine of the CONTROLLER, which is used to specify the conditions for issuing flight commands that control the vertical motion of the aircraft (Figure 5).

The guidance system has four High-Level Modes:

- Minimum Time: engaged to minimize time during a landing,
- Minimum Fuel: engaged to minimize fuel during a landing,
- Take Off/Go Around: engaged when a landing is missed, and
- Standby : engaged when no other vertical high-level mode is active.

These modes control the aircraft via the Intermediate Modes:

- Altitude: capture and hold a particular altitude, and
- Glideslope: track and capture the glideslope, which is a line through space relative to the ground.

The Intermediate Modes achieve their goals by interacting with the Low-Level Modes, which can be initiated only by the guidance system (that is, the pilot cannot directly instruct the guidance system to enter one of these modes). The three guidance-system-initiated modes are climb, level flight, and descend.

The appropriate vertical operational mode is determined using inputs from the environment (such as the Reference Flight Path Table and input from the sensors) as well as the aircraft model (see Figure 4). The Aircraft Model includes process state information, such as the position of the gears and flaps.

Mode Control Panel. The MODE CONTROL PANEL, which models pilot inputs to the guidance system, is composed of four parallel state machines that model the control, lateral and vertical mode buttons that the pilot selects on the MCP, and other input data. Again for space reasons, we describe only the vertical mode selections.

There are six buttons on the MCP that the pilot can select to control the vertical modes of the aircraft, as described above: minimum-time, minimum-fuel, take off/go around, standby, altitude, and glideslope .

The first four buttons represent the pilot selection of a high-level vertical mode, while the last two represent selection of an intermediate vertical mode.

Displays. The final component of the guidance system is the Displays. We model the information on two of the aircraft displays: the MCP and the Primary Flight Display. The annunciators on the MCP indicate the current control, lateral and vertical modes. The Primary Flight Display models the relationship between current data values, such as current altitude, and their desired values, such as reference altitude. Pilots use this information to monitor the behavior of the guidance system during automatic flight or to control the guidance system during manual flight.

Transition: $\boxed{\text{Not Active}} \longrightarrow \boxed{\text{Active}}$

Location: Controller $\mapsto$ Vertical-Modes $\mapsto$ Intermediate-Mode $\mapsto$ Altitude

Trigger Event: MCP-Selection

Condition:

OR

MCP-Selection = Altitude	T	F	F	F
MCP-Selection = Glideslope	F	T	F	F
Glideslope-Position IN STATE Below-Glideslope	.	T	.	.
MCP-Selection = Minimum-Time	F	F	T	F
MCP-Selection = Minimum-Fuel	F	F	F	T

Output Action: MCP-Light-Altitude-Button

Figure 3: A definition of the transition from state NOT ACTIVE to state ACTIVE for the ALTITUDE state machine, which is an INTERMEDIATE MODE within the VERTICAL MODES of the CONTROLLER (see Figure 5). The transition can take place only when the trigger event MCP-Selection occurs and either the pilot selects the Altitude, Minimum Time or Minimum Fuel buttons on the Mode Control Panel, or the pilot selects the Glideslope button on the Mode Control Panel and the plane is below the glideslope. If the transition is taken, the Altitude button on the MCP is lit.

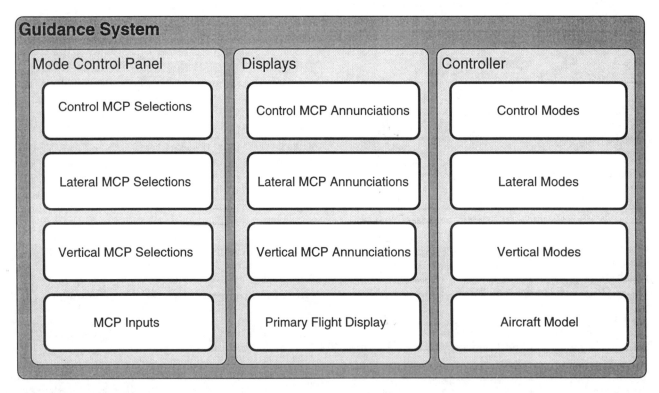

Figure 4: At a high-level of abstraction, the guidance system model consists of three parallel state machines, which themselves are composed of parallel state machines. The MODE CONTROL PANEL models the operator's input to the guidance system, which includes the control, lateral and vertical operating mode selections along with the MCP inputs such as the reference speed or reference altitude. DISPLAYS models the control, lateral and vertical modes annunciated on the MCP, and the data values displayed on the Primary Flight Display. The CONTROLLER models the logic that determines the control, lateral and vertical operating modes of the guidance system. The control logic is defined using a model of the aircraft along with the controller operating modes.

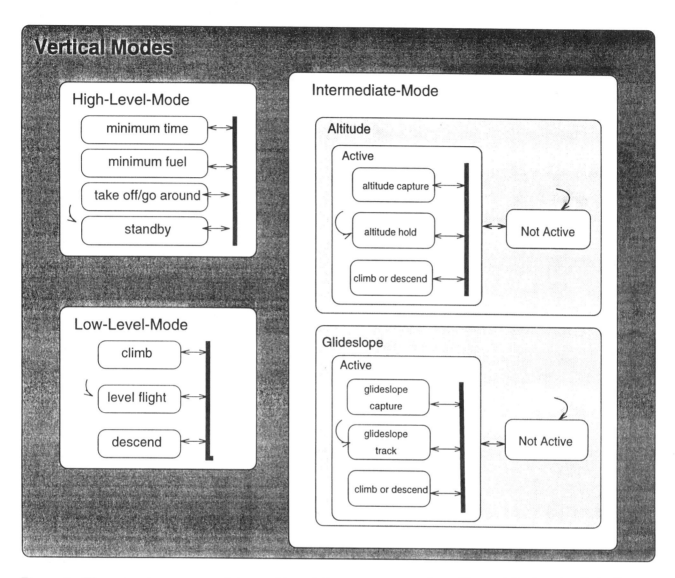

Figure 5: The VERTICAL MODES state machine of the CONTROLLER from Figure 4 further detailed. It consists of three parallel state machines: HIGH-LEVEL-MODE modeling the high-level vertical flight control modes; INTERMEDIATE-MODE modeling modes that can be initiated by either the pilot or the guidance system; and LOW-LEVEL-MODE modeling modes that can only be initiated by the guidance system.

Formal Methods for V&V of partial specifications: An experience report

Steve Easterbrook and John Callahan
{steve,callahan}@cs.wvu.edu
NASA/West Virginia University Software IV&V Facility
100 University Drive
Fairmont, WV 26554

Abstract

This paper describes our work exploring the suitability of formal specification methods for independent verification and validation (IV&V) of software specifications for large, safety critical systems. An IV&V contractor often has to perform rapid analysis on incomplete specifications, with no control over how those specifications are represented. Lightweight formal methods show significant promise in this context, as they offer a way of uncovering major errors, without the burden of full proofs of correctness. We describe an experiment in the application of the method SCR to testing for consistency properties of a partial model of the requirements for Fault Detection Isolation and Recovery on the space station. We conclude that the insights gained from formalizing a specification is valuable, and it is the process of formalization, rather than the end product that is important. It was only necessary to build enough of the formal model to test the properties in which we were interested. Maintenance of fidelity between multiple representations of the same requirements (as they evolve) is still a problem, and deserves further study.

1 Introduction

Requirements engineering methods typically provide a set of notations for expressing software specifications, together with tools for checking properties of specifications, such as completeness and consistency. In general, such methods demand a full commitment. It is assumed that the method will be used to construct a complete specification, which will then act as a baseline for subsequent development phases. However, to validate and verify large specifications for safety-critical real-time systems, it is sensible to apply a number of different methods, to overcome weaknesses and biases of each individual method. For example, a formal method might be used to model a critical portion of an informal specification, to check safety and liveness properties of that portion. In order to manage the application of multiple methods, it is necessary to develop and maintain alternative representations of partial specifications, and to express the relationships between them.

This paper describes some preliminary work on the use of formal specification as a tool for Independent Verification and Validation (IV&V). Our intention is to use formal methods not as a part of the development process itself, but as a 'shadow' activity, performed by an independent team of experts. Our long-term expectation is that this approach will turn out to be a less painful way of introducing formal methods into well-established, large-scale software development processes.

There are a number of questions that need to be addressed before formal methods can be used in this way. Most published case studies of formal methods have focussed on the use of a formal specification as a baseline from which design and code can be verified [3]. In contrast, we have been applying formal methods for intermittent "spot checks" to test for errors as the requirements evolve. The term "lightweight formal methods" has been used to describe this approach [15]. In this context, the the formal specification is dispensable – what is important are the insights gained from *the process of* formalizing partial views of the requirements and from validating properties of the resulting models. However, it is still necessary to demonstrate fidelity between the original (informal) specification, and the formal model. Furthermore, iterative application of this approach can be greatly facilitated if the relationships between the partial views are captured.

The context for this work is the development of software for the International Space Station (ISS) project. Boeing Space and Defense Group Houston (Prime) is responsible for supervising the overall development and integration of International Space Station software. There are three Product Groups (PGs), McDonnell Douglas Aerospace, Rockwell Aerospace - Rocketdyne and Boeing Space and Defense Group Huntsville, who are developing several key Computer Software Configuration Items (CSCIs). There are also several International Partners (IPs) including Russia, Japan, Canada, and the European Space Agency, who are developing software that will need to be incorporated into ISS. With over 45 flight computers and an estimated 1.1 million source lines of flight code, the potential problems are considerable. Software IV&V is currently being performed by Intermetrics, under an interim contract. The Intermetrics team is based at Fairmont, W.Va., with personnel stationed in Houston and Huntsville in order to interact with the develop-

160

ment teams.

In section 2, we outline the IV&V process, and discuss the aspects of this process that hinder effective IV&V. With this as background, the remainder of the paper focuses on the use of methods and tools within this process. We present two experiments in the use of formal specification. For these we used a combination of AND/OR tables [8], and the Software Cost Reduction (SCR) approach [9]. The first experiment involved the translation of a portion of the Fault Detection, Isolation and Recovery (FDIR) specification into a formal notation. This experiment confirmed that the natural language used in the Software Requirements Specification (SRS) documents is inherently ambiguous, and that the task of generating formal specifications from this documentation is fraught with difficulty. In the second experiment, we applied an automated consistency checking tool, to test some formal properties of the specification. Although this experiment demonstrated that important disjointness properties did not hold, the results did not add any more value to the analysis. The first experiment had already demonstrated that the way in which these requirements were expressed was a problem. The errors found in the second experiment were attributable to the same problem.

Application of formal methods in this context was not always easy. The informal specification from which we derived our models did not permit an easy translation into a state-based model. We encountered severe problems in demonstrating fidelity, and providing traceability between the two. Section 5 discusses these problems, and sketches out further work aimed at eliciting relationships between partial specifications by extracting information from fine-grained process capture.

We conclude that in an IV&V context, the analytical benefits offered by formal methods have to be weighed against the effort needed to maintain fidelity between a formal model and the informal specification used by the development team. An IV&V team needs to be able to perform partial analyses on partial specifications, without being tied to any one formalism. The analysis carried out must be sufficient to reveal important problems, as opposed to surface defects. Further analysis is a waste of effort until these problems have been fixed. This conclusion implies a change of perspective for the use of formal methods: while the specification is still evolving it is important to identify quickly any major defects; it is not necessary to perform a complete analysis. Tools that are geared towards finding and characterizing such problems (E.g. SCR* [10], Nitpick [11], etc.) are more useful than tools geared towards proving correctness (E.g. theorem provers).

2 The IV&V Process

For *Independent* Verification and Validation (IV&V), the software customer hires a separate contractor to analyze the products and process of the software development contractor. This analysis is performed in parallel with the development process, throughout the software lifecycle, and in no way replaces in-house V&V. IV&V is applied in high-cost and safety-critical projects to overcome analysis bias and reduce development risk. The customer relies on the IV&V contractor as an informed, unbiased advocate to assess the status of a project's schedule, cost, and the viability of its product during development. In full IV&V, the IV&V contractor has managerial, financial and technical independence, and reports to the customer, not the developer. Most importantly, the IV&V contractor should be engaged as early as possible in the project: studies have shown that IV&V has the biggest impact in the early phases, especially in the requirements phase [13].

An example IV&V activity is the analysis of specifications on the Space Station project. An SRS is written by the relevant development contractor for each Software Configuration Item (CSCI). These are written in natural language, and follow the format of DOD-STD-2167A. The IV&V contractor periodically receives copies of the SRS documents, in various stages of completion. These are analyzed for technical integrity by the IV&V contractor, in order to identify any requirements problems and risks. The kind of analysis performed will vary according to the level and the type of specification, and will cover issues such as clarity, testability, traceability, consistency and completeness. If problems are identified, the IV&V contractor may recommend that either the requirements be rewritten, or the problem be tracked through subsequent phases.

Performing IV&V on large projects is far from straightforward. Problems faced by the IV&V contractor include:

resource allocation – A complete, detailed analysis of the entire system is infeasible. Effort has to be allocated so as to maximize effectiveness. For example, a criticality and risk analysis might be performed to determine which components need the most scrutiny. Timing is also a factor; effort needs to be allocated at the right points in the development of a product (e.g. a document), so that the product is mature enough to be analyzed, but not so mature that it cannot be changed.

short timescales – To be most effective, IV&V reports are needed as quickly as possible. There is always a delay between the delivery of an interim product to the IV&V team, and the completion of analysis of that product. During this time, the development process continues. Hence, if IV&V analysis takes too long, the results might be available too late to be useful. In general, the earlier an error is reported, the cheaper it is to correct.

lack of access – Contact between the development team and the IV&V team is difficult to manage. The IV&V team needs to maintain independence, whilst ensuring they obtain enough information from the developers to do their job. From the developers' point of view, interaction with the IV&V team represents a cost overhead, which can interfere with project deadlines. Inevitably, the

IV&V contractor has less access to the development team than is ideal.

evolving products – Documentation from the development team is usually made available to the IV&V contractor in draft form, to facilitate early analysis. The drawback is that documents may be revised while the IV&V team is analyzing them, making the results of the analysis irrelevant before it is finished.

reporting the right problems – The IV&V contractor has, by necessity, considerable discretion over the kinds of analysis to perform on different products. It also has discretion over which problems to report. It is vital to the effective use of IV&V that the IV&V contractor prioritizes the problems it identifies. If too many trivial problems are reported, this may swamp the communication channels with the developer and the customer.

lack of voice – The IV&V contractor may have difficulty in getting its message across, especially when the development contractor disputes IV&V's assessment. Often, problems found by IV&V have cost and schedule implications, and in such circumstances the customer may be more willing to listen to assurances from the developer. The effectiveness of IV&V then depends on having a high-placed advocate within the customer organization.

Despite these problems, IV&V has been shown to be a cost-effective means of improving the quality of the software product, and providing extra assurance for high-cost, safety-critical projects [12]. In addition to providing analysis of project artifacts (e.g. requirements, code, test plans), the presence of IV&V in the lifecycle also has a positive effect on the quality of the software. Our work suggests that the *interaction* between the IV&V and development teams drives improvements in both products and processes. This effect, however, is difficult to capture and quantify.

3 Methods and Tools in IV&V

An important aspect of IV&V is the choice of the right methods and tools. Ideally, an IV&V contractor will have access to all the tools used by the development team, including the ability to share all project databases. However, the IV&V team also needs to supplement these with additional methods and tools, to address any gaps or weaknesses in the coverage of the developer's tools. These additional tools need to complement the developer's tools, so that interoperability does not become a problem. The use of these additional tools is an important factor in ensuring that IV&V is truly independent.

It is often the case that the use of a particular method or tool by the IV&V team leads to the adoption of that method or tool by the developers. In part this is due to the 'watchdog effect': if the developers know that their product will be analyzed in a particular way, it is in their interest to perform the analysis

themselves before releasing it. If this seems to be a rather negative reason to adopt a technique, there is also a positive aspect. Because the IV&V team is out of the critical path for the software development effort, they have more scope for experimentation with new techniques than the developers [1]. Hence, in some ways the IV&V team can play a role as a proving ground for new techniques, and can come to be an agent of process improvement. For these reasons, we believe that IV&V offers a practical route through which formal methods may be introduced into projects that would otherwise not be able to adopt them.

There are still problems to be overcome whenever the IV&V team adopts a tool that is not used by the developers. Compatibility with the developers' tools is important. For example, if the IV&V team uses a formal specification tool, the informal specification delivered by the developers will need to be translated into the formal specification language not just once, but each time the developers produce a new draft. Any problems identified by using the tool must be traced back to the informal specification, before they can be reported. There must be a reasonable assurance that the formal specification remains faithful to the original, otherwise any analysis performed on it is worthless. Hence, keeping track of the relationship between the formal and informal specifications is vital.

4 Experiments with formal methods

Having described the role that an IV&V contractor plays in the software process, and outlined the issues involved in the selection of tools and techniques for IV&V, we now present our work on the use of formal methods in the IV&V of requirements specifications. We performed two experiments. The first was a formalisation of individual requirements statements into a tabular form, to improve clarity. The second was the development of a formal model of these requirements, which was then tested for consistency.

Currently, the development contractors on the Space Station project use natural language specifications extensively. We are working with the IV&V team to explore how formal methods can enhance the kinds of analysis they perform on the developer's informal specifications. Here, we will report our work with the Fault Detection, Isolation and Recovery requirements for the main command and control bus. An example requirement is given in figure 1.

Our initial interest in formal methods was twofold. First, it was clear that the informal specifications were hard to understand, and would benefit from a clearer representation. We needed a notation that was both precise and easy to read. Leveson's AND/OR tables [8] provided us with a solution. During the development of the RSML specifications for TCAS II, Leveson adopted these AND/OR tables in preference to predicate calculus, as they were readable by a wide range of people. This tabular representation was well suited to the Space Station FDIR requirements (see table 1), as it mapped directly onto the individual requirements statements.

Second, we needed a way to verify that the specified functionality was internally consistent. For the FDIR

(2.16.3.f) While acting as the bus controller, the C&C MDM CSCI shall set the e,c,w, indicator identified in Table 3.2.16-II for the corresponding RT to "failed" and set the failure status to "failed" for all RT's on the bus upon detection of transaction errors of selected messages to RTs whose 1553 FDIR is not inhibited in two consecutive processing frames within 100 millisec of detection of the second transaction error if; a backup BC is available, the BC has been switched in the last 20 sec, the SPD card reset capability is inhibited, or the SPD card has been reset in the last 10 major (10-second) frames, and either:

1. the transaction errors are from multiple RT's, the current channel has been reset within the last major frame, or

2. the transaction errors are from multiple RT's, the bus channel's reset capability is inhibited, and the current channel has not been reset within the last major frame.

Figure 1: An example of a level 3 requirement for FDIR of the Command and Control bus for Space Station. This requirement specifies the circumstances under which all remote terminals (RTs) on the bus should be switched to their backups.

requirements, this meant checking that the conditions specified for each recovery action were mutually exclusive, and that the requirements covered all possible conditions. Hand checking these properties would have been hard, so we sought a tool to help. We examined several tools, before selecting SCR* [10]. SCR offered two important advantages. First, the notation was primarily tabular, which appeared to be an important aid to readability. Second, the tool had automated checking for properties such as coverage and disjointess of a state based model [9]. In addition, this tool did not require us to build a complete formal model of the Bus FDIR functionality in order to check these properties.

4.1 Experiment 1: Translation

Our first experiment concerned the translation of requirements like that shown in Figure 1 into a formal notation. Leveson's AND/OR tables allowed us to represent arbitrary combinations of conjunctions and disjunctions without ambiguity, and in a form that was clearly readable. Table 1 shows the tabular form of the requirement in Figure 1.

For the IV&V team, this was a significant improvement in readability. More importantly, the process of producing the tables ensured that the analysts fully understood the requirement. This benefit is very important for IV&V. In many cases, just reading a specification is insufficient to really appreciate the detail. Short of repeating the development process from scratch, it can be hard for the IV&V analyst to understand a specification in the same way that its authors understand it. Translating it into a table, however, proved to be a valuable clarification process.

There was, unfortunately, a problem. Translation of a single requirement, like the one above, was not a straightforward task. Translation of this requirement took several attempts until we were happy with the table, and even then we were not convinced that it was right.

We conducted an experiment to investigate the problem. We gave the English language version to

four different people, all of whom had some experience of representing requirements using tables, and asked them to produce the tabular form. Two of these people were domain experts, and two were not. We were interested in exploring the scope for misinterpretation of the requirements from the point of view of both domain experts who write such requirements, and other stakeholders, such as the programmer who would have to implement them.

We received four different answers. These differed in both the number of conditions identified (i.e. number of rows in the table) and the number of combinations under which the function would be activated (i.e. columns in the table). The version shown in Table 1 is a synthesis of the four answers, representing what we currently believe is the intended interpretation.

The differences in the responses show that the original requirement is riddled with ambiguities. For example, the mixture of 'ands' and 'ors' in the requirement is a problem because, unlike programming languages, English does not have any standard precedence rules. It is not clear how to scope the various subclauses, either. For example, the timing condition 'within 100 millisec...' could refer to the inhibition of the FDIR, or to one or both of the required setting operations. With a little domain knowledge, it is possible to eliminate some interpretations, but this is by no means a trivial task, and there is no guarantee that everyone who needs to read this requirement will get it right.

The experiment demonstrated three important results. Firstly, the tabular forms were very helpful in resolving misunderstandings. For example, it would be difficult to discover that our four subjects had different interpretations of the original requirement without asking them to re-write it. By re-writing it in tabular form, we could identify exactly where the disagreements lay, and then take each discrepancy in turn and discuss what we thought the most likely interpretation was. From this, we were able to synthesize a 'best' interpretation. Obtaining individual translations and comparing them was more effective in identifying differences in our understandings than our initial

		OR			
	C&C MDM acting as the bus controller	T	T	T	T
	Detection of transaction errors in two consecutive processing frames	T	T	T	T
	errors are on selected messages	T	T	T	T
	the RT's 1553 FDIR is not inhibited	T	T	T	T
	A backup BC is available	T	T	T	T
A	The BC has been switched in the last 20 seconds	T	T	T	T
N	The SPD card reset capability is inhibited	T	T	.	.
D	The SPD card has been reset in the last 10 major (10 second) frames	.	.	T	.
	The transaction errors are from multiple RTs	T	T	T	T
	The current channel has been reset within the last major frame	T	F	T	F
	The bus channel's reset capability is inhibited	.	T	.	T

Table 1: A Leveson-style table for requirement 2.16.3.f. This table summarizes the conditional part of the requirement in Figure 1, showing four combinations of conditions (the four columns) under which the specified action should be carried out).

attempts to work together to produce a single translation. This confirms a hypothesis described in [5], that negotiating requirements conflicts is more effective if we start with a precise description of each person's individual viewpoint. Note that our final version was different from all four of the individual versions, implying that if the final version is correct, all four individual attempts were wrong!

This leads to the second result, which is that translation of informal requirements into a formal notation is error prone. All four of our subjects had some experience of using such tables, so the problem lies not in the correct use of the notation, but in the interpretation of the informal statement of requirements. The requirement we used in the experiment is perhaps an extreme example, given its rather convoluted English. However, there is enough scope for misinterpretation in the process of formalizing the requirements to cause us to worry about the fidelity of our formal models.

The third result is that the whole process was remarkably good at identifying ambiguities in the original specification. By producing different interpretations and comparing them, we were able to identify a systematic pattern of ambiguities in the way the English language requirements were written. Hence, even if the IV&V team fail to persuade the development team to adopt a tabular notation, they can at least help them to correct the ambiguities in the English.

In fact, the development contractors have used the tabular notation occasionally, in the most recent versions of the specifications. Initially, they resisted the IV&V team's requests to adopt a tabular notation, largely because of schedule constraints. They have now begun to use the notation for revisions of the specifications, especially in areas where reviewers had had problems with readability. We regard this as a small but important process improvement, inspired by the IV&V team.

4.2 Experiment 2: Analysis of Partial Specifications

Our second goal was to check some of the properties of the FDIR specification that could not be checked by hand. One of the important validity checks for these requirements is that an action is specified for each possible combination of failure conditions. Another check is that no combination of conditions has conflicting actions specified for it. We refer to these as coverage and disjointness checks, respectively [14].

In practice, there were two approaches that IV&V could take to verify such properties. They could obtain the development team's failure model, validate this model, and then verify the requirements against the model. Or they could generate their own behavioral model of the requirements as described, verify that it is internally consistent, and then validate this against their understanding of the system. The latter approach was chosen, partly because the IV&V team has had difficulty obtaining the original models on which the specification is based, and partly because the latter approach was more likely to overcome analysis bias.

We chose SCR as an appropriate model to perform these analyses for a number of reasons. First, the tabular notation used in SCR maps onto the AND/OR tables we had already generated in a fairly systematic way. Each AND/OR table represents a single row in a mode transition table in SCR. Second, there was a tool (SCR*) available for checking SCR specifications which included both coverage and disjointness tests, and which had a simulator built in for animating the complete state-based model. A model checker was being added. Furthermore, the consistency checker in the SCR* tool provides counter-examples whenever an inconsistency is found. Our early experiments with a theorem prover (PVS [14]) were abandoned because when a proof failed, it took too long to discover the problem. The provision of counter-examples is impor-

164

tant in tracing problems back to the informal specification, and in convincing the development team that there really is a problem.

The first step was to produce an SCR model of the specified FDIR behavior. At this stage we had six AND/OR tables, similar to the one shown in Table 1, representing the six paragraphs, a to f, of section 2.16.3 of the requirements. Each paragraph isolates one failure mode, and specifies an appropriate action. We merged these into a single table, modeling each failure mode as a separate SCR mode (Table 2).

Merging the AND/OR tables to produce Table 2 was not straightforward. Although there were a number of conditions common to several of the tables, the wording varied, and it was not always obvious whether similar sounding phrases actually referred to the same condition, due to inconsistencies in the use of terminology. For example the condition "the bus has been switched in the major (10-second) frame" appeared in one paragraph, and "the bus has been switched in the last major frame" appeared in another. We initially assumed these to be identical. However, this led to an inconsistency in the table. In fact the former refers to the *current* frame, while the latter refers to the *previous* frame. There were numerous places where we had to make assumptions to proceed, and we carefully recorded these as annotations to the original text, to be checked with the developers.

The modes we have identified are not present explicitly in the informal specification. Our modes correspond intuitively to failure modes, but might not be a particularly good choice for simulation or model checking purposes, because they really express output events rather than states. However, they suit our purpose, as the table in this form can be checked directly for coverage and disjointness without completing the model. In fact, the complete model would be complicated: a clock would be needed to implement the bus processing frames, together with several timers to keep track of historical state. Even then, SCR cannot (currently) represent timing conditions on the required functions.

Having created the table, we then checked it for coverage and disjointness. Not surprisingly, the table is not disjoint: in fact there is an overlap between every possible pair of rows. Analysis of the counter-examples provided by the SCR* tool indicates a systematic under-specification of the conditions. The original model of the FDIR system was a procedural model with an explicit order on the checks that need to be performed. The specification does not have this explicit ordering, and the described conditions do not adequately express this ordering. However, this result was not a surprise: the IV&V team had already submitted a report suggesting that the ordering be made explicit in the specification.

While we were producing this analysis, a new draft of the specification was released. The section specifying Bus FDIR requirements had been re-written, partly due to issues raised by the IV&V team, both before and after our first experiment. The new version is much clearer (but does not use our tables). It is also much simpler: several failure modes and at least half the conditions expressed in Table 2 have been removed, and the disjointness problem described above has been corrected.

Hence our formal analysis was redundant before it was complete. In practice, it would have been possible to perform the analysis much earlier: we delayed the work until a full release of the SCR* tool was available. However, we can now apply the same technique to other parts of the specifications, and expect that in some cases it will identify new problems, while in others it will supply concrete evidence of known problems. Once the requirements are stable, we plan to build a complete model of the FDIR subsystem, and use a model checker to study its behavior under repeated and intermittent fault conditions.

5 Discussion

We have described our on-going work with formal methods as a tool for an Independent V&V team to perform analysis of software requirements. Our initial results are encouraging: the translation process was extremely valuable in identifying ambiguities and improving our understanding of the specification. In this process, a number of errors were found. Analysis of a partial formal specification demonstrated an important error in the specification, and appears to be a powerful means of gaining maximal results from minimal effort. We constructed just enough of a model to test the properties we were interested in, without any further commitment to the method.

However, our experiments have revealed two related problems: it is hard to guarantee fidelity between informal and formal specifications, and it is hard to manage consistency between partial specifications expressed in different notations.

Although the major finding of our formal analysis is valid, we are not confident that the partial model is faithful to the version of the developer's specification on which it is based. This fidelity issue is more of a problem in IV&V than in development. A formal model developed by the IV&V team cannot replace the informal specification. The IV&V team must therefore either persuade the developers to adopt formal notations themselves, or take care to maintain fidelity between the developers' informal specifications and their own formal models. With the current state of practice, wholesale adoption of formal methods by the developers on an existing project is unlikely [4].

The fidelity problem is important to IV&V because the formal models developed by IV&V are produced for the purposes of checking the developer's specifications. The models are only useful for this purpose if they are accurate representations of the developer's specifications. Also, when analysis of the formal models reveals problems in the specifications, these problems must be traced back to the informal specification before they can be reported.

Although the fidelity problem seriously affects the utility of any formal analysis performed by the IV&V team, we should point out that it does not affect all the benefits of formal specification. The process of translating pieces of the informal specification into a formal notation has benefit not just for the analysis

Current Mode	Conditions											Next Mode
	errors in two cons. frames	bus swch'd last frame	bus switch inhibit	bus swch'd this frame	backup BC avail.	BC swch'd in last 20 sec	card reset inhibit	card reset last 10 frames	errors from mult. RTs	channel reset last frame	channel reset inhibit	
Normal	@T	-	-	F	-	-	-	-	-	-	-	switch buses
	@T	-	T	F	-	-	-	-	-	-	F	reset the channel
	@T	T	-	F	-	-	-	-	-	-	F	
	@T	-	-	-	-	-	F	F	T	T	-	reset the card
	@T	-	-	-	-	-	F	F	T	F	T	
	@T	T	-	-	-	-	-	-	F	T	-	switch RT to backup
	@T	F	T	-	-	-	-	-	F	T	-	
	@T	T	-	-	-	-	-	-	F	F	T	
	@T	F	T	-	-	-	-	-	F	F	T	
	@T	-	-	-	T	F	T	-	T	T	-	switch BC to backup
	@T	-	-	-	T	F	T	-	T	F	T	
	@T	-	-	-	T	F	-	T	T	T	-	
	@T	-	-	-	T	F	-	T	T	F	T	
	@T	-	-	-	T	T	T	-	T	T	-	switch all RTs
	@T	-	-	-	T	T	T	-	T	F	T	
	@T	-	-	-	T	T	-	T	T	T	-	
	@T	-	-	-	T	T	-	T	T	F	T	

Table 2: An SCR Mode transition table. Each of the central columns represents a condition, showing whether it should be true or false; '-' means "don't care"; '@T' indicates a trigger condition for the mode transition. The four columns of table 1 correspond to the last four rows of this table. The semantics of SCR require this table to represent a function, so that the disjunction of all the rows covers all possible conditions (coverage), and the conjunction of any two rows is false (disjointness).

that it leads to, but also for the removal of ambiguities and for improved understanding. For this benefit, it is the *process* of formalization, rather than the end product that is important.

The fidelity problem is really a special case of a more general problem: management of consistency between partial specifications expressed in different notations. For instance, the AND/OR tables have a clear relationship with the SCR mode tables, but if we make a correction to one of the AND/OR tables, it is fairly tedious to identify the corresponding correction in the SCR tables. Similarly, each time the developers issue a new informal specification, we need to update our tabular representations. Although it may seem that the use of both AND/OR tables and SCR models together would compound this problem, the opposite is true. The AND/OR tables mapped clearly onto the textual requirements, while the relationship between the AND/OR tables and the SCR model was relatively straight forward. Therefore, the use of AND/OR tables as an intermediate representation reduced the traceability gap, and made it easier to keep the formal model up to date. There remains, however, a significant bookkeeping problem.

There is a growing body of work on handling inconsistency in specifications. Our previous work demonstrated how to delay the resolution of inconsistency, and provided a generic framework for expressing consistency relationships [6]. Other work has taken consistency checking further, making use of semantic models underlying a method to determine what consistency rules are needed and how to operationalize them. For example, Heitmeyer's work with consistency checking in SCR [9] uses the semantics of SCR to define a series of consistency rules ranging from simple syntactic checks (e.g. that all names are unique) to sophisticated properties of tables (e.g. coverage and disjointness). Similarly, Leveson's work on consistency checking in RSML [8] uses the semantics of the statechart formalism to determine a set of consistency rules that can be tested, tractably, using a high level abstract model. In both these approaches, the completeness of the formal specifications is important, and consistency checking is seen as part of the process of obtaining a complete, consistent specification.

Unfortunately, these approaches do not help with consistency checking between partial specifications expressed in different notations. Because the IV&V process is concurrent with and complementary to the development process, there is an unusually large amount of flexibility in how a formal method can be used. There is no need to make a commitment to any one formal notation, just as there is no need to develop complete specifications. In fact, the aim of the IV&V agent is not to perform complete analyses, but to do just enough analysis to check specific aspects of the software. Development of complete formal models is therefore unnecessary and may be counter-productive. For example, in our second experiment, the limited analysis we performed on a partial model was sufficient to reveal a major problem; the existence of this problem meant that any further effort to complete the model would have been wasted.

While the use of partial specifications offers greater flexibility in the use of methods and tools, it also means that we do not have a well-defined method from which to generate a set of consistency relationships. There are implicit consistency relationships between the assorted partial specifications drawn from different methods, but there is no overall 'method' to to tell us what these relationships are. Actually, there is a method: the problem is that it is implicit, and to some extent is generated on the fly. For example, there is a method for generating SCR mode tables from the AND/OR tables, but the method was not defined before we did it. With some effort, we could formalize this method, and define semantic relationships between the two types of table. However, this effort will only be worthwhile if we intend to re-use the method extensively. In the meantime, we would like to have tools to help us keep track of consistency relationships in our opportunistic use of partial specifications.

In our previous work defining consistency relationships between viewpoints, we assumed that the majority of such rules are defined by the method [6]. The viewpoints framework explicitly supports the process of method definition, in which, among other things, the inter-viewpoint relationships are defined. Hence the general problem of defining arbitrary relationships between any two notations is avoided. However, we also recognized that some consistency relationships could not be defined in this way, and gave the example of a user-defined synonym relationship between two different labels. We also outlined an approach to discovering such relationships through low level process monitoring. We now regard this type of consistency relationship as vital to any approach involving partial specifications.

Without a method to define *a priori* consistency relationships, we are forced to discover the relationships as the work proceeds. In fact this is not as hard as it sounds. By recording low level actions on the partial specifications, we begin to build up a fine-grained process model, which can provide information about consistency relationships. For example, by observing cut and paste operations during the creation of our AND/OR tables and our SCR mode tables, it is possible to determine the relationship between rows in the AND/OR tables and rows in the mode table. In the weakest case, this will provide us with a simple traceability link. In fact, we believe we can do better than this. There is enough information in the edit actions not just to identify traceability links, but to define the relationship expressed by the link. For example, it should be possible to determine enough information to define a consistency rule that can automatically check that each column of the AND/OR table is consistent with its corresponding row in the mode table. We plan to explore this avenue further, by capturing and analyzing this kind of process information.

6 Conclusions

This paper has described our initial work in the use of formal methods in an IV&V project. We have discussed how the demands placed on methods and tools in IV&V are different from their use in a development context. We have also discussed how IV&V can act as a process improvement agent, and hence can be a fruitful way of introducing formal methods into large projects.

As with all potential uses of a new method, any extra effort needed to use the method must be more than offset by the benefits it brings. Use of a method in IV&V is no different. We can divide the benefits of using a formal method such as SCR into two areas:

1. The process of translating portions of a specification into a tabular notation helps to detect ambiguities and increase readability, even if the translation is only partial. The process can also be used to catch misunderstandings, thus increasing the confidence that the IV&V team is interpreting the specification correctly. The process of having several analysts produce their own tabular translations was particularly useful in this respect. Differences in the tables they produced allowed us to pinpoint exactly what the disagreement was about.

2. The resulting tables can be analyzed for attributes such as coverage and disjointness. This is a substantial contribution to the IV&V team's efforts to check the technical integrity of the specifications. Such attributes are particularly hard to analyze from the informal specifications. Most importantly, this analysis can be conducted without the need to build complete models.

The problems we encountered in applying formal methods were as follows:

1. The process of translating into a formal notation is error-prone. Only by duplicating the translation effort were we able to discover just how much scope there is for misinterpretation. Luckily, the resulting tables are very readable. Therefore it is much easier to compare different tables than it is to compare different versions of the informal specification.

2. For IV&V, fidelity and traceability between the informal and formal specifications is difficult to guarantee. The value of any analysis carried out by IV&V on the formal model is entirely dependent on how faithful the formal model is to the developer's informal specification. The IV&V's formal model can not be used in place of the informal specifications produced by the developers.

3. Opportunistic use of partial specifications means that there is not a well-defined method from which to derive consistency rules. Maintenance of consistency in our partial specifications became a real problem.

The problems of consistency checking in partial specifications written in different notations is important enough to warrant more attention. We plan to study the problem in more detail by developing a set of tools based on the ViewPoint framework [7], which will allow us to model relationships between partial specifications written by different people. We are also exploring how this problem relates to that of linking test case scenarios to requirements [2]. Finally, we are continuing the experiments described in this paper by examining how model checking can be used to validate the specifications.

Acknowledgments

Our thanks are due to Chuck Neppach and Dan McCaugherty for many interesting discussions of the work presented here, and to Frank Schneider, Edward Addy, John Hinkle, George Sabolish, Todd Montgomery and Butch Neal for detailed comments on earlier drafts of this paper. This work is supported by NASA Cooperative Research Agreement NCCW-0040.

References

[1] V. Basili. The experience factory and its relationship to other improvement paradigms. In *Proceedings of the 4th European Software Engineering Conference, Garmish-Partenkirchen, Germany,* September 1993.

[2] J. Callahan and T. Montgomery. An approach to verification and validation of a reliable multicast protocol. In *Proceedings of the ACM International Symposium on Software Testing and Analysis (ISSTA)*, January 1996.

[3] D. Craigen, S. L. Gerhart, and T. Ralston. Formal methods reality check: Industrial usage. *IEEE Transactions on Software Engineering*, 21(2):90–98, 1995.

[4] D. H. Craigen, S. L. Gerhart, and T. J. Ralston. An international survey of industrial applications of formal methods, vol 1: Purpose, approach, analysis and conclusions. Technical Report NRL/FR/5546–93-9581, Naval Research Laboratory, 1993.

[5] S. M. Easterbrook. Handling conflict between domain descriptions with computer supported negotiation. *Knowledge Acquisition: An International Journal*, 3(4):255–289, 1991.

[6] S. M. Easterbrook and B. A. Nuseibeh. Managing inconsistencies in evolving specifications. In *Second IEEE International Symposium on Requirements Engineering*, pages 48–55, March 1995.

[7] S. M. Easterbrook and B. A. Nuseibeh. Using viewpoints for inconsistency management. *BCS/IEE Software Engineering Journal*, 11(1), January 1996.

[8] M. Heimdahl and N. Leveson. Completeness and consistency analysis of state-based requirements. In *Proceedings of the 17th International Conference on Software Engineering*, pages 3–14, April 1995.

[9] C. Heitemeyer, B. Labaw, and D. Kiskis. Consistency checking of scr-style requirements specifications. In *Second IEEE International Symposium on Requirements Engineering*, pages 56–63, March 1995.

[10] C. Heitmeyer, A. Bull, C. Gasarch, and B. Labaw. Scr*: A toolset for specifying and analyzing requirements. In *Tenth Annual Conference on Computer Assurance (COMPASS '95)*, pages 109–122, June 1995.

[11] D. Jackson and C. A. Damon. Elements of style: Analysing a software design feature with a counterexample detector. In *International Symposium on Software Testing and Analysis (ISSTA)*, pages 239–249, January 1996.

[12] Jet Propulsion Lab. Cost-effectiveness of software independent verification and validation. Technical Report NASA RTOP 323-51-72, NASA JPL, October 1985.

[13] R. O. Lewis. *Independent Verification and Validation: A Lifecycle Engineering Process for Quality Software*. J. Wiley & Sons, 1992.

[14] S. Owre, J. Rushby, and N. Shankar. Analysing tabular and state-transition specifications in pvs. Technical Report CSL-95-12, Computer Science Laboratory, SRI International, 1995.

[15] H. Saiedain, J. P. Bowen, R. W. Butler, D. L. Dill, R. L. Glass, A. Hall, M. G. Hinchey, C. M. Holloway, D. Jackson, C. B. Jones, M. J. Lutz, D. L. Parnas, J. Rushby, J. Wing, and P. Zave. An invitation to formal methods. *IEEE Computer*, 29(4):16–30, 1996.

Extended Requirements Traceability: Results of an Industrial Case Study

Orlena Gotel & Anthony Finkelstein
Department of Computer Science
City University
London EC1V 0HB
[olly|acwf]@soi.city.ac.uk

Abstract

Contribution structures offer a way to model the network of people who have participated in the requirements engineering process. They further provide the opportunity to extend conventional forms of artifact-based requirements traceability with the traceability of contributing personnel. In this paper, we describe a case study that investigated the modelling and use of contribution structures in an industrial project. In particular, we demonstrate how they made it possible to answer previously unanswerable questions about the human source(s) of requirements. In so doing, we argue that this information addresses problems currently attributed to inadequate requirements traceability.

1: Introduction

The inability to answer questions regarding the human source(s) of requirements information has been found to result in claims of *requirements traceability* problems [5]. An approach to address this problem, based on modelling the *contribution structure* underlying requirements, was presented in [6]. This paper describes a case study designed to evaluate, through demonstration, whether use of the approach helps answer these outstanding questions and, in so doing, alleviates an important class of requirements traceability problems. The case study is based on a real industrial project.

In Section 2, we explain what requirements traceability is and describe the underlying reason for long-term requirements traceability problems. We provide examples of the kind of questions that are problematic for practitioners to answer as a consequence. We then outline an approach to address this fundamental problem and summarise how it is anticipated to provide answers to these questions. In Section 3, we describe the case study material we gathered and used to validate our claim. Since

the approach did not exist at the outset of the case study project, its requirements did not drive the data gathering and its use was not on the project's critical path. In Section 4, we demonstrate how the approach was applied to this data in a post-hoc manner. This application means we are only in a position to validate the feasibility of the approach and the usefulness of the information it provides, say to reveal information about the project's evolution and assist the maintenance process, in a subjective and historical manner. We do this in Section 5, where we show how this information makes it possible to answer questions regarding *involvement*, *responsibility*, *ramifications*, *change notification* and *working relationships*. Based on our experiences and practitioner comments, we highlight some outstanding issues and make recommendations in Section 6.

2: Contribution structures for traceability

In this section, we describe what requirements traceability is, why it is important and what the problems with it are. We then outline an approach to address a fundamental problem that currently makes it difficult to recover information about the human source(s) of requirements information.

2.1: Requirements traceability

Requirements traceability refers to the ability to describe and follow the life of a requirement in both a forwards and backwards direction (i.e., from its origins, through its development and specification, to its subsequent deployment and use, and through periods of on-going refinement and iteration in any of these phases). It is considered a primary technique to help with many project-related activities, like ensuring that systems and software conform to their changing requirements, but is commonly cited as a problem area by practitioners.

169

Although the number of tools that claim to support requirements traceability is growing, some more recent ones being described in [10, 11, 13, 14], the schemes that need to be established prior to their use have received rather less attention. With few exceptions, examples being the requirements traceability models of the U.S. DoD [8, 9] and the requirements traceability meta models arising from the NATURE project [12], endeavours to improve the potential for requirements traceability have mostly involved uncovering and recording as much information as possible about the requirements engineering process, then linking it in interesting ways for trace retrieval. This can lead to an unwieldy mass of unstructured and unusable data without some *a priori* discrimination concerning the type of requirements information that practitioners are likely to need and for what purposes.

Following an empirical study reported in [5], we argued that the most fundamental information to record for relieving *long-term* requirements traceability problems was that which identified the *human source(s)* of requirements information. We found that, what are perceived to be requirements traceability problems tend to arise when practitioners are unable to answer questions about the personnel who had been involved in the production and refinement of requirements. This is because people are often considered the ultimate baseline whenever requirements need to be re-examined or re-worked. Examples of these problematic questions include:

(1) Who has been involved in the production of this requirement and how?

(2) Who was originally responsible for this requirement, who is currently responsible for it and at what points in its life has this responsibility changed hands?

(3) At what points in this requirement's life have the working arrangements of all involved been changed? Accordingly, within the remit of which groups do decisions about this requirement lie?

(4) Who needs to be involved in, or informed of, any changes proposed to this requirement?

(5) What are the ramifications, regarding loss of project knowledge, if a specific individual or group leaves?

2.2: Contribution structures

We described an approach to address this more focal problem in [6]. The approach is based on modelling the contribution structure underlying requirements. This model reflects the network of people who have contributed to the artifacts produced in the requirements engineering process. In [6], we also described how the approach can be implemented and gave scenarios of use. Formalisation of the approach and the inferences it supports can be found in

[4]. We only summarise the main steps of the approach in Figure 1 and below.

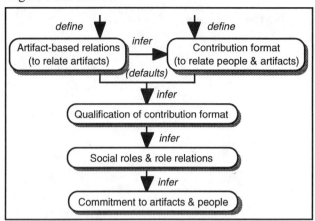

Figure 1: Main steps of the approach.

Working through Figure 1, minimal semantics are given to the artifact-based relations ordinarily put in place for requirements traceability. For example, based on the notion of *communicative function*, an artifact can either *reference* or *adopt* the content of a linked artifact, depending on whether or not their content overlaps. A record of the people who contributed to an artifact's production is maintained in its *contribution format*. For example, based on Goffman's work on the nature of participation in social encounters [3], this can delineate the *principal*, *author* and *documentor* of an artifact. These categories were chosen for their analytic potential to infer details about the social roles, role relations and commitments of those involved. The information provided from these steps makes it possible to extend conventional forms of *artifact-based* requirements traceability with a form of *personnel-based* requirements traceability.

3: Case study

In this section, we give details of the company, project and participants of the case study. We describe the data we gathered and our method for so doing.

3.1: Project

The project came from a communications company employing about twenty-five people. The company runs many projects concurrently, providing solutions to various communications-related problems. The objective of this particular project was to supply a dedicated communications service to complement a customer's disaster recovery programme. The project was initiated in February 1992 and went live at the end of March 1992.

In August 1992, the idea of developing a generic service for other customers was discussed. Six versions of a requirements and design specification were drawn up throughout September 1992. These were then abandoned until the end of October 1992, when new staff were employed to develop and market the service. Following much staff turn-over, the generic service did not go live until February 1994. Between October 1992 and February 1994, the specification evolved into an operational service, an operations manual and a high-level manager's guide. Since February 1994, the generic service and its documentation has undergone continuous modification to account for the requirements of new customers.

Most of the artifacts produced during the project were informal and paper-based. All that remains within the company today is an early specification, an up-to-date operations manual, an up-to-date manager's guide, customer contracts and miscellaneous correspondence. Requirements traceability has not been maintained. Those still involved in the project are no longer aware of from where or from whom the various aspects of the current service have been derived. Some problems have resulted from this loss of information but, because the project is restricted in scope, and because the team has been small and exhibited some staff continuity, these have not been critical to its maintainability and success.

3.2: Data gathered

The work that occurred from the initial discussion about providing a generic service, through to the sixth version of the requirements and design specification, was followed closely. We observed all the meetings that took place, made notes, took audio recordings and collected photocopies of any tangible artifacts produced. We also participated in some aspects of the process. During this time, a detailed picture of what had happened when developing the initial customer-specific service was reconstructed with those who had been involved. From the end of October 1992, we maintained a record of the main artifacts produced due to this specification. We also maintained a record of the people involved in the production and distribution of these artifacts.

One hundred and sixty-six main artifacts were produced in the project. These relate to four main phases:
(1) Development of the customer-specific service (twenty-three artifacts between February and March 1992).
(2) Development of the baseline for the generic service (sixty-five artifacts from August to September 1992).
(3) Development of the initial generic service (thirty-nine artifacts from October 1992 to July 1993).

(4) Extension of the generic service to address new requirements (thirty-nine artifacts from September 1993 to June 1995).

For the purposes of the case study, our definition of "artifact" applied to single physical documents. This was to promote identification and to enable us to examine the viability of the approach at a coarse level of granularity before introducing further complexity.

Fifty-eight people contributed directly to the project. These included individuals and groups from within the company and from outside. To maintain confidentiality, we use alphabetic identifiers when we refer to these individuals and groups in the remainder of this paper.

4: Application of approach

In this section, we outline how the approach was applied. Based on the data we had gathered, key project participants were tasked to reconstruct the main artifact-based relations and to give them some semantics. They were also tasked to reconstruct the contribution format for each artifact, prompted by contextual material. We then applied the last three steps of the approach to examine what could be inferred about the project and its social roles, role relations and people's commitments.

4.1: Artifact-based relations

For each project phase, its artifacts were numbered according to production order. The temporal relations between them was then clarified, based on [1]. The coarse flow-down of information and influence amongst these artifacts was also established. These orderings for the artifacts produced in phase one are shown in Figure 2.

Table 1 shows how semantics were assigned to these relations. From the original reason provided for the relation by participants, the nature of the relation was categorised according to classifications of *cohesion* and *coherence* [2]. Based on this classification, its broad communicative function was identified as either *referencing* or *adopting*. Although the more detailed semantics have implications for selective traceability, consensus was found difficult to establish at that level, whilst easier to agree at the coarser level.

Figure 2a highlights the adopts relations of Table 1. Since these tend to capture *parent-child* or *predecessor-successor* relations, they provide for what we regard as conventional forms of artifact-based requirements traceability. Figure 2b goes on to illustrate how the references relations of Table 1 provide additional contextual information that is often not integrated and used for requirements traceability purposes.

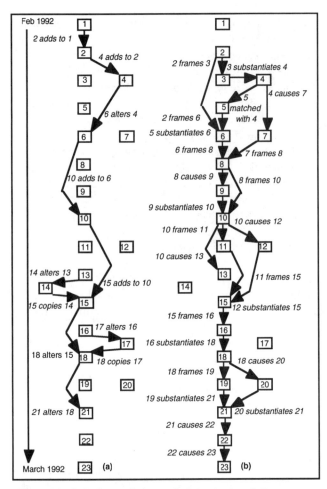

Figure 2: Relations of phase 1: (a) adopts relations - arrows suggest flow-down of content; (b) references relations - arrows suggest direction of influence.

4.2: Contribution format

The contribution format of each artifact was established to indicate the individuals and/or groups who contributed in the capacities of principal, author and documentor. The contribution formats of some of the artifacts produced in phase one are shown in Table 2.

4.3: Qualification

The capacities of each contribution format were qualified to provide more details about contributions and contributors. As an example, the authorial capacity was qualified to indicate the levels and types of dependency upon other authors, providing a citation-like network. Following the resulting authorial trails, we are able to see: how each progressive author made use of previous people's contributions; which authors produce the most

original artifacts; which authors use their own or another's contributions the most often; whose contributions get referenced with the greatest frequency; and so on. We can also begin to assess the influence of a person's authored contributions on the surrounding body of artifacts and on the project as a whole. Such details can help identify those to notify following different types of change or those to contact regarding different types of query.

(1) Informal description of relation given by practitioner	(2) Relation in terms of cohesion & coherence	(3) Broad communicative function
2 qualifies 1	2 adds to 1	2 adopts 1
2 is the reason for 3	2 frames 3	3 references 2
4 defines 2	4 adds to 2	4 adopts 2
2 is the reason for 6	2 frames 6	6 references 2
3 assists with 4	3 substantiates 4	4 references 3
5 is compared with 4	5 matched with 4	5 references 4
6 refines 4	6 alters 4	6 adopts 4
5 assists with 6	5 substantiates 6	6 references 5
7 responds to 4	4 causes 7	4 references 7
6 is the reason for 8	6 frames 8	8 references 6
7 is background for 8	7 frames 8	8 references 7
9 is a result of 8	8 causes 9	9 references 8
9 assists with 10	9 substantiates 10	10 references 9
10 elaborates 6	10 adds to 6	10 adopts 6
8 is background for 10	8 frames 10	10 references 8
10 is reason for 11	10 frames 11	11 references 10
12 replies to 10	10 causes 12	12 references 10
13 replies to 10	10 causes 13	13 references 10
15 extends 10	15 adds to 10	15 adopts 10

Table 1: Semantics for the relations of phase 1.

Artifact	Principal	Author	Documentor
1	BH	BI	AW
2	AT	BB={AW/AV/AT/AR/AX/AU}	BB={AW/AV/AT/AR/AX/AU}
3	AA	AA/AE	AA
4	AA	AA/AE	AU
5	AA	AA/AQ/AP/BB={AW/AV/AT/AR/AX/AU}	AA
6	AA	AA/AT	AA
7	BH	BI	BL
8	AA	AA/AE	AA
9	AA	AA/BB={AW/AV/AT/AR/AX/AU}	AA
10	AA	AA	AA

Table 2: Contribution formats for artifacts 1 to 10. AA/AE means person AA & person AE were joint contributors in the given capacity. BB refers to a group, so its members are given in curly brackets.

4.4: Social roles & role relations

The social roles that people assume when contributing to artifacts can be inferred from the information gathered so far. For instance: if a person is both the principal and author of an artifact, they can be said to be its *devisor*; if they are solely the documentor, they can be said to be its *relayer*. The ensuing role relations between people when they contribute jointly to artifacts, say as a *devisor/relayer* pair, reveals more about the underlying contribution structure. Not only can we see whom has collaborated with whom, we can also see how they collaborated and whether these role relations have varied or been sustained throughout a project.

To explain the use of such information, we compare the social roles of two of the project leaders. AI was the project leader when artifacts 99 to 127 were produced and a contributor to twenty-two of these. AJ was the project leader when artifacts 128 to 162 were produced and a contributor to twenty-six of these. Their social roles when contributing to these artifacts, as well as their role relations to collaborators, are shown in Tables 3 and 4.

Social role of AI	On how many artifacts?	How many on own?	Social roles of people who collaborate with AI & number of times
True author (i.e., contributes as P, A & D)	16	13	BM=true author (x1) AA/AE/AD=ghost author (x1) AA/AE/AD/AT=ghost author (x1)
Nominal author (i.e., P & D)	2	0	AA/AE/AD/AG=ghost author (x1) BB/AP/AQ=ghost author (x1)
Representative (i.e., A & D)	2	0	AD=sponsor (x2)
Ghost author (i.e., A)	2	0	AD=sponsor & BO=relayer (x2)

Table 3: Social roles & role relations for AI.

Social role of AJ	On how many artifacts?	How many on own?	Social roles of people who collaborate with AJ & number of times
True author (i.e., P, A & D)	7	2	BM=true author (x1) AE=ghost author (x 4)
Ghost author (i.e., A)	9	0	AD=sponsor & AP/AS=relayer (x1) AD=sponsor & AP=relayer (x5) AD=sponsor & AR=relayer (x2) AD=sponsor, AF=ghost author & AO=relayer (x1)
Devisor (i.e., P & A)	9	0	AE=ghost author & AW=relayer (x2) AE=ghost author & BQ=relayer (x1) AL=relayer (x4) AM=relayer (x2)
Sponsor (i.e., P)	1	0	AE=true author (x1)

Table 4: Social roles & role relations for AJ.

From these tables, we can see that AI worked on his own on over half the artifacts he contributed to, else he worked with small groups of people. As he worked

largely as a *true author*, he was evidently a self-sufficient documentor. Although details delineating the type and content of artifacts have not been included in these tables, it is noteworthy that AD tended to collaborate with AI as a *sponsor* when dealing with customer-related artifacts. In contrast to AI, we can see that AJ worked rarely on his own and collaborated mainly with one or two others. He had a strong dependency on AE as his *ghost author* when they worked together and on many other people as *relayers*, the latter hinting at the need for secretarial support. It is noteworthy here that AD was ultimately responsible for about a third of the artifacts that AJ had contributed to.

There could be many reasons for the subtle differences in how these two people with the same job description worked in the project. AI did not close any sales and focused on developing a marketable service. In contrast, AJ focused on selling what AI had developed and only made subsequent additions to it to account for new customer requirements. Notably, it was with such additions that AE collaborated with AJ as ghost author. Since AE had also collaborated with AI as ghost author during the earlier project phase, this collaboration obviously served to maintain some continuity.

4.5: Commitment

Table 5 indicates the kind of information that can be inferred about the commitments of project contributors, both to artifacts and to other people. We can see that, as AP has predominantly been a relayer (i.e., purely a documentor), she is mainly responsible for physical artifacts. She is only responsible for the content of artifacts when collaborating with others. She has never been responsible for their ultimate effect (i.e., a principal). We can also see the people AP is committed to through their collaboration on artifacts. For instance, we can identify AD and AJ as those with whom AP has collaborated the most often, as well as identify the number and type of artifacts on which they collaborated. By extension, we can examine those people that AP is committed to due to the artifact-based relations that situate her contributions in the wider network of artifacts.

The intersection and difference between commitments can uncover much interesting information. For example, we can identify: which people have collaborated with specified others the most or least often; which people are committed to the same set of other people; which people have collaborated with customers; which people are committed to the same type of artifacts and for the same aspects; which people have contributed to those artifacts that are the initial sources of requirements; and so on.

Artifacts AP contributes to	Aspect of artifact committed to (is the commitment shared with others?)	Other contributors & number of artifacts on which collaborate
5	Content (as one of many contributors)	AD (x6)
22	Physical (on own)	AJ (x6)
31	Physical (on own)	AQ (x4)
41	Content (as one of two) Physical (on own)	AT (x4)
85	Content (as one of two)	AU (x4)
96	Content (as one of many) Physical (as one of two)	AW (x3)
100	Content (as one of many)	AV (x3)
111	Content (as one of many)	AR (x3)
139	Physical (as one of two)	AX (x3)
140	Physical (on own)	AC (x2)
148	Physical (on own)	AD (x2)
151	Physical (on own)	AJ (x2)
154	Physical (on own)	AQ (x1)
155	Physical (on own)	AT (x1)

Table 5: AP's artifact & collaborator commitment stores.

5: Results & discussion

In this section, we demonstrate how the questions of Section 2 can be addressed. We also mention other forms of analysis the approach makes possible. The reader is referred to [4] for a more detailed description and a thorough evaluation.

5.1: Involvement

Who has been involved in the production of this requirement and how?

One of the requirements in version two of the requirements and design specification, artifact 49, led to much investigation and many artifacts that later became redundant. It was a requirement pursued throughout phase two of the project and cited in all six versions of the specification. Once dropped in phase three, its impact only surfaced over time. The resulting problems could have been alleviated with knowledge of its original source and of those who had pushed for its concern.

Following application of the approach, this requirement was traced back to artifact 27. Note that, we define an "original contribution" to be one that does not depend upon other artifacts for its existence; we do not attempt to measure degrees of originality here. The contribution format at the source shows that AA was writing requirements in the name of a collective. Having delineated the contribution format of internal components, AX can be identified as the member who originated this particular requirement. Knowledge of the source makes it possible to recover AX's original intention, one that was actually misconstrued by AA in the project. Furthermore,

we can see how this misconstrued requirement pervaded subsequent artifacts, due to AA's backing and no later recourse to AX. We can also see which people ended up doing the most redundant work as a consequence. Notably, it was a requirement that dominated many of AA's early and individual contributions.

5.2: Responsibility

Who was originally responsible for this requirement, who is currently responsible for it and at what points in its life has this responsibility changed hands?

Phase three of the project saw the introduction of a manager's guide, its latest version being artifact 160 in phase four. Table 6 shows a subset of the information gathered relating to this artifact and its earlier versions.

Manager's guide	Artifact 160	Artifact 150	Artifact 138	Artifact 125	Artifact 118
Version	5	4	3	2	1
Principal	AJ	AJ	AJ	AI	AI
Author	AJ	AJ	AJ	AI	AI
Document	AM	AL	AL	AI	AI
Adopts relations	Adds to 150	Adds to 138	Adds to 125	Alters 118	None
References relations	Matched with 159	Matched with 149	Matched with 137	Matched with 124	(a) Matched with 115 (b) Framed by 108
Principal (of referenced artifact)	AJ	AJ	AJ	AI	(a)AI (b)AA
Author (of referenced artifact)	AJ	AJ	AJ	AI	(a)AI (b)AA
Doc (of referenced artifact)	AM	AL	AL	AI	(a)AI (b)AA

Table 6: Changes in responsibility for manager's guide.

From Table 6, we can see the transition between AI's original work on the guide and AJ's later work on it. We can also see that AJ only made additions to what AI originally produced. The working arrangements also changed from AI working on his own to AJ working in conjunction with one other person doing the physical documentation. Therefore, although AI was originally responsible for all aspects of the guide, AJ is now responsible for its content and effect, whilst AM is now responsible for all physical aspects of the document. Table 6 further shows that the guide has been aligned with versions of the operations manual throughout its evolution, these being artifacts 159, 149, 137, 124 and 115. The only other artifact with which the first version of the guide is related is artifact 108. Inspection of this artifact reveals that AA, as its true author, was originally responsible for the idea to develop this guide.

5.3: Working arrangement & remit

At what points in this requirement's life have the working arrangements of all involved been changed? Accordingly, within the remit of which groups do decisions about this requirement lie?

In Figure 3, we depict the contributors to the formal versions of the requirements and design specification produced in phase two. From this, we can see that any decisions about the later versions of the specification lie with AC, AA, AE and AG. However, decisions relating to its earlier versions lie with different subsets of this group. Notably, we can see that AE provides continuity through the evolution of the specification, since he remains its sole documentor and one of its authors.

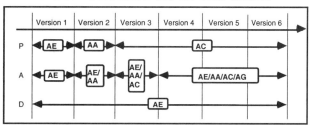

Figure 3: Changing decision making authority amongst members of the group contributing to the requirements & design specification.

In Table 7, we delineate the social roles of the contributors to the different versions of this specification. From this, we can see the subtle transformation in the role relation between AE and AA as other people became involved. We can also see how the role relations between all those involved became stable with version four. With such information about how group members come together, including how their interrelations change, we can begin to examine the impact of changing working arrangements on different attributes of an evolving artifact, like its attention to technical detail and so forth.

Requirements & design specification	Social roles & role relations of direct contributors
Version 1 (artifact 46)	AE=true author
Version 2 (artifact 49)	AA=devisor, AE=representative
Version 3 (artifact 61)	AC=devisor, AA=ghost author, AE=representative
Version 4 (artifact 74)	AC=devisor, AA/AG=ghost author, AE=representative
Version 5 (artifact 84)	AC=devisor, AA/AG=ghost author, AE=representative
Version 6 (artifact 88)	AC=devisor, AA/AG=ghost author, AE=representative

Table 7: Working arrangements of those contributing to the requirements & design specification.

By extending the analysis of this specification into phase three (not shown above), we are able to see that the ultimate responsibility for the specification passed from AC to AH once AC left the project. Interestingly, it did not pass back to one of those who had held this responsibility earlier on. Whilst AH held this position when contributing to the evolution of the specification, no further contributions were made to it by members of the original team. They only reassembled when AI took over AH's position in phase four. This information helps to explain why development of the specification proceeded successfully in phase four, but was compounded by problems and misunderstandings in phase three.

5.4: Change notification

Who needs to be involved in, or informed of, any changes proposed to this requirement?

Changes were not made to the content of the operations manual after AI left the project in phase three. As of version three, artifact 124, each new version saw the introduction of a new section to add novel features to the generic service. Had a change been proposed to the section introduced in version six, artifact 159, a section that described a new electronic mailbox service to be implemented, we would be able to identify all those who contributed, so able to check who would need to be involved in the change process. Similarly, we would be able to identify all those who made subsequent use of this service in later work, so able to check who would need to be informed of changes. These trails are shown in Figure 4. Only crude details are provided about artifact content in the figure. Trace visualisation is an on-going research issue that is not explored here.

In examining those involved in the production path of the mailbox service, we can see that it arose following a request from a specific customer, CF, in correspondence captured in artifact 152. We can also use these trails to see that the requirement for a mailbox service was raised earlier in artifact 114, a list of requirements drawn from all the customer correspondence received in phase three. In particular, this requirement had been noted by customer BX, subsequently documented formally in artifact 122, then reported more fully in artifact 127. This report was used as background material when the requirement for the service surfaced again later.

In examining those involved in the usage path of artifact 159, we can see that it is adopted by artifact 166, and referenced by artifacts 160 and 161. If internal links were present from the section on the mailbox service in artifact 159 to artifact 166, we could see we would need to inform AT and AW of any change. Where project policy is to inform the authors of any artifacts referencing ones that are to be changed, we would also be able to see the need to inform AJ and AF. With knowledge of such trails, different types of change or change proposal can be dealt with in the most desirable way, on a project-specific basis, with automatic notification.

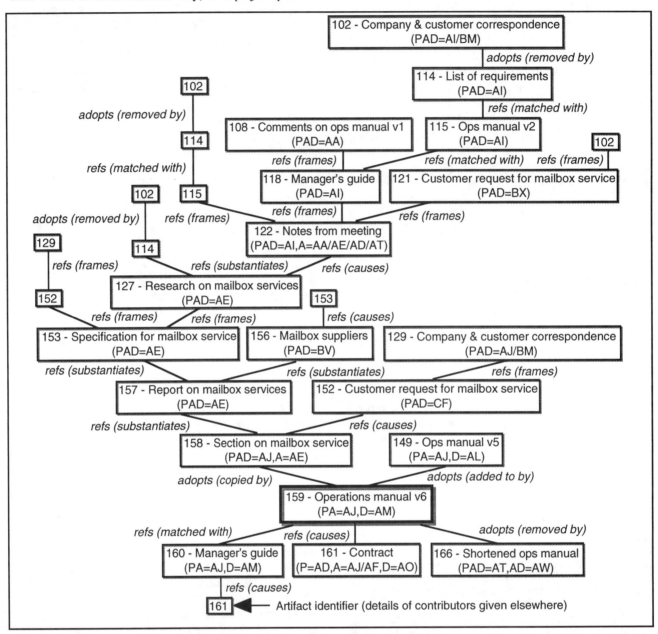

Figure 4: Who needs to be involved in, & informed of, any changes proposed to the new section of the operations manual introduced in version six, artifact 159. (PA=AJ,D=AM means person AJ is principal & author, person AM is documentor.)

5.5: Ramification

What are the ramifications, regarding loss of project knowledge, if a specific individual or group leaves?

AC left the project at the end of phase two. Before his departure, we can examine which of AC's contributions are unused by other people, so those that the other project participants are probably not aware of. In this way, we can ensure that his outstanding commitments are passed on and not lost, identify alternative points of contact for AC's contributions, so smooth staff turn-over. We list AC's contributions and collaborators in Table 8. We also list those artifacts that adopt or reference AC's contributions in this table to examine their contributors in turn.

Artifact	Other contributors (AC's collaborators)	Adopted by (artifacts)	Referenced by (artifacts)
26	None	30	28
28	AA=true author	30	29
30	None	34/35/36	31/32/33/39
31	AP=relayer	None	32/33
34	AA=rep, AE/AG=ghost	None	47
35	AA/AE/AG=ghost	None	50
36	AA/AG=ghost, AE=rep	45	44
37	AA=rep, AE/AG=ghost	None	47
38	AA/AG=ghost, AE=rep	None	44
50	None	56	54/55/85
51	AA=rep, AE/AG=ghost	None	55
52	AA/AE/AG=ghost	58	56
53	AA/AG=ghost, AE=rep	58	None
56	AE=ghost	None	58/73/85
58	AE=ghost	60/61	59
59	AE=ghost	None	None
61	AA=ghost, AE=rep	63/64/65	None
63	AA=rep, AE/AG=ghost	None	68
64	AA/AE/AG=ghost	None	None
65	AA/AG=ghost, AE=rep	74	None
73	AT/AU=ghost	75/76/77	80/87
74	AA/AG=ghost, AE=rep	75/76/77	None
75	AA=rep, AE/AG=ghost	None	78/79
76	AA/AE/AG=ghost	81	78/80
77	AA/AG=ghost, AE=rep	81/83/84	78
78	AA/AG=ghost, AE=rep	82	81
80	None	81	85
81	AE=rep	83/84	82/85
82	AE=rep	None	84
84	AA/AG=ghost, AE=rep	88	None
85	AQ/AP=ghost	None	86
86	AQ=ghost	None	None
87	AT=ghost	None	None
88	AA/AG=ghost, AE=rep	None	None

Table 8: AC's legacy. Where AC is a contributor to the artifacts cited in columns three & four, its identifier is given in bold. (rep = representative, ghost = ghost author.)

By inspection of Table 8, we can see which of AC's contributions are not used in any way by distinct others. Firstly, we can see that AE must be aware of AC's individual contribution in artifact 80. This is because he adopted its content in artifact 81, when working in conjunction with AC, in both an authorial and documenting capacity. We can also see that, since AQ, AP and AT are relatively minor players in the project, we might need to alert the key players to artifacts 85, 86 and 87. We can thereby signal which of AC's artifacts are still pending approval for integration into the critical path.

As we can see who has contributed with AC, and in what role relations, we can pass on this information if there are later queries about any of his contributions. If a new person is to take over AC's commitments, we can identify AC's long-term, transitory and current collaborators for contact purposes. By indicating those who have made use of AC's contributions, especially in conjunction with AC himself, we can identify those who

are likely to have had additional communication with AC concerning his artifacts. Potentially, these people can act as replacement contact points.

5.6: Further analyses

It becomes possible to carry out other forms of analysis as a by-product of the approach. These can provide much value-added information. For example, the number of contributors to each artifact in a project can highlight phases of group activity and those artifacts perhaps more prone to later query. Similarly, the number and type of contribution made by specific individuals or groups in a project can highlight its driving forces and its stable backbone. Although premature to generalise, interesting future work would be to consider the health of a project in terms of contribution and contributor *profiles*.

6: Conclusions

Members of the company we studied agreed that the data we revealed about the contribution structure underlying the project rang true. It identified: the right people to help rectify matters where problems of misunderstanding surfaced; those to involve in requirements change; how to handle staff turn-over; amongst other things. In particular, it provided information about social roles and role relations that could not have been determined from the company's organisational chart or work allocation timetables. This information was considered invaluable to inform how work could be allocated in future projects and to entertain the notion of requirements reuse.

However, although we were fortunate to have access to high-quality material, the case study has some limitations for demonstrating and evaluating our approach: requirements traceability was not practiced in the organisation studied; the development philosophy was informal and unstructured. A different perspective would no doubt arise in those organisations with some form of requirements traceability or document control already in place, or by those currently experiencing problems caused by inadequate requirements traceability. Similarly, by those organisations running larger projects involving many people and artifacts, or by those with explicit process improvement agendas. A summary of the main issues that arose during the case study, concerning the use of the approach and the information it provides, are given in Table 9. These suggest areas for further research.

Drawing from this case study, we suggest that the approach is practical and feasible. Furthermore, it need not be overly labour-intensive if introduced in a suitable setting and in an appropriate manner. For instance, if introduced into organisations that already practice some form of requirements traceability, incrementally and as an extension to current requirements traceability schemes. Even with crude extensions distinguishing basic types of artifact-based relation and contribution, it becomes possible to trace those involved in different aspects of a project and to reveal their working relations. This provides for a more comprehensive form of requirements traceability that is able to answer many problematic and outstanding questions. Eventually, were such information gathered across projects and organisations, it would become possible to investigate how the organisation of the requirements engineering process itself impacts practice. This information could be used as a basis for process and quality improvement programmes.

Main issues concerning use of the approach	Main issues concerning use of the information the approach provides
Whose job is it to record contributors & to insert artifact-based relations? How much is it feasible to do automatically?	The time to analyse & act upon the data has implications for use during a project. How to make its use transparent in activities like change management?
Balancing the granularity & semantics of artifacts & relations against the complexity of the contribution structure modelled & the traceability provided.	Overwhelming analytical opportunities for organisational, project & workflow analyses. What information can best inform practice in particular organisations & projects?
A need to account for how an "author" actually contributes when there are many authors.	Sensitivity of information indicates a need to re-examine organisational cultures & to introduce use policies.
When should details of the undocumented events that influence an artifact, like informal interactions, be captured & how?	A need to take care in analysis & generalisation. Does a large number of contributions really indicate productivity, quality, centrality, e.t.c.?
How to balance the work involved versus the benefits reaped? How to ensure commitment? e.t.c.	Integration with other forms of organisational modelling (e.g., how could it be used in the context of the Actor Dependency model [15]?)
A need to expand the social dimension. How to account for artifact distribution details, so we can examine who contributes as a consequence?	No metrics provided. Is it a real advance over current practice? Is it cost-effective in providing answers to personnel questions during a project?

Table 9: Outstanding issues & research directions.

Acknowledgements

The authors acknowledge the comments and assistance of colleagues, particularly David Michael, Wolfgang Emmerich, Stephen Morris and George Spanoudakis. They would also like to thank the company of the case study. Finally, thanks to Steve Fickas, Eric Yu and the anonymous referees for their recommendations.

References

[1] Allen, J. F. Maintaining Knowledge about Temporal Intervals, *Communications of the ACM*, Volume 26 (November 1983), pp. 832-843.

[2] De Beaugrande, R. A. and Dressler, W. U. *Introduction to Text Linguistics*, Longman (1981).

[3] Goffman, E. Footing, *Semiotica*, Volume 25 (1979), pp. 1-29.

[4] Gotel, O. C. Z. *Contribution Structures for Requirements Traceability*, Ph.D. Thesis, Imperial College of Science, Technology and Medicine, University of London (August 1995).

[5] Gotel, O. C. Z. and Finkelstein, A. C. W. An Analysis of the Requirements Traceability Problem, *Proceedings of the IEEE International Conference on Requirements Engineering*, IEEE Computer Society Press, Colorado Springs, Colorado (April 1994), pp. 94-101.

[6] Gotel, O. C. Z. and Finkelstein, A. C. W. Contribution Structures, *Proceedings of the Second IEEE International Symposium on Requirements Engineering*, IEEE Computer Society Press, York, U.K. (March 1995), pp. 100-107.

[7] Gotel, O. C. Z. and Finkelstein, A. C. W. Revisiting Requirements Production, *Software Engineering Journal*, Volume 11 (May 1996), pp. 166-182.

[8] Harrington, G. A. and Rondeau, K. M. *An Investigation of Requirements Traceability to Support Systems Development*, Naval Postgraduate School, Monterey, California (September 1993).

[9] Laubengayer, R. C. and Spearman, J. S. *A Model of Pre-Requirements Specification (pre-RS) Traceability in the Department of Defense*, Naval Postgraduate School, Monterey, California (June 1994).

[10] Macfarlane, I. A. and Reilly, I. Requirements Traceability in an Integrated Development Environment, *Proceedings of the Second IEEE International Symposium on Requirements Engineering*, IEEE Computer Society Press, York, U.K. (March 1995), pp. 116-123.

[11] Pinheiro, F. A. C. and Goguen, J. A. An Object-Oriented Tool for Tracing Requirements, *IEEE Software*, Volume 13 (March 1996), pp. 52-64.

[12] Pohl, K. PRO-ART: Enabling Requirements Pre-Traceability, *Proceedings of the Second IEEE International Conference on Requirements Engineering*, IEEE Computer Society Press, Colorado Springs, Colorado (April 1996), pp. 76-84.

[13] Structured Software Systems Limited. *Cradle: Systems Engineering Guide*, Document RM/CRY/006/01, Issue 1, Product Version 1.8X, 3SL, Barrow-in-Furness, Cumbria, U.K. (February 1995).

[14] TD Technologies, Inc. *SLATE: System Level Automation Tool for Engineers*, Marketing Literature, http://www.slate.tdtech.com (1995).

[15] Yu, E. S. K. and Mylopoulos, J. Understanding "Why" in Software Process Modelling, Analysis, and Design, *Proceedings of the Sixteenth International Conference on Software Engineering*, IEEE Computer Society Press, Sorrento, Italy (May 1994), pp. 159-168.

Session 8B

Workshop

Organizer

Alistair Sutcliffe
City University of London, UK

"Exploring Scenarios in Requirements Engineering"

Workshop
Exploring Scenarios in Requirements Engineering

Alistair Sutcliffe

Centre for HCI Design

City University

Northampton Square

A.G.Sutcliffe@uk.ac.city

1 Introduction

Use of examples, scenes, narrative descriptions of contexts, mock-ups and prototypes have attracted considerable attention in Requirements Engineering, Human Computer Interaction and Information Systems communities. Loosely all these ideas can be called scenario based approaches, although exact definitions are not easy beyond saying these approaches emphasise some description of the real world. Experience seems to tell us that people react to 'real things' and that this helps clarifying requirements. Indeed the widespread acceptance of prototyping in system development points to the effectiveness of scenario based approaches. However, we have little understanding about how scenarios should be constructed, little hard evidence about their effectiveness and even less idea about why they work. This workshop will explore some of these issues underlying scenario based RE.

One of the earliest demonstration of effective scenario based design was the IBM voice message system for the Los Angeles Olympics [3]. Since then scenarios have come in a variety of shapes and forms. Interpretation of scenarios seems to depend on their usage and how they are generated. In the HCI community scenarios have been proposed as detailed descriptions of a usage context so design decisions can be reasoned about [2] or small-scale examples of an existing product which are used to anchor discussion about different design theories [9]. In software engineering 'use cases' have been developed as informal narrative description of use, responsibilities and services within object oriented design [5]. Scenarios in the information systems community have evolved the concept of a rich picture which gives the social and environmental setting of a required system so arguments can be developed about the impact of introducing technology, and the matching between user requirements and task support provided by the system [6]. Finally, in Requirements Engineering, scenario scripts have been proposed as test data for checking dependencies between a requirements specification and the users/environment in which it will have to function [7].

I have been interested in how scenarios can be used with other techniques such as design rationale to explore requirements [8]. An important influence has been Colin Pott's inquiry cycle [7], which is one of the few examples of giving methical advice about how to use scenarios. More recently, I have been taking a broader look at what constitutes scenario based approaches and why they should be effective with colleagues on the EU Long Term Research project CREWS (appropriately enough titled Cooperative Requirements Engineering with Scenarios). The following ideas are a result of our collaboration.

2 A Framework of Issues

Presentation and Representation. Scenarios have been described in variety of media. Narrative text is probably most common, with several formatted variants, such as scripts and event histories. Other media have been used to amplify scenarios, typically still image for diagrams, sketches and photographs of the application and its environment, and video to illustrate a system context and to record design scenarios being discussed in meetings. Scenarios may also be presented as implemented software system in prototypes. mock-ups and concept demonstrators developed in multimedia authoring tools such as Director.

What is represented in scenarios varies from concrete descriptions of reality to designed systems which are then run as simulations to present a future vision of how the system will behave. An example is the traffic light scenario used in the ARIES system to simulate operation of a prototype design [1]. At a more general level simulations of system operation in logical terms have also been described as scenarios, e.g. animated Petri nets, but this seems to be stretching the definition a little far. Representation may either be in natural, informal media (e.g. text, picture, video) or in more formal abstractions- diagrams, conceptual models or formal specification languages.

In summary, scenarios may be presented in a variety of media, either natural language text and images or in symbolic form as diagrams, formal notations, etc. Furthermore the content which is represented may be just facts recorded in a medium or a designed artefact that delivers its own presentation as a simulation.

Content. Contents may not be definable in terms of concise semantics if the scenario contains facts pertinent to the real world expressed in text or images.

Content may be described at different levels of modelling concern, e.g. social, cognitive (individual user), or technical system. The content of scenarios needs to be accessible so their scope can be judged, for instance typical contents are

- Organisational information: structures of companies, groups, departments, membership of people (agents) in groups, geographical and topographical information about the world of interest

- Stakeholder information: characteristics of people, their views, opinions roles, responsibilities, aims and objectives

- Behavioural information: tasks, activities, actions, events, what people and the system does

- Objects: entities, data, information, attributes, etc.

- Episodes: context information of events, situations, history of past events, setting for future imagined events

Content tends to delineate "scenarios in the wide" which describe chunks of the world including the social level (e.g., [6]) from "scenarios in the narrow" which contain more detailed description of behaviour and event dependencies of a system where the design intent has already been established [4]. At one extreme, scenarios become large scale models of the world, for instance war games used by the military and scenario exercises employed in training for the emergency services. In some cases such scenario models can be implemented, as in virtual reality war games. One distinction which may be drawn is between concrete and more abstract descriptions, where instance scenarios concentrate on details of individual agents, events, stories and episodes with little or no abstraction while type scenarios describe facts in categories and abstractions derived from experience.

This distinction does categorise many examples, but in some cases part of the knowledge may be expressed at the type level (e.g. events) while the rest is considered at the instance level, e.g., agents (John, Mary, etc.).

Process. Scenarios are used in a variety of different activities in RE. Frequently they are used as a means of situating thought for requirements elaboration, and as the starting points for more detailed modelling, as in the use case tradition [5]. Validation is a common motivation for using scenarios, such as cross checking dependencies between scenario scripts and functions in a requirements specification in the Inquiry Cycle [7]. Another use, less frequently reported in the literature is to anchor negotiation by use of scenarios or their use in explanation, marketing presentations, etc. in which the scenario is used to bring the proposed system to life for the audience.

An interesting intersection of process and content is when a scenarios becomes part of the Requirements specification and possibly vice versa. Creating scenarios usually entails acquiring domain knowledge. If this knowledge is abstracted it becomes part of a model. The enterprise modelling tradition may be seen as creating generic models from scenarios. At some point in the RE process part of a model will be elected as a Requirements specification leaving the rest as 'outside the machine boundary'. Of course as analysis progresses this boundary may shift either way, which sets up interesting relationship between domain models, scenarios and requirements specifications. Clearly scenarios at the instance level are always outside the boundary of the designed system, however for a scenario at the type level, the situation is not so clear.

Process also intersects with representation. Scenarios represented in prototype form are used in validation by walkthrough inspections. Simulations are also used for validation, however, the role of the scenario is subtlety different. Where formal constraints can be applied simulation scenarios converge with verification of requirements specifications.

Clearly scenarios have many different interpretation and uses, but as yet there is little consensus about what constitute a scenario based approach and how RE may be improve by such approaches.

3   Objectives of the Workshop

The workshop will explore the different concepts of scenarios which have been proposed to establish if there is a common view about what constitutes a scenario and then ask a series of questions about how they may be effectively used:

- How can we create or model scenarios? Where is the boundary between instance level and generic information? What are the subtypes of scenarios ranging from descriptions of real-world scenes to simulations and prototypes of future systems?

- Where can scenarios be used in RE, how do scenario like descriptions function in communication, and how can interaction between the stakeholders in RE sessions with (or within) scenarios be improved?

- Contributions, experience reports, tales of success and failure will be sought, so we can survey the current state of the art (and practice) in scenario based RE, look forward to the research challenges, and propose ways of technology transfer of the knowledge already present in the RE community.

References

[1] K. M. Brenner et al. Utilising Scenarios in the Software Development Process. IFIP WG 8.1 Working Conference on Information Systems Development Process. December, 1992.

[2] J. M. Carroll. The Scenario Perspective on System Development. In *Scenario-Based Design: Envisioning Work and Technology in System Development*, ed. J.M. Carroll, 1995.

[3] J. D. Gould. Utilising Scenarios in the Software Development Process. IFIP WG 8.1 Working Conference on

Information Systems Development Process. December, 1987.

[4] M. Jackson *Software Requirements and Specifications.* Addison Wesley, 1995.

[5] I. Jacobson. The Use-Case Construct in Object-Oriented Software Engineering. In *Scenario-Based Design: Envisioning Work and Technology in System Development*, ed. J.M. Carroll, 1995.

[6] M. Kyng. Creating Contexts for Design. In *Scenario-Based Design: Envisioning Work and Technology in System Development*, ed. J.M. Carroll, 1995.

[7] C. Potts et al. Inquiry-Based Requirements Analysis. *IEEE Software.* pp. 21-32, 1994.

[8] A. G. Sutcliffe. Requirements Rationales: Integrating Approaches to Requirements Analysis. In *Proceedings of Designing Interactive Systems, DIS'95*, ed. G. M. Olson and S. Schuon. pp. 33-42, ACM Press, 1995.

[9] R. M. Young and P. Barnard. The Use of Scenarios in Human-Computer Interaction Research: Turbocharging the Tortoise of Cumulative Science. *CHI + GI 87 Human Factors in Computing Systems and Graphics Interface.* Toronto, 1987.

Session 9

Keynote Address

Speaker

David Harel
The Weizmann Institute of Science, Israel

*"Will I Be Pretty, Will I Be Rich? Some Thoughts on Theory vs. Practice
in Software Engineering"*

Will I be Pretty, Will I be Rich?
Some Thoughts on Theory vs. Practice in Systems Engineering

David Harel
The Weizmann Institute of Science
Rehovot, Israel

"The mathematician's patterns, like the
painter's or the poet's, must be *beautiful*; ...
there is no permanent place in the world for
ugly mathematics."

(G. H. Hardy [H, p. 25])

(at a cocktail party)   A: "I'm writing a best-seller."
                         B: "Short of money, eh?"

(Cartoon in the *New Yorker*)

1 Preamble

This is a very short summary of a talk presented at the Third International Symposium on Requirements Engineering (RE '97).[1] The talk attempts to put forward some thoughts on theoretical vs. applied research in the specification and design of reactive, highly concurrent systems. By its very nature, such a talk is bound to be disorganized, rambling, non-self-contained, and extremely subjective. It is; and the written summary you are reading is even worse, since it not only omits the details of the examples used in the talk, but also lacks the intonations, facial gestures and hand-waving that are part and parcel of talks that have little technical content.

Oh well. So be it.

2 A 3-way Classification

This is a conference on requirements engineering. It concerns real systems, and tries to deal with the very real problems that arise in their development. This talk deals with theory — theoretical computer science, to be specific. In theoretical research, the methods and tools are mathematical but the theory is often geared towards particular kinds of real systems. Thus, there are conferences dedicated to the theory of programming languages (the Principles of Programming Languages conference, POPL), to the theory of database systems (Principles of Database Systems, PODS), to

the theory of concurrent systems (the CONCUR conference), to the theory of distributed systems (Principles of Distributed Computing, PODC), and more.

What sort of theory is done in such application-oriented meetings? Should applied people take notice of this work? Should the general theory community take an interest in the applications that lead to it? What sort of theory is done by theoreticians in general, and why? How about the converse questions: Should theoreticians peddle their merchandise to other theoreticians? Are they doing enough to serve the needs of the real-world practitioners?

At the heart of the talk is an attempt to clarify some of the issues behind these questions, by dividing the research carried out by theoreticians into three kinds, which will be referred to as Type 1, Type 2 and Type 3 theory.[2]

Type 1 theory concerns true foundations and principles. It should be robust, deep and of fundamental nature, and should explain, generalize and enlighten. Such are the basics of computability and complexity theory, for example, as they emerge from the work of Church, Turing, Rabin, Cook and many others.

Type 2 theory responds directly to the needs arising in applications. It should be pragmatic and specific, molding itself to fit the requirements posed by real-world difficulties, and it should result in things that work and can actually be used. Such is the fast Fourier transform, for example, or those parts of the theory of context-free languages that lead to efficient compilation techniques.

Type 3 is theory for the sake of theory (TST). It should be mathematically elegant, yet difficult and clever, and should be of interest to other theoreticians. Much of the work theoreticians do is of this type.

The borderlines between these are fuzzy, and as time goes by migration often take place: Some Type 3 results and techniques eventually become Type 1, and sometimes — but more rarely — Type 3 work becomes applicable, thus turning into Type 2.

---

[1] Similar talks, of the same title and with a very similar written summary, were given at the 13th ACM Symp. on Principles of Database Systems (PODS) in 1994, and at the 6th Int. Conf. on Concurrency Theory (CONCUR) in 1995.

[2] While our interest here is mainly in theory carried out in conjunction with practical fields of computer science, such as concurrent systems and programming, many of the points made can be modified to apply to theory in general. Also, the 3-way classification proposed here is somewhat different from the one proposed by Raghavan [R] for general STOC/FOCS theory.

TST is legitimate and desirable, and not only because it might get upgraded. It is absolutely essential to the well-being and substance of a scientific community. Even so, most theoreticians will never admit to doing Type 3 work.[3]

One difference between Types 1 and 2 on the one hand and Type 3 on the other is in the judges. While the quality of TST is inevitably determined by theoreticians, the ultimate test of both Type 1 and Type 2 is in the opinions of real-world people, such as systems engineers and programmers. A non-applicable piece of work can be considered by theoreticians to be Type 1, but it cannot fully deserve that label unless engineers and programmers can be made to appreciate its virtues too. Otherwise, there are exactly two possibilities, (i) the theory is bad, or (ii) it is TST (in which case it might be excellent, but the applied guys couldn't really have known).

3 Did Hoare and Milner do Theory?

This part of the talk, the main one, is dedicated to illustrating the points with examples. A sample question to ponder is this: Were Hoare and Milner in their pioneering work on CSP [Ho] and CCS [M] doing theory, and, if so, was it Type 1, 2 or 3?

Some of the examples discussed in the talk include the following: Database topics, such as Codd's work [C1, C2], query-by-example [Z], datalog (cf. [U]), and computable queries [CH]; state-based formalisms for specifying system behavior, such as Petri nets [Re] and statecharts [Ha1], including their practicality, and results on their relative expressive power and succinctness [RS, MF, EZ, DH]; recent work on adapting statecharts to the object-oriented paradigm; theoretical and practical aspects of verifying finite-state systems by executable specifications (see, e.g., [Ha2]), or the recently proposed methods based on BDD's [B, B+]. Also mentioned is a research spinoff into drawing graphs nicely [DaH, HS].

4 Post-Ramble

There is also a message in all this. Subjective, and perhaps trivial, but here it is anyway.

A typical theoretician wants his/her work to end up being Type 1. But setting out in advance with this in mind is usually pointless. Taking the theoretician's perspective here, we can try to *aim* in the direction of Type 1, by being collective and general. We should avoid overly specialized theories, ones that seem to apply only to a special case of some special language,

---

[3] Hardy, the great number theorist, was a notable exception, stating, in the famous passage from [H, p. 90]: "I have never done anything 'useful'. No discovery of mine has made, or is likely to make, directly or indirectly, for good or ill, the least difference to the amenity of the world. [...] I have just one chance of escaping a verdict of complete triviality, that I may be judged to have created something worth creating." He was wrong, of course, as any modern-day cryptographer will tell you.

model or approach. We should seek results that are as generic and as all-encompassing as possible. Robustness is the name of the game. And we should always keep in mind that the essence of true Type 1 must be appreciable by non-theoreticians too, and it is our responsibility to expose and elucidate it.

As to Type 2, while theory people are by no means obliged to produce applicable work, some of us really want to. If we are interested in *actively* carrying out Type 2 work, we should get out there and become involved. We should take a real interest, listen attentively to what the real-world people ask for, and study their thought-patterns and work-habits. Only then can we try to see if there are ways we can help. The problems arising out there are usually much harder than we tend to think. Riches don't come easy. Doing our work in isolation and then trying to impose our ideas on the real world is bound to fail. If engineers and programmers do not find it beneficial to use the result of an application-oriented research effort — for whatever reasons — that piece of research is probably quite useless. We should be humble; they are the absolute judges.

So much for us theoreticians. What can be said here to the practitioners, e.g., of the kind participating in this conference?

Well, as far as Type 2 theory goes, simply don't give in. Be demanding; be pedantic, or even idiosyncratic. Explain and justify your problems and needs to the theoreticians. Let them in on your whims and fancies; you might just turn lucky. But be patient, since most theoreticians cannot muster the down-to-earth attitude an engineer needs in order to function well in the face of real-world problems. Some of us can't even program!

When it comes to Type 1 work, the practitioners should be the ones to show an interest. Theory can be more than just pretty mathematics. Some of it is deep, sweeping and fundamental. It will usually not be of direct help in your daily work, but it very often addresses truly basic issues, capturing phenomena that are at the heart of the field — that field in which your real-world work is done. Be open. Listen to it. It might not be quite as way-out as you think.

References

[B]   R.E. Bryant, "Graph Based Algorithms for Boolean Function Manipulation", *IEEE Trans. on Computers* C-**35**:8 (1986), 677–691.

[B+]   J.R. Burch, E.M. Clarke, K.L. McMillan, D.L. Dill and J. Hwang, "Symbolic Model Checking: 10^{20} States and Beyond", *Inf. and Comput.* **98** (1992), 142–170.

[CH]   A.K. Chandra and D. Harel, "Computable Queries for Relational Data Bases", *J. Comput. Syst. Sci.* **21** (1980), 156–178.

[C1]   E.F. Codd, "A Relational Model of Data for Large Shared Data Banks", *Comm. Assoc. Comput. Mach.* **13**:6 (1970), 377–387.

[C2]   E.F. Codd, "Relational Completeness of Data Base Sublanguages", In *Data Base Systems* (Rustin, ed.), Prentice-Hall, Englewood Cliffs, N.J., 1972.

[DaH]   R. Davidson and D. Harel, "Drawing Graphs Nicely Using Simulated Annealing", *ACM Trans. on Graphics*, in press.

[DH]   D. Drusinsky and D. Harel, "On the Power of Bounded Concurrency I: Finite Automata", *J. Assoc. Comput. Mach.* **41** (1994), 517–539. (Preliminary version appeared in *Proc. Concurrency '88*, LNCS 335, Springer-Verlag, New York, 1988, pp. 74–103.)

[EZ]   A. Ehrenfeucht and P. Zeiger, "Complexity Measures for Regular Expressions", *J. Comput. Sys. Sci.* **12** (1976), 134–146.

[H]   G.H. Hardy, *A Mathematician's Apology*, Cambridge Univ. Press, 1940.

[Ha1]   D. Harel, "Statecharts: A Visual Formalism for Complex Systems", *Sci. Comput. Prog.* **8** (1987), 231–274.

[Ha2]   D. Harel, "Biting the Silver Bullet: Toward a Brighter Future for System Development", *Computer* (Jan. 1992), 8–20.

[HS]   D. Harel and M. Sardas, "Randomized Graph Drawing with Heavy–Duty Preprocessing", *J. Visual Lang. and Comput.*, **6** (1995), 233–253. (Also, *Proc. Workshop on Advanced Visual Interfaces*, ACM Press, New York, 1994, pp. 19–33.)

[Ho]   Hoare C.A.R, "Communicating Sequential Processes", *Comm. Assoc. Comput. Mach.* **21**, (1978), 666–677.

[MF]   A.R. Meyer and M.J. Fischer, "Economy of Description by Automata, Grammars, and Formal Systems", *Proc. 12th IEEE Symp. on Switching and Automata Theory*, 1971, pp. 188–191.

[M]   Milner, R., *A Calculus of Communicating Systems*, Lecture Notes in Computer Science, Vol. 94, Springer-Verlag, New York, 1980.

[RS]   M.O. Rabin and D. Scott, "Finite Automata and Their Decision Problems", *IBM J. Res.* **3** (1959), 115–125.

[R]   P. Raghavan, Electronic mail contribution to a debate on the future of theory, Feb. 17, 1994.

[Re]   W. Reisig, *Petri Nets: An Introduction*, Springer-Verlag, Berlin, 1985.

[U]   J.D. Ullman, *Principles of Database and Knowledge-Base Systems*, Vols. I and II, Computer Science Press, 1988.

[Z]   M.M. Zloof, "Query-by-Example: A Data Base Language", *IBM Systems J.* **16** (1977), 324–343.

Session 10

Languages and Tools

Auditdraw: Generating Audits the FAST Way
Neeraj K. Gupta, Lalita J. Jagadeesan, Eleftherios E. Koutsofios, and David M. Weiss

It has often been said that there is leverage to be gained in devising software development techniques specific to coherent families of applications. This paper reports on how one family of applications was addressed using an approach that is interesting because it includes a broad range of family-specific software development artifacts, which include an application-oriented language, a tool-set, and a process for domain analysis, requirements specification, and code generation. The contribution of the paper is not in the particulars of the artifacts and process for the one reported family, but rather in the methodological insights that can be generalized to other families.
— *Sol Greenspan*

The Integrated Specification and Analysis of Functional, Temporal, and Resource Requirements
Hanene Ben-Abdallah, Insup Lee, and Young Si Kim

The requirements for an industrial process-control system often concern time and physical resources, because these are important concepts in the environment that the system will be controlling. This paper illustrates the potential benefits of special-purpose requirements languages. A language with built-in concepts of time and resources is used to specify conveniently both the requirements for and the design of a process-control system. The design is then proved to satisfy the requirements.
— *Pamela Zave*

Generating Provably Consistent Code from Hierarchical State Machines
David J. Keenan and Mats P.E Heimdahl

Requirements specification languages emphasize readability, understandability, ease of use, and analyzability. Ultimately, however, production quality code is the desired end-product. It would be great if code could be generated automatically from specifications in such languages! It would be ideal if the generation were done in a way which gave assurance to the exact correspondence between specification and code!! This paper shows promising results in exactly this direction.
— *Martin Feather*

Auditdraw: Generating Audits the FAST Way

Neeraj K. Gupta[1], Lalita Jategaonkar Jagadeesan[2], Eleftherios E. Koutsofios[3]
and David M. Weiss[2]

Abstract

Through a research/development collaboration, we have applied the FAST domain engineering process to the audits software in Lucent Technologies' 5ESS telephone switching system. Our collaboration has developed an application-oriented language, toolset, and accompanying process for specifying the requirements and generating the code for the 5ESS audits software. We describe the FAST process, our language, and the expected benefits of this project.

Keywords

requirements specification, requirements elicitation, software reuse, domain analysis, domain engineering, telecommunications, application-oriented languages, application generators, process, software engineering, requirements engineering

1 Introduction

Industrial software engineers continually face the question of how to produce their software faster, at lower cost, with more features. One approach to answering this question is to organize software into families, identify the requirements for the family, parameterize the requirements for individual family members, and to reuse assets within and across families to generate family members rapidly [9]. We describe here a collaboration between researchers and developers to apply a process, called FAST, that embodies such an approach. Our target family was the set of programs that audit the database in Lucent Technologies' 5ESS®telephone switching system to ensure that the switch operates reliably. Each member of the family is known as an audit. The FAST process guided us to develop a set of reusable assets that enable us to generate the C code for audits that is currently developed manually.

In particular, our collaboration has developed the following:

- an application-oriented language, called Auditdraw, designed especially for specifying the requirements and generating the code for the 5ESS audits software,

- a set of tools to help audits developers use Auditdraw, and

- a new process for developing audits using Auditdraw.

We expect that Auditdraw and its toolset and process will significantly increase productivity and significantly decrease cost and interval in audits software development in the 5ESS switch; we note, however, that it is not yet in production use.

Section 2 of this paper gives a brief description of Lucent Technologies' 5ESS switch and its audits software, section 3 describes the FAST process, and section 4 describes how we applied it to the audits domain. Section 5 contains our conclusions.

2 Lucent Technologies' 5ESS Telephone Switching System and Audits

Lucent Technologies' 5ESS telephone switching system [8] provides highly reliable telecommunications services; one of the key factors in ensuring system integrity and stability is the presence of reliable data.

In particular, audits [4] are programs that help ensure fault-tolerance of relational databases. In the 5ESS telephone switching system, these programs perform run-time checks on the consistency of data entities, and perform the appropriate corrections on data when an inconsistency is detected. For example, in a 5ESS switch, the status of units connected to the switch is maintained in the switch's database. Some units are arranged in hierarchies, and units that are related to each other by the hierarchy must have consistent status, e.g., if a parent unit is out of service, then all of its children must either be out of service or in a transient state.

---

[1]Independent Consultant, with Lucent Technologies 5ESS Software Development Organization, ngupta@dt2k.com

[2]Software Production Research Dept., Bell Laboratories, 1000 E. Warrenville Rd., Naperville, IL 60566 (USA) {lalita,weiss}@bell-labs.com

[3]Network Services Research Center, AT&T Labs – Research, 600 Mountain Ave, Murray Hill, NJ 07974 (USA) ek@research.att.com

A group of software developers are responsible for the 5ESS audits software. The requirements for audits come from developers from other subsystems, and are written in English. These requirements are typically given after the data design for a new or modified feature has been completed, and they specify the data entities that need to be audited, how to access those data entities, and what consistency checks to perform between different data entities. Requirements capture is followed by the construction of an audit design, and this design is subsequently reviewed at a high level design meeting and then a low level design meeting. Based on this design, the audit developers program the audit in C; this program is then reviewed in a formal inspection. The errors found during the code inspection are corrected and then the testing process begins. The first phase consists of unit testing, where the audits developers test the individual C functions. The second phase consists of integration testing, where other developers test all of the C code comprising the new feature – including code from other subsystems – by executing scenarios based on the requirements. This process is depicted in Figure 1.

With the encouragement of both research and development management, we formed a collaboration between researchers in the Bell Labs Software Production Research Department and developers in the 5ESS audits group to improve the audits development process. The collaboration has been applying the FAST process to the audits domain.

3 The FAST Process

The FAST process assumes that most software development is redevelopment, and that software production can be organized around families of systems to avoid much of the rework typically involved in redevelopment.[1] The goal of FAST is to provide a systematic approach to analyzing potential families and to develop facilities for efficient production of family members. Key to the process is finding the requirements for the family and appropriate abstractions for representing them, creating a language for specifying the requirements of individual family members, and then translating specifications of family members into deliverable software. Put another way, FAST is a systematic process for family-oriented, abstraction, specification, and translation.

FAST has two subprocesses, as shown in Figure 2.

- Defining the requirements for the family and developing a set of reusable assets for producing family members. This subprocess is known as domain engineering. We call its early phases domain analysis.

- Using the assets to produce family members, primarily by generation. This subprocess is known as application engineering.

The two subprocesses are connected by feedback loops to guide the evolution of the family and its assets.

3.1 Defining the Family and Developing the Reusable Assets

Defining the family means identifying the requirements for potential family members, characterizing what they have in common, and how they differ. For example, every audit must check for the existence of data before it attempts to access the value of the data. Furthermore, such a check must be done within the same time segment as when the data are accessed. All members of the family of 5ESS audits must obey this rule. On the other hand, the particular data to be accessed, the way in which the check is made, and the way in which the data are accessed vary over reasonably well-defined sets, and there are certain combinations that are not included in the family.

Just as one may organize requirements for single systems in a variety of ways, one may organize requirements for families in the same variety of ways. As an example, one might use categories such as interfaces to devices, interfaces to external systems, and behavior. Within each category (or further sub-categories), one may describe what's true for all family members (commonalities), what varies among family members (variabilities), and what the range of variability is. Section 3.1.1 describes the process that we use for eliciting such requirements and the artifact that results.

Note that no matter whether one prefers object-oriented approaches, functional approaches, or other approaches, one must decide what the potential family members are. We believe that this is equivalent to predicting what kinds of requirements changes are likely to occur during the lifetime of the family, and is of crucial importance in developing family members.

3.1.1 Eliciting Requirements: The Commonality Analysis

We use a process called commonality analysis to elicit the requirements for a family. A commonality analysis

---

[1]FAST is a variant of Synthesis, which is described in [2, 10, 11]. The primary differences are that FAST does not have a separate activity for bounding a domain, and uses a much more structured approach for defining a family than does Synthesis. In addition, FAST relies more heavily on compiler-building technology than does Synthesis.

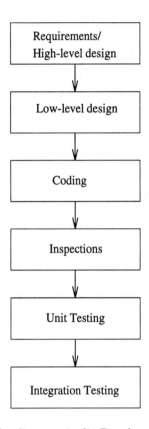

Figure 1: The Current Audit Development Process

is a structured, moderated discussion among a group of domain experts. Its result is a document, also called a commonality analysis [1], whose key parts include

1. A dictionary of terms used in discussing the family.

2. A list of assumptions that are true for all family members; these assumptions are known as commonalities and are requirements that every member of the family must meet. An example of a commonality is the requirement that every audit must check for the existence of data before it attempts to access the value of the data and that the check must be done within the same time segment as the access.

3. A list of assumptions about what can vary among family members; these assumptions are known as variabilities and are requirements that distinguish among family members. The way in which data are accessed by an audit is a variability.

4. A list of parameters that define, for each variability, the possible set of values that it can have and the time at which a value must be fixed when specifying a family member. The possible data access methods used by audits form the set of values for the variability described in the preceding paragraph. For a given audit, the access methods it uses must be declared when the audit is specified.

5. A list of issues that arose during the course of the analysis and, for each issue, a brief discussion of its resolution.

As much as possible, we use standard forms for expressing the terms, commonalities, variabilities, and parameters of variation. Except for the parameters of variation the standard forms are just structured prose. One example is commonalities that take the form "There is a mechanism for ..." An instance might be: "There is a fixed set of mechanisms that an audit may use for accessing data." For parameters of variation we use a table that includes, as appropriate, mathematical descriptions of the value spaces of variabilities. Where standard forms do not fit, we use free text for describing terms, commonalities, and variabilities.

A commonality analysis document for a family provides the basis for designing a specification language

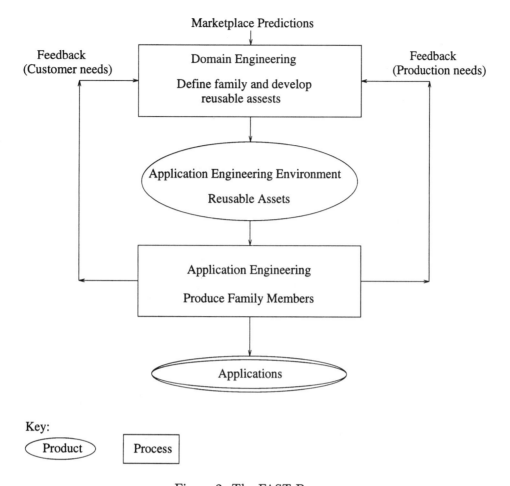

Figure 2: The FAST Process

and other reusable assets for the family.

The commonality analysis process is organized into phases that are designed to elicit terms, commonalities, variabilities, and parameters of variation by consensus from a group of 5-10 domain experts. Early phases of the process concentrate on gaining agreement among the domain experts on the objectives of the analysis and on the boundaries of the discussion. The intermediate stages of the analysis focus on gaining consensus for the definitions of commonly-used, important terms, for commonalities, and for variabilities. Later stages focus on parameterizing the variabilities and on reviewing the results of the analysis for completeness, consistency, and readability.

All stages of the commonality analysis are guided by a moderator who understands the FAST process, the role of commonality analysis within the FAST process, and the development culture in which the process is being used. Moderators have considerable discretion in adapting the process to different groups, but rarely change the structure of the artifact, i.e., we are flexible about the structure of the process, but inflexible about the structure of the document. A more detailed description of the commonality analysis process and artifact can be found in [12].

3.1.2 Reusable Assets

The reusable assets for a family consist of all the procedures, tools, and artifacts needed to produce family members, known in FAST as an application engineering environment. For example, a language for specifying family members and a translator for generating C code from a specification in the language are typically included in the environment. Those who use the environment follow a process specified by its developers. For 5ESS audits, the Auditdraw language and its translator form the initial environment. As the family of audits evolves, the environment will also, as the translator is enhanced and new tools are added to the environment.

3.2 Generating Family Members

The application engineering environment is designed to help its users to generate members of the family very rapidly. Much of its effectiveness depends on how accurately requirements for potential family members were predicted during domain analysis. When predictions about what family members will be needed in the future are accurate, the environment will be very effective. For this reason, a key input to the family definition process is predictions about marketplace trends.

Key to the environment is a well-designed language for specifying requirements. Its users should be able to specify particular family members just by specifying the variations considered during the definition of the family. For example, they should be able to specify for audits the data to be accessed by an audit. The language should allow them to do so in a way natural to the family, i.e., using the abstractions, such as data item fetch, that are used to define the family. The environment should provide them with facilities for verifying the choices they have made, e.g., verifying that all values for a particular data item have been checked.

The environment embodies both the process for creating family members envisioned during the definition of the family and the tools, procedures, and artifacts needed to carry out that process. Its users create a model of the family member that they would like to produce and then generate the family member. For 5ESS audits, the model is a specification expressed in Auditdraw. Generation of the family member is accomplished by supplying the specification to the Auditdraw translator, which performs completeness and consistency checks and generates the appropriate code.

3.3 Applicability of FAST

The FAST process is worth applying when the cost of domain engineering is repaid by the decrease in cost and development time for future family members, i.e., when the domain engineering cost can be amortized over the family members that are produced with the results of domain engineering. Such repayment occurs in the following situations:

- When a system will exist in many variations over a long period of time,

- When there is considerable time and effort being devoted to making continual changes to a system,

- When there are many customers for a system, each of whom wants the system customized for his or her purposes,

- When it is important to produce variations on a system quickly.

Much of our experience in applying FAST has been in legacy systems that are still in demand, where there is a reservoir of knowledge about the system, and where change to the system has become slow and costly compared with marketplace demands. Users of FAST often view it as a way to gain a competitive advantage in speed and cost.

We usually apply FAST by seeking a domain within a large, legacy system where there is frequent change occurring at relatively large cost. Such a domain is often an isolatable section of the system where the changes can be encapsulated, and where a group of software developers has responsibility for making the changes. Section 4 describes the application of FAST to such a domain within the 5ESS software. Although this is a typical application of FAST, we also believe it will work wherever developers are able to make informed decisions about family requirements.

3.4 Organizing FAST Applications

In addition to performing a commonality analysis, the FAST domain engineering process includes activities for designing and implementing the application engineering environment, and the application engineering process for using the environment to produce applications. A detailed description of these activities is beyond the scope of this paper.

As shown in Figure 2., we perform domain engineering and application engineering iteratively, reanalyzing, refining, and improving the environment as necessary. For the early iterations, we generally establish a collaboration of researchers and software developers to develop the initial version(s) of the environment. For the commonality analysis, the moderator is frequently a researcher teamed with 5-10 domain experts. For language design and implementation activities, the team is often composed of one researcher and two or three developers.

As prototypes of the environment become available, more developers are added to the team as testers. As the environment becomes ready for production, a wider set of developers is trained in its use, and researchers take a decreasing role in further environment development. We expect that the domain experts will become the owners of the environment and continue its development based on the feedback they get from its use and from forecasts of marketplace needs. Most of the domains that we are currently engineering are still in their first iteration of domain engineering.

Our experience, which is still very limited, indicates that the resources needed to develop the first version of an application engineering environment suitable for production use is less than five staff years of effort. We consider these domains to be in the first major iteration of the domain engineering-application engineering cycle shown in Figure 2. We currently have approximately ten domains somewhere in their first major iteration.

4 FAST and Audits

4.1 Domain Analysis and Application-Oriented Language Development

The audits software development group and several researchers collaborated to perform a domain analysis for the 5ESS audits domain, the first step in the FAST process. Thus, the requirements described in the commonality analysis [1] have been validated by a large group of audits experts.

For the language development phase, a member of the 5ESS audits software development organization visited the Software Production Research Department for approximately one year and a half. This interaction enabled a very fruitful and crucial exchange of ideas and concerns between research and development. Prior to our collaboration, there had been a prototypical rule-based language developed in the audits software group, on which we have capitalized.

Based on the domain analysis and this previous language prototype, we developed the Auditdraw language [3], designed especially for specifying the requirements for the 5ESS audits software. In the course of our work, we discovered that audit requirements can be very naturally represented as a form of decision trees, in which a decision involves the retrieval of a data entity from a database, and the comparison of its value to the value of some other data. Since some sub-trees may be identical, we have generalized these decision trees to directed acyclic graphs. The leaves of these graphs represent reports to be generated and corrective actions to be taken. Since these graphs correspond to audit requirements, the decisions, reports, and actions all have precisely specified behavior.

Auditdraw is a language for specifying such decision graphs. In the graphical view, the programmer interactively draws the graph on either a workstation or a PC; he/she uses a graphical interface designed especially for audit graphs, which is built on the graphical layout tool "dotty" [5, 6]. In addition to the graphical specification, the programmer also writes a simple companion declaration section that specifies the name of the audit and its interfaces, the data to be audited, and the data access methods to be used. The graph and declaration section are then together automatically translated into semantically-equivalent executable code.

A simple Auditdraw graphical specification is given in Figure 3. This specification gives an Auditdraw representation of a small piece of an audit in the 5ESS telephone switching system, which checks the consistency of parent and child circuits appearing in hierarchical circuits. Specifically, the states of a parent

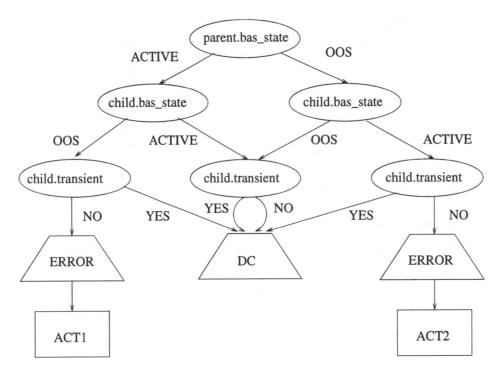

Figure 3: An Auditdraw Example

circuit and its child circuit should be identical: they should either both be active (ACTIVE) or both be out of service (OOS). If not, then an ERROR report should be generated and some corrective action (ACT1 or ACT2) taken. The only exception is when the child circuit is in a transient state, in which case any inconsistencies should be ignored and a "don't care" (DC) report should be generated. In our sample specification, the ovals represent the data entities to be checked (and the application of their associated retrieval method) and the arcs represent transitions. An arc is traversed if the value of the data entity in the source oval of the arc matches the value written on the arc; control is then passed to the target oval of the arc. Trapezoids represent error reports to be generated and boxes represent actions to be taken. So, for example, if the parent state is ACTIVE, the child state is OOS and the child is not in a transient state, then the ERROR report will be generated and the action ACT1 performed.

As illustrated above, Auditdraw specifications have a precisely defined semantics that model the behavior of audits. This enables static analysis – such as completeness checking and optimization – to be performed. For example, Figure 4 depicts an optimization of the Auditdraw graph shown in Figure 3; the first check performed in the optimized version is whether or not the child circuit is currently in a transient state.

If so, the graph is exited immediately after generating a DC report.

Two views are supported by Auditdraw: a graphical view, described above, and a rule-based view. In the rule-based view, the programmer explicitly writes every maximal path of the graph in a textual format. These two views are interchangeable: a specification written in the rule-based view can be automatically translated into the graphical view, and vice-versa. Both of these views can be automatically translated into semantically-equivalent executable code.

4.2 Current Status and Future Plans

The commonality analysis for the Audits domain [1] was completed in May, 1994. A majority of software developers from the 5ESS Audits software development group and several members of the Software Production Research Department participated in this analysis.

The design of (both views of) the Auditdraw language, and the development of the graphical toolset and code generator was completed in July, 1995. This toolset is currently under trial in the Audits development group. As part of this trial, we have developed an Auditdraw graph specifying an actual 5ESS software audit; this graph consists of 56 nodes and 113 edges. The C code automatically generated from this graph using Auditdraw consists of approximately 600 lines.

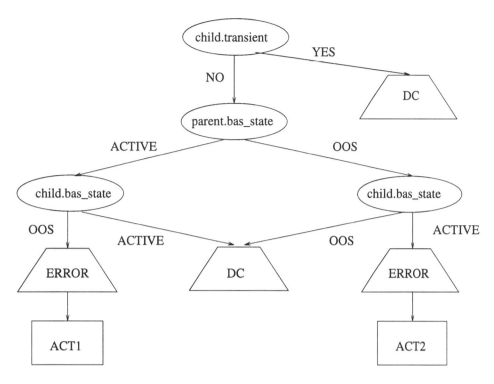

Figure 4: An Auditdraw Optimization Example

Since a formally-defined language and toolset now exist for the graphical view of Auditdraw, we plan to build on and extend this toolset after the completion of the trial. In particular, we plan to develop a toolset for the rule-based version of Auditdraw, as well as a debugger, optimizer, and "Auditdiff" tool for comparing the graphs/text of different audits specified using Auditdraw; this latter tool should be especially useful in the maintenance of audits. We also plan to develop a training course for Auditdraw.

4.3 Expected Benefits

The 5ESS audits software development group has begun a trial of the Auditdraw language and toolset. In this trial, several 5ESS audits will be specified using Auditdraw, and the resulting executable code will be tested in the 5ESS production environments.

Since Auditdraw is a high-level language designed especially for specifying audits, we believe that audits written in Auditdraw will be faster and easier to write and maintain. In particular, audit requirements can be specified directly in Auditdraw, and executable code can be automatically generated. We expect that this will significantly reduce coding errors, and that the resulting audits will be of a higher quality. Furthermore, automatic code generation from high-level specifications eliminates the need for several phases in the current development process: namely, the low-

level design, coding, code inspection and unit testing phases. The Auditdraw process is shown in Figure 5; the reader should contrast this with the current audit development process shown in Figure 1.

As a side benefit of this streamlined process, we expect that audits will be available to other subsystems earlier, aiding in the debugging of those subsystems.

Thus, we believe that the Auditdraw language and toolset will significantly increase productivity and significantly decrease cost and interval in audits software development in the 5ESS switch. While Auditdraw is specific to the audits subsystem in 5ESS, it is also readily adaptable to other platforms. For example, the generated code could be re-targeted to a different language. More generally, Auditdraw is generic to databases; namely, it can provide fault-tolerance for a very large variety of data. We are currently investigating possible applications to other Lucent Technologies switching platforms.

5 Conclusions

Our application of FAST to the audits domain should significantly improve the efficiency of the audits development process. We believe this is a result primarily of the following factors:

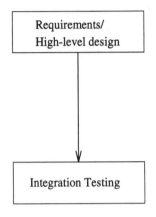

Figure 5: The Auditdraw Process

- *Family-oriented viewpoint*

  The FAST process is based on identifying the common requirements for a family of systems, parameterizing the requirements for individual family members, and finding the appropriate abstractions for easily expressing the commonalities and variabilities in requirements among family members. The commonality analysis process helps to ensure that these requirements and the abstractions remain correct and suitably expressive as the family of systems evolves.

- *Generation of code from requirements specifications*

  Requirements expressed in these abstractions are automatically translated into executable code. This streamlines the development process and ensures that the software satisfies the requirements.

- *Maintainability of software*

  Since the abstractions for expressing requirements are tailored for a particular family, in this case the 5ESS audits software, changes in the requirements are easily expressed by modifications in the requirements specifications. Furthermore, maintainability of the software in the face of changes in requirements is greatly aided by the automatic generation of code.

- *Static analysis of requirements specifications*

  The formally-defined nature of the graphical representation makes it possible to analyze statically the audits requirements for consistency and completeness, relative to the data that is being audited.

- *Reuse*

  Artifacts of the requirements engineering process can be reused in the following ways:

  Abstractions hold over large subsets of audits family members. In particular, commonalities are re-used in all the members of the family. For example, every 5ESS audit must check for the existence of data before it attempts to access the value of the data. Similarly, variabilities typically hold across large subsets of family members: for example, many audits share the same data access operations.

  The language translator reuses these abstractions for many audit designs, since the commonalities and variabilities are built into the language. For example, existence checks and appropriate data access operations are automatically inserted into audit designs during code generation.

  Abstractions are often shared between families. For example, we fully expect that many abstractions common to 5ESS audits will also be shared by audits in other switching platforms. Thus, Auditdraw and its process can be tailored to a variety of fault-tolerant databases.

  In addition, we believe that the collaborative nature of the FAST projects, i.e., researchers and software developers working together to use the FAST process to create reusable assets, will greatly improve the chances for success. Not only did each type of collaborator bring specialized knowledge needed for the domain analysis and domain implementation, but we expect that transfer of the new technology that resulted from the collaboration will be greatly eased by having the technology users be part of the process.

Acknowledgments

We thank the 5ESS Audits group for many helpful discussions on this project, and Chris Ramming and Curt Tuckey for many helpful discussions on the language and toolset. We are grateful to Bob Colby, Cy Rubald, and Mary Zajac for their vigorous support of this collaboration, and Joe Paule and Eric Sumner, Jr. for their vision in initiating it.

References

[1] G. Babu, M. Baron, A. Charles, J. D'Mello, D. Ebright, N. Gupta, L. Jagadeesan, S. Ozdemir, S. Patel, J. Paule, M. Phreykz, R. Trygar, and D. Weiss. Commonality analysis for audits. Technical report, AT&T Bell Laboratories, May 1994. Internal Document.

[2] G.H. Jr. Campbell, S.R. Faulk, and D.M. Weiss. Introduction to synthesis. Technical Report INTRO-SYNTHESIS-PROCESS-90019-N, Software Productivity Consortium, 1990.

[3] N.K. Gupta, L.J. Jagadeesan, E.E. Koutsofios, and D.M. Weiss. User's guide for auditdraw. Technical report, AT&T Bell Laboratories, May 1995.

[4] G. Haugk, F.M. Lax, R.D. Royer, and J.R. Williams. The 5ESS(TM) switching system: Maintenance capabilities. *AT&T Technical Journal*, 64(6 part 2):1385–1416, July-August 1985.

[5] E.E. Koutsofios and S.C. North. Viewing graphs with Dotty. Technical Report 59113-930120-04TM, AT&T Bell Laboratories, 1993.

[6] E.E. Koutsofios and S.C. North. Applications of graph visualization. In *Graphics Interface '94, Banff, Alberta*, pages 235–245, 1994.

[7] R.C.T. Lai and D.M. Weiss. A formal model of the FAST process. Technical Report Bell Labs Technical Memorandum, BL0112650-950707-30TM, AT&T Bell Laboratories, July 1995.

[8] K.E. Martersteck and A.E. Spencer. Introduction to the 5ESS(TM) switching system. *AT&T Technical Journal*, 64(6 part 2):1305–1314, July-August 1985.

[9] D.L. Parnas. On the design and development of program families. *IEEE Transactions on Software Engineering*, SE-2:1–9, March 1976.

[10] Synthesis guidebook, volume i, methodology definition. Technical Report SPC-91122-MC, v. 01.00.02, Software Productivity Consortium, December 1991.

[11] Synthesis guidebook, volume ii, case studies. Technical Report SPC-91122-MC, v. 01.00.02, Software Productivity Consortium, December 1991.

[12] D.M. Weiss. Defining families: The commonality analysis. Submitted to IEEE Transactions on Software Engineering, July 1996.

The Integrated Specification and Analysis of Functional, Temporal, and Resource Requirements*

Hanêne Ben-Abdallah and Insup Lee
Department of Computer and Information Science
University of Pennsylvania
Philadelphia, PA 19104

Young Si Kim
Electronics and Telecommunications Research Institute
P.O.Box 106 Yusong
Taejon, Korea

email: {hanene@saul, lee@central, yskim@saul}.cis.upenn.edu

Abstract

The Graphical Communicating Shared Resources, GCSR, is a specification language with a precise, operational semantics for the specification and analysis of real-time systems. GCSR allows a designer to integrate the functional and temporal requirements of a real-time system along with its run-time resource requirements. The integration is orthogonal in the sense that it produces system models that are easy to modify, e.g., to reflect different resource requirements, allocations and scheduling disciplines. In addition, it renders the verification of resource related requirements natural and straightforward. The formal semantics of GCSR allows the simulation of a system model and the thorough verification of system requirements through equivalence checking and state space exploration. This paper reviews GCSR and reports our experience with the production cell case study.

1 Introduction

The timed behavior of a real-time system is affected not only by the time its components take to execute and synchronize, but also by delays introduced due to the scheduling of tasks that compete for shared resources. Most current real-time formalisms adequately capture delays due to component synchronization, e.g., Statecharts [5], Modechart [7] and timed extensions of the classic untimed process algebras CSP and CCS [3, 4, 6, 11, 13, 12]. These formalisms, however, abstract out resource-specific details. This motivated the Communicating Shared Resources (CSR) paradigm [8, 9] to provide a formalism where the run-time resource requirements of a real-time system can be specified together with its functional and temporal requirements. The integration of the three types of requirements allows designers to consider resource-induced constraints early in the development cycle, explore alternate resource allocations, and to eliminate unimplementable design alternatives without expensive prototyping.

Within the CSR paradigm, the Algebra of Communicating Shared Resources (ACSR) [8] and the Graphical Communicating Shared Resources (GCSR) [2, 1] have been developed. ACSR is a timed process algebra and GCSR is a graphical language. The novelty of these formalisms relative to existing real-time formalisms is their representation of resources and priority. Without an explicit notion of resources, the specification of resource-bound systems requires that some artificial means be used to model resource requirements, such as defining processes to represent resources. Models that lack explicit priorities require that a process be created for the sole purpose of arbitrating priorities and implementing preemption. Providing explicit notions of resources and priority within the CSR formalisms results in specifications that are close analogues of the systems they model and that are easier to modify to reflect different resource allocations and scheduling disciplines.

In this paper, we use the GCSR language to illustrate the specification and analysis of real-time systems within the CSR paradigm. The graphical syn-

*This research was supported in part by NSF CCR-9415346, AFOSR F49620-95-1-0508, ARO DAAH04-95-1-0092, NSF-STC-SBR-8920230 and a grant from ETRI. The views and conclusions contained herein are those of the authors and should not be interpreted as necessarily representing the official policies or endorsements, either expressed or implied, of the Air Force Office of Scientific Research or the U.S. Government.

198

tax of GCSR adopts the intuitive notions of edges and nodes in control flow diagrams and produces modular, hierarchical and thus scalable specifications. The semantics of GCSR is described as a labeled transition system either directly or indirectly through a translation to the process algebra ACSR. The GCSR-ACSR connection makes it possible for the graphical language to benefit from well-founded process algebraic analysis techniques: automated equivalence checking, state space exploration, testing and execution. These analysis techniques are used to verify whether a GCSR design of a system satisfies its requirements. Furthermore, to facilitate the design within the CSR paradigm, we have developed a toolset, PARAGON[1], that assists in the graphical and textual (for ACSR) description of real-time system models and that supports the above verification techniques.

To experiment with the expressiveness of the GCSR language and to evaluate the benefits of integrating the resource requirements with the functional and temporal requirements, we modeled and analyzed the "Production Cell" case study [10]. This example represents a realistic, industrial real-time application, where safety requirements are essential and can be met by the application of formal methods. For this application, the GCSR modularity allowed us to distinctively model the two main agents in the system: the environment (i.e., the cell's physical machines) and the software system (i.e., software controllers) [14]. The graphical syntax of GCSR facilitated the visualization of the system components and their resource and communication dependencies. In addition, the explicit specification of the resources allowed us to express in a natural way the violations of safety requirements pertinent to system resources. Furthermore, analysis techniques provided within the CSR paradigm are suitable for verifying that our solution satisfies all of the requirements. In this paper, we illustrate the GCSR language and analysis techniques applied to parts of the production cell example.

Paper organization. The next section informally describes the production cell as reported in [10]. Section 3 reviews the GCSR language. Section 4 presents parts of our design solution for the production cell modeled in GCSR. Section 5 illustrates the analysis techniques applied to the presented design solution. Section 6 summarizes our evaluation in terms of the suitability of the GCSR language for requirements engineering of real-time systems in the class of the production cell.

---

[1]PARAGON stands for Process-algebraic Analysis of Real-time Applications with Graphics-Oriented Notation.

2   The Production Cell Example

Figure 1 depicts the top level view of the production cell as described in [10]. The system has two conveyer belts that are connected by a traveling crane, a positioning table, a two-armed robot, and a press. In addition, the system is equipped with a set of sensors (e.g., photoelectric cells) and actuators. The main functionality of the system is to forge metal plates in the press. This is accomplished through the following cycle of operation: the deposit belt conveys unforged metal plates towards the traveling crane which moves each metal plate to the feed belt; the feed belt conveys metal plates towards the the elevating rotary table which rotates to position a metal plate at the robot arm's reach; the robot extends its first arm, picks up the metal plate, retracts its first arm, rotates to position its first arm in front of the press, extends its first arm, places the plate in the press, and retracts its first arm away from the press; the press closes, forges the metal plate, and opens again; finally, the robot extends its second arm, picks up the forged metal plate, retracts its second arm, and rotates to position its second arm on the deposit belt where it unloads the metal plate. To achieve its functionality autonomously, the system receives sensory information and controls the behavior of the machines through a set of actuators.

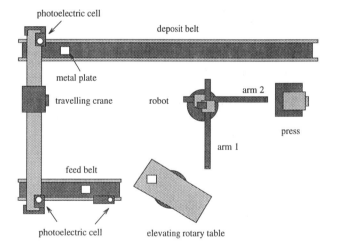

Figure 1: Top View of the Production Cell

The behavior of the system must satisfy two types of requirements. One is safety requirements, e.g., the system must limit the motion of a machine within its allowed ranges, disallow a machine from colliding with another, and disallow a metal plate to be dropped in unsafe areas. The second is liveness requirements, e.g., the system must guarantee that each metal plate on the feed belt eventually arrives at the end of the deposit belt after it has been forged at the press.

The goal of the case study is to develop a software *controller* for the production cell. The controller basically serves as a system monitor and scheduler. It collects sensory information from the set of sensors in the system and instructs the machines through the set of actuators to react in such a way that the system realizes its predefined functionality.

3 The GCSR Language

The GCSR language is based on the view that a real-time system consists of a set of communicating components, called *processes*, that execute on a finite set of serially shared resources and synchronize with one another through communication channels. The use of shared resources is represented by timed *actions*, and synchronization is supported by instantaneous *events*. The execution of an action is assumed to take nonzero time units with respect to a global clock, and to consume a set of resources during that time. The execution of an action is subject to the availability of the resources it uses. Contention for resources is arbitrated according to the priorities of competing actions; priorities are static, i.e., fixed and are drawn from the set of natural numbers. To ensure uniform progress of time, processes execute actions synchronously. Time can be either dense or discrete; however, we consider discrete time only to simplify the description of the GCSR semantics and for implementation reasons.

Unlike an action, the execution of an event is instantaneous and never consumes any resource. Processes execute events asynchronously except when two processes synchronize through matching event names, i.e., channels. Graphically, a GCSR process is represented by a finite set of *nodes* that are connected with directed *edges*. Figure 3 shows the graphical GCSR objects. Before describing the details of these symbols, we first introduce a simple example to illustrate the intuition behind the GCSR objects.

3.1 Example

Figure 2 shows the high level structure of the robot and parts of a model for the robot component of the production cell. The robot consists of two components: a controller, *RController_3*, and a machine model, *RModel_3*. The two components execute in parallel, exclusively use the set of resources $C = \{rbMotor\}$, and privately synchronize their activities through the set of communication events

$$R = \{rLeft, rRight, rStop, rAtPr, rAtDb, rAtRt,$$
$$ra1Extend, ra1Out, ra1Retract, ra1In, ra1Stop,$$
$$ra2Extend, ra2Out, ra2Retract, ra2In, ra2Stop\}.$$

The model for the robot consists of three components that describe the behavior of the hardware components of the robot. The process *RArm1_3* describes the behavior of the first robot arm. Initially, it behaves like the process *Wait*, which basically waits by consuming time but no resources. The execution of the *Wait* process can be interrupted by the reception of two events from the controller. One, event $(ra1Retract?, 1)$, is the reception of the event named *ra1Retract* at priority level 1 and which instructs it to retract the first arm; the second is $(ra1Extend?, 1)$ which instructs it to extend the first arm. The process *RArm1_3* does not have any preference between the reception of either event; the choice between them is non-deterministic.

We assumed that the robot arm is initially retracted. Thus, when the event $(ra1Retracr?, 1)$ is received, the execution of the *Wait* process is immediately aborted and the event $(ra1Crash!, 1)$ is instantaneously sent to indicate that the robot arm has moved beyond its limits which is a safety violation. Afterwards, the execution flow in the process *RArm1_3* enters the filled box, which indicates a deadlocked state. The semantics of the GCSR language ensures that if a component enters a deadlock state, so is the whole system.

The second possible behavior of the process *RArm1_3* from the initial node is to receive the event $(ra1Extend?, 1)$; in this case the motor resource, *rbMotor*, is used at priority level 1 to extend the first robot arm. The extension action takes *Trh1* time units to finish, after which the process *RArm1_3* has two possible behaviors. One is to instantaneously send the event $(ra1Out!, 2)$ to the controller to inform it that the arm is fully extended; another is to send instantaneously the event $(ra1Crash!, 1)$ to indicate that the robot arm may move beyond its limits. The communication with the controller has a higher priority than crashing, which is the desirable behavior.

If the first robot arm manages to communicate with the controller, it again has two possible behaviors: either immediately receive the event $(ra1Stop?, 2)$ to stop the extension, or again send the crash signaling event. If the first robot arm successfully receives the stop event, it idles until it receives signals from the controller, and so on so forth.

3.2 GCSR Nodes and Edges

Figure 3 (a) shows the graphical symbols for the GCSR nodes. The *Resource* attribute of a time-consuming node is a set of (resource, priority) pairs, with the restriction that each resource is listed at most once; this enforces the notion of serial resource usage. The *Name* attribute of a reference node refers to the name of a GCSR process. The *Restrict* and *Close* attributes of a compound node are sets of event names and resource names, respectively.

The motivation for various node symbols in GCSR

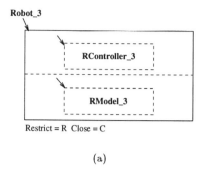

(a)

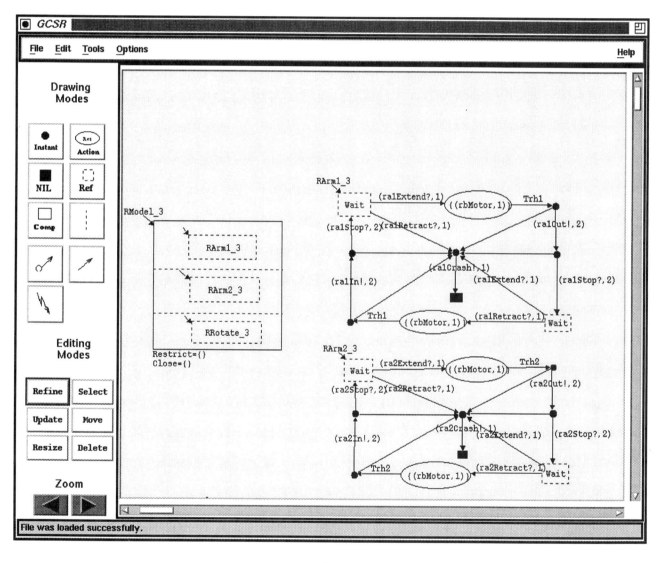

(b)

Figure 2: (a) High Level Structure of the Robot

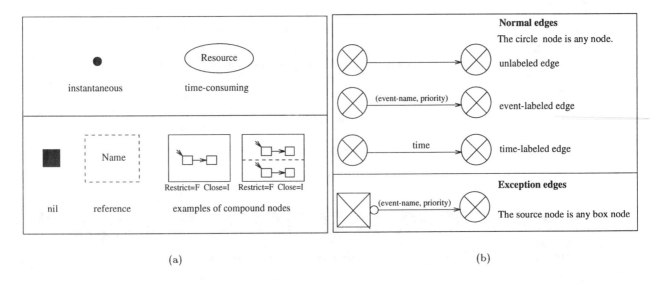

(a)

(b)

Figure 3: (a) GCSR nodes; (b) GCSR edges

is a succinct and scalable representation of the different system activities and components. The instantaneous node requires that no delay be allowed before the next activity. In contrast, the time-consuming node describes a time consuming activity. Furthermore, the Resource attribute of a time-consuming node explicitly describes the required resources for the system activity, which makes it easy to modify any resource requirement to reflect different resource allocation and scheduling disciplines.

The *nil* node describes a halting process, i.e., end of system execution. The reference node allows the decomposition of a large specification into subspecifications which eases the visual structuring of such a specification. On the other hand, the compound node visually distinguishes a system action from a system component. It is essential in supporting scalable and modular specifications since it allows a designer to: 1) group GCSR processes into a higher level entity, 2) connect several GCSR processes that are executed sequentially, and 3) reflect the fact that system components execute in parallel.

In addition to a structural modularity, compound nodes also provide for semantic modularity by encapsulating dependencies through their Restrict and Close attributes. The Restrict attribute identifies a set of events that are visible only among the GCSR processes inside the node; the Close attribute identifies a set of resources that are reserved for the nested GCSR processes, even if their time consuming actions do not explicitly request them.

GCSR nodes can be connected with edges to de-

scribe sequential execution. GCSR offers four types of edges shown in Figure 3 (b). The distinct symbols for the first three types of edges and an exception edge are motivated by the desire to support a structured, hierarchical specification in which edges do not cross node boundaries and to graphically distinguish two types of control flow: one that is externally controlled by an interacting process and one that is triggered internally through voluntary release of control by raising an exception. The second type of control flow is described by an exception edge. Control moves to the destination node of an exception edge when the process of the source node executes an exception event that labels the exception edge. The transfer of control through an exception edge allows synchronization between a process inside a compound node with an outside node and thus emulates a transition between nodes at different levels of nesting.

3.3 Informal Semantics

Intuitively, the behavior of a GCSR process consists of a sequence of execution steps each of which represents either a communication event or a time and resource consuming action. A communication event can be either a receive or send operation, respectively designated by the symbols "?" and "!" in front of the event name.

For example, in Figure 4 (a), once execution reaches the instantaneous node, the system *sends* the event named *e* at priority 1 and then execution moves instantaneously to the target node of the event-labeled edge from where execution continues. On the other hand, in Figure 4 (b) execution remains in the time-consuming

202

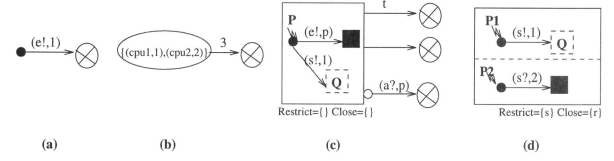

Figure 4: GCSR samples of partial specifications

node for 3 time units while using simultaneously the *cpu*1 resource at priority level 1 and the *cpu*2 resource at priority level 2; after three time units in the source node, execution moves to the target node of the time-labeled edge.

In addition to this basic notion of sequential execution in GCSR, several processes can be combined through compound nodes to describe a large system where processes execute in sequence or in parallel. For example, in Figure 4 (c), once execution reaches the compound node, the GCSR process P is executed for at most t time units, after which execution moves to the target node of the time-labeled edge. The execution of P can also be terminated in two other ways. One way is through an interrupt; this is represented by the unlabeled edge, which can be taken any moment during the execution of P. When this happens, the execution of P is aborted and execution moves to the target node. Another way of terminating the execution of P is through an exception raised by P sending the event a. At this time, the execution moves to the target node of the exception edge which is labeled with the *receive* event a at priority p. The difference is that, in the interrupt case the enabling of the edge is caused by a process other than P, whereas in the exception case the enabling of the edge is caused by the process P itself.

GCSR processes can also be combined in parallel by nesting them inside a compound node. When control reaches a compound node, it simultaneously enters all the initial nodes of its nested GCSR processes. Control can move through event-labeled edges in different nested processes in an interleaved fashion for unrestricted events, but synchronously in any two nested processes for restricted events. In addition, if control spends time in one nested process, then control in all the remaining nested processes must be in time-consuming nodes where it can spend time. This forces synchronous time passage between parallel processes.

Furthermore, since resources are assumed to be serial, the set of resources used in all the time-consuming nodes that simultaneously have control must be disjoint. For example, in Figure 4 (d) the processes $P1$ and $P2$ execute in parallel. Furthermore, $P1$ and $P2$ communicate privately through the event named s and use the resource r exclusively; that is, if the compound node is combined in parallel with another GCSR process, this latter can not communicate with $P1$ and $P2$ through the event s and will not have access to the resource r unless it requires it at a higher priority.

The overall behavior of a GCSR process can be formally described as a labeled transition system where each transition represents an execution step with the label being a communication event or a time and resource consuming action. As seen in the previous examples, there might be several transitions that are simultaneously possible. The selection among them is done via a notion of priority; see [2, 8] for details.

A second way of defining the execution steps of a GCSR process is through a translation to the timed process algebra ACSR which also has an operational semantics. Each GCSR process is translated to an ACSR process. The translation is consistent with the direct semantics in the sense that the corresponding ACSR process has an equivalent labeled transition system to the one generated directly from the GCSR process [2, 1]. The translation between GCSR and ACSR makes it possible for GCSR to benefit from the analysis techniques developed within the process algebraic setting [9]: simulation, state space exploration, and equivalence checking. ACSR offers several notions of equivalence that are based on prioritized bisimulations over labeled transition systems. The congruent notions of equivalence allow the modular verification of a system: a large system can be analyzed by analyzing its components individually.

4 A Design for the Production Cell

Our design produces a distributed controller for the whole production cell. In this design, the desirable behavior is achieved through arbitration of the usage of shared resources and synchronization through the set of sensors and actuators.

To make the design solution easier to understand, we reflect the hardware structure of the system. We model each machine as a *process* and the overall system as the machine processes running in parallel. We also distinguish between the machine model (i.e., environment) and its controller (i.e., software): each machine is represented by a *model* process whose desirable behavior is dictated by a *controller* process. In addition, to facilitate specification and analysis, we produce a solution in a hierarchical and modular fashion; that is, we describe the system components at different levels of abstraction where each level show more details than another and augments it in the support of the safety and liveness requirements. Due to space limitations, we only present the detailed specification of parts of the production to illustrate the main points. For a complete description of the case study refer to [1].

Design assumptions. Our design uses two types of communication events: those that represent information from the actuators and sensors, and those that we added to synchronize between the distributed components of the controller, e.g., event *prLoaded* which the robot controller sends to the press controller after loading the press. The actuators send digital signals to start or stop an action of electric motors and electromagnets in the system. Each actuator signal is therefore represented by an event in a straight forward way—e.g., event *prDownward* signals the press to start moving down and event *prStop* signals the press to stop its motion. Since the controllers use the actuators to dictate the machine behavior, the controller processes are in charge of sending the actuator events to the machine models.

The sensors report either digital values (i.e., from switches and photoelectric cells), or continuous values (i.e., from potentiometers.) Since our formalism does not support data values, we represent both types of sensory information with events whose occurrences describe the critical values in the state of the monitored machine. For example, to describe how far the first robot arm is extracted, we use the events *ra1Out* and *ra1In* to describe the facts that the first arm is completely extracted and completely retracted, respectively; intermediate positions are ignored. Since sensory information report the machine status, the machine models are in charge of sending the sensory events to the controllers.

Resource usage and arbitration is part of the environment being modeled in order to fomalize the requirements. Our design represents several dedicated resources such as electric motors, as well as all those resources shared between two components in the system. For example, the robot uses exclusively a motor, resource *rbMotor*, to rotate, and the robot and press share usage of the *press* resource. The shared resources are used to detect potential collisions between the sharing components.

The original description of the production cell [10] did not specify any timing requirement, despite the fact that this is a *real-time application*. We therefore introduced our timing assumptions which represent two types of activities in the system. One is to carry out a system function, e.g., "extend the first robot arm", the second is to reserve shared resources before using them, e.g., "the robot must reserve the press before unloading a forged metal plate". The reservation activity allowed us to avoid using unnecessary software communication events between the various parts of the controller. In addition, the notion of priority allowed us to ensure that a resource reservation has a lower priority than its usage and thus guarantee progress in the system.

4.1 System Structure

The high level structure of the production cell is specified by the following ACSR process

$$ProdCell \stackrel{\text{def}}{=} [(\ Press \| Robot \| ElevRotTable \| \\ FeedBelt \| Crane \| DepositBelt\ ) \setminus F]_I$$

where the operator $P_1 \| P_2$ denotes the parallel execution of processes, the operator $P \setminus F$ restricts communication events listed in F, and the operator $[P]_I$ close the resources listed in I.

In the process *ProdCell* the set of restricted events represent all of the communication events exchanged between the machines

$$F = \{\ crCanLoad,\ crUnloaded,\ rtCanLoad, \\ rtUnloaded,\ prUnloaded,\ prLoaded\ \}$$

and the closed resources represent all of the resources in the production cell, e.g., the robot motor *rbMotor* and the *press* resource. We next describe the detailed design of the robot and press components; we omit other components due to space limitations.

4.2 Robot

Figure 2 shows the structure of the robot and parts of the detailed description of the robot model. We note that in the robot model (and all remaining machine models) we adopted the following strategy to reflect

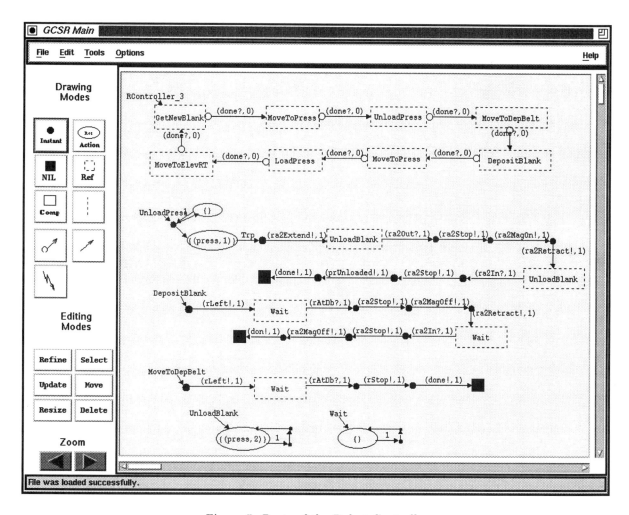

Figure 5: Parts of the Robot Controller

accurately the relationship between a machine and its controller: The robot model only constrains when to send sensory events in order to reflect an accurate state of the machine. However, it is ready to accept all actuator events from the controller in all states. A safety violation due to receiving a wrong event is marked by sending the special crash events and a system deadlock. A safe controller, on the other hand, must constrain when to send specific actuator events to avoid any safety violation.

Figure 5 shows parts of the components in a detailed specification of the robot controller. The controller, process *RController_3*, recursively executes a set of tasks in sequence. By inspecting the controller, it is easy to see that our design enforces the following safety requirements: When the robot must unload the press, process *UnloadPress* of Figure 5, it extends its second arm only when the resource *press* is available. As we see shortly, the *press* resource can be reserved only when it is not in use, i.e., for pressing or position-

ing the press plates. During the reservation step, the robot tries to acquire the *press* at priority level 1. The moment it manages to reserve the *press*, it starts using it at priority level 2 and in a non-preemptive mode, as described by the process *UnloadBlank*. In addition, when the second robot arm is extending or retracting near the press, the press does not move. This is guaranteed by holding the resource *press* in the process *UnloadBlank* which executes between the actuator events *ra2Extend* and *ra2Stop*. Similar restrictions are imposed on the first arm of the robot when it operates at the press.

4.3 Press

Figures 6 and 7, respectively, show the detailed description of the controller and model processes for the press. Note that in the process *PController_3*, the press controller holds the *press* resource at priority level 2 between the time it signals the press model to move down and the time it signals it to stop moving. This disallows the robot from acquiring the *press* during the

205

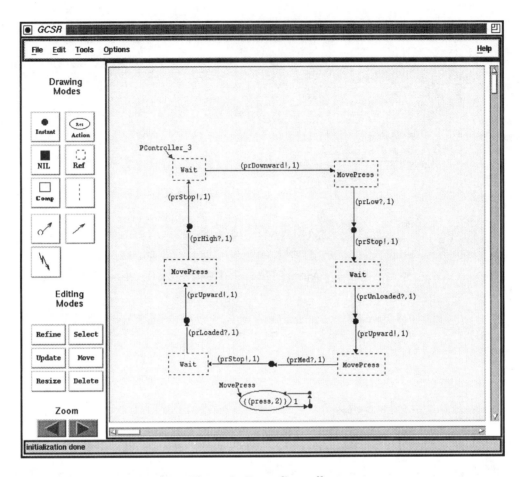

Figure 6: Press Controller

reservation step and, therefore, avoids potential collisions between a robot arm extending towards the press and the mobile plate of the press. In addition, when the press is forging a metal, the *press* resource is used by the press model, process *PModel_3*, at the highest priority level and in a non-preemptive mode.

The press model, process *PModel_3*, marks potential out of range motion of the press with a crash event, $(prCrash!, 1)$. As we see in the next section, testing the occurrence of this event indicates the out of range, and thus unsafe, motion of the press.

5 Design Analysis

According to the paper by Zave and Jackson [14], the requirements are desired properties of the environment, e.g., the robot and press machine models. If the design of a controller is satisfactory, then the properties of the design, together with the known or assumed properties of the environment, will imply the desired properties of the environment [14].

We use three basic techniques to verify that our design satisfies the safety and liveness requirements:

equivalence checking, testing, and deadlock detection. Certain requirements are easy to express as an abstract specification that can be inspected for correctness– e.g., safety in the robot. When possible, we describe the requirements and verify that a machine model together with its controller design is equivalent to the requirements. The equivalence proves that our design ensures that the system satisfies the requirements, under the assumptions made about the machine models. For those properties where a requirements specification is not easy to construct, we use testing to verify the correctness of our design. A tester either confirms the requirements or shows a requirement violation. In addition, our design solution uses deadlock to model out of range motion and machine collisions. We verify these safety violations by searching for deadlocked states in our design solution.

To make the analysis of our design solution manageable, we exploit its modularity. When a safety requirement involves two or more machines, e.g., machine collisions, we verify the correctness of the relevant machines grouped together. Our analysis was facilitated

206

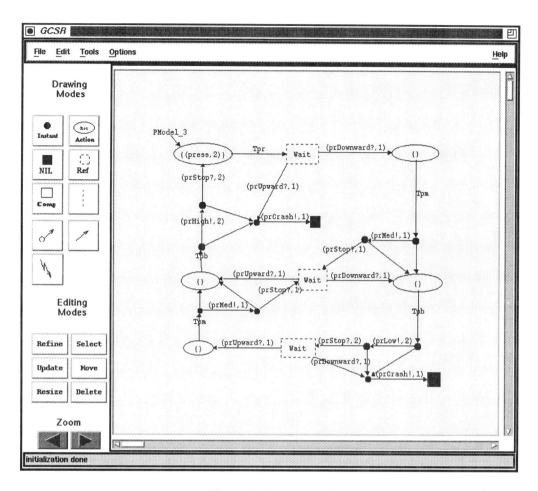

Figure 7: Press Model

by the automated translation of GCSR specifications to ACSR processes and the PARAGON toolset which supports the verification of systems modeled in ACSR.

Due to space limitations, we next illustrate the above analysis techniques in a few instances; the complete analysis can be found in [1].

Safety analysis: Part 1. As we saw in the design of the robot and press, our design solution uses special *crash* events to mark potential out of range motions and uses resource related deadlocks to detect collisions between machines.

We verified this part of the safety requirements by developing tester processes that deadlock when they detect the occurrences of a crash event. This procedure can be done either on the whole system or separately on the machines involved as illustrated below:

$$RobotSafetyTest \stackrel{\text{def}}{=} \quad \emptyset : RobotSafetyTest$$
$$+ (ra1Crash, 1).NIL$$
$$+ (ra2Crash, 1).NIL$$
$$+ (rbCrash, 1).NIL \quad .$$

The tester either idles (action $\emptyset$) or receives one of the crash events and becomes the NIL process which is deadlocked. We used the automated GCSR to ACSR translation and PARAGON to verify that our detailed design for the robot running in parallel with the process *RobotSafetyTest* is deadlock free for a fixed set of timing assumptions; more specifically, the process

$$(Robot\_3 \parallel RobotSafetyTest) \backslash F$$

where $F = \{prCrash, ra1Crash, ra2Crash, rbCrash\}$ does not have any deadlocked state. (PARAGON provides a command that allows the user to examine statistics of the labeled transition system of a process, e.g., number of states, number of transitions, and number of deadlocked states.)

Safety analysis: Part 2. A robot can drop a metal plate in two unsafe areas: between the elevating rotary table and the press, and between the press and the deposit belt. The first safety violation is detected by the occurrence of the event *ra1MagOff* before the event *rAtPr*. The second safety violation is detected by the occurrence of the event *ra2MagOff* before the event *rAtDb*.

It is straightforward to model each of these requirements as an abstract process that bascially executes the relevant events in the correct order and idles in between. We then abstract details from the design of the robot controller by the event and resource name abstraction operators in PARAGON to hide irrelevant events and resources. Finally, we use the automated equivalence checking in PARAGON to verify that the process describing the safety requirement is equivalent to the abstracted design of the robot controller.

In addition, one advantage of the graphical notation, is that we can inspect the flow of control within the GCSR description of the robot controller (Figure 5) to verify that the relevant events indeed occur in the correct order. In Figure 5, we can easily see that the process *RController_3*, the controller executes the subprocess in charge of moving the robot to the deposit belt, process *MoveToDepBelt*, before the subprocess in charge of unloading the metal plate on the deposit belt, process *UnloadBlank*. Thus the event *rAtDb* which is received by the subprocess *MoveToDepBelt* is received before the event *ra1MagOff* is sent in the subprocess *UnloadBlank*.

Another safety requirement for the press is that it can close only when no robot arm is positioned inside. In our design, a violation of this safety requirement is described by the simultaneous use of the *press* resource by the robot and the press. Since resources can be use in a serial mode only, such a situation leads to a deadlock in our design and more precisely in the process

$$[\, (\, RController\_3 \| Press\_3 \,) \backslash \{ prUnloaded, \, prLoaded \} \\ ]_{\{press, \, rMotor, \, db\}} \quad .$$

Note that since only the robot controller uses the *press*, and not the robot model, we only needed to verify the requirements for the robot controller; this is an advantage of the modular semantics of the ACSR and GCSR formalisms.

6 Conclusion

We have presented Graphical Communicating Share Resources (GCSR). The GCSR language allows the formal specification of the functional, temporal, and the resource requirements of a real-time system. The graphical notation of GCSR helps to visualize the overall structure of the system, its component dependencies (in terms of resources and communication), and the flow of control within each component. The syntax of GCSR produces modular and hierarchical descriptions, which is essential in the case of large-scale systems. In addition, the GCSR-ACSR connection allows

GCSR to benefit from well established analysis techniques for process algebras. The operational semantics of GCSR allows the execution of a specification to explore sample behaviors, which enhances the customers' understanding of the requirements.

As the production cell example illustrated, GCSR can be used both as a requirements language and a design language. Zave and Jackson [14] outline two necessary expressive capabilities in a requirements language: distinction between three categories of actions, and expression of constraints on actions in all categories. The three action categories result from viewing the system during the requirements phase as composed of at least two agents: the environment and the software system—which can be composed of several agents. The three action categories are: 1) environment-controlled shared, 2) environment-controlled unshared, and 3) software-controlled shared. An action is environment (software) controlled if the environment (software) performs it. An action is shared between the environment and software system if both agents can jointly observe it.

Modularity in GCSR allows us to describe distinctively the environment agents (e.g., machine models) and the software system agents (e.g., software controller). In addition, the notions of restricted events and closed resources make it easy to derive syntactically (or visually) the actions shared between the environment and the software system. The restricted events represent the interface between the environment and its software system and are therefore shared. Unrestricted events, on the other hand, are unshared—e.g., the crash events which the machine models produce to mark safety violations are not received by the machine controllers. Also, in our formalism, the environment and software system can reserve a set of resources to share exclusively through the concept of resource closure. Since resources can be used in serial, the environment and software system implicitly share them. When either agent fails to acquire a resource, it can conclude that the other agent is using it at a higher priority level. In terms of action control, the GCSR syntax does not enforce any division. It is the user's responsibility to make the distinction. For communication events, as shown in the production cell, one can adopt the convention that a sender is the controller, which makes it easy to derive syntactically event control information. However, a similar concept cannot be used to define the control of resource-consuming actions since most resources are shared and used interchangeably between the environment and software system.

The second necessary feature of a requirements language is to express constraints on actions in all categories. Constraints in a real-time system are either functional (e.g., action a can only happen in a particular state) or temporal (e.g., action a has duration t and deadline d.) In GCSR, the functional constraints can be modeled through the notions of interrupt, exception, and sequential and parallel execution. The temporal constraints can be modeled through the notion of the time consuming actions and temporal scope of execution. GCSR allows the specification of an additional type of constraint: prioritized resource usage which affects the timed behavior of a real-time system. The notions of resources and priorities in GCSR produce a model that encompasses run-time resource constraints, thus, a model that is more dependable and realistic. In particular, the notions of resources renders more intuitive the analysis of resource related safety requirements; for example, to verify the absence of a collision between the robot and the press, we verified that our specification does not have a deadlock due to an attempt by the robot and the press to use simultaneously the *press* resource.

Acknowledgement. The authors gratefully acknowledge the comments and suggestions made by the referees, as well as Pamela Zave, in improving the quality of this paper.

References

[1] H. Ben-Abdallah. *Graphical Communicating Shared Resources: a Language for the Specification, Refinement, and Analysis of Real-Time Systems.* PhD thesis, Department of Computer and Information Science, The University of Pennsylvania, Philadelphia, PA 19104, August 1996.

[2] H. Ben-Abdallah, I. Lee, and J.Y. Choi. A graphical language with formal semantics for the specification and analysis of real-time systems. In *IEEE Proceedings of Real-Time Systems Symposium (RTSS' 95)*, Pisa, Italy, December 1995.

[3] J. Davies and S. Schneider. An Introduction to Timed CSP. Technical Report PRG-75, Oxford University Computing Laboratory, Programming Research Group, August 1989.

[4] J. Davies and S. Schneider. An Introduction to Timed CSP. Technical Report PRG-75, Oxford University Computing Laboratory, UK, August 1989.

[5] D. Harel. Statecharts: A visual formalism for complex systems. *Science of Computer Programming*, 8:231–274, 1987.

[6] M. Hennessy and T. Regan. A Temporal Process Algebra. Technical Report 2/90, Univ. of Sussex, UK, April 1990.

[7] F. Jahanian and A. K. Mok. Modechart: A specification language for real-time systems. *IEEE Transactions on Software Engineering (to appear)*, November 1989. IBM Tech Report RC 15140.

[8] I. Lee, P. Brémond-Grégoire, and R. Gerber. A Process Algebraic Approach to the Specification and Analysis of Resource-Bound Real-Time Systems. *Proceedings of the IEEE*, pages 158–171, Jan 1994.

[9] I. Lee, H. Ben-Abdallah, and J.Y. Choi. A process algebraic method for the specification and analysis of real-time systems. In C. Heitmeyer and D. Mandrioli, editors, *Formal Methods for Real-Time Computing*, chapter 7. John Wiley & Sons, Chichester, January 1996.

[10] C. Lewerentz and T. Linder, editors. *Formal Development of Reactive Systems: Case Study Production Cell*, volume 891 of *Lecture Notes in Computer Science*. Springer-Verlag, 1995.

[11] F. Moller and C. Tofts. A Temporal Calculus of Communicating Systems. In *Proc. of CONCUR '90*, pages 401–415. LNCS 458, Springer Verlag, August 1990.

[12] X. Nicollin and J. Sifakis. The Algebra of Timed Processes ATP: Theory and Application. Technical Report RT-C26, Institut National Polytechnique De Grenoble, November 1991.

[13] W. Yi. CCS + Time = An Interleaving Model for Real Time Systems. In *Proc. of Int. Conf. on Automata, Languages and Programming*, July 1991.

[14] P. Zave and M. Jackson. Four dark corners of requirements engineering. To appear in *ACM Transactions on Software Engineering and Methodology*. January 1997.

Generating Code from Hierarchical State-Based Requirements

Mats P.E. Heimdahl

University of Minnesota, Institute of Technology

Department of Computer Science, 4-192 EE/CS Bldg.

Minneapolis, MN 55455

heimdahl@cs.umn.edu

David J. Keenan

Hughes Information Technology Systems

16800 E. Centretech Pkwy, Bldg. 485, M/S 5M-82

Denver, CO 80011-9046

dkeenan@redwood.dn.hac.com

1 Introduction

Computer software is playing an increasingly important role in safety-critical embedded computer systems, where incorrect operation of the software could lead to loss of life, substantial material or environmental damage, or large monetary losses. Although software is a powerful and flexible tool for industry, these very advantages have contributed to a corresponding increase in system complexity. It is becoming clear that the power software can bring to a system can also undermine the ability of the analyst to comprehend, and consequently control, the system's behavior.

In a previous investigation, the Irvine Safety Research Group, under the leadership of Dr. Nancy Leveson, developed a requirements specification language called the Requirements State Machine Language (RSML) suitable for the specification of safety-critical control embedded systems [14, 15]. To make RSML suitable as a requirements specification language usable by all stake holders in a specification effort, the syntax and semantics were developed with readability, understandability, and ease of use in mind. The usefulness of the language was demonstrated through the successful development of a requirements specification for a large commercial avionics system called TCAS II (Traffic alert and Collision Avoidance System II) [14, 15]. Furthermore, we have developed a collection of automated analysis procedures that check an RSML specification for desirable properties such as completeness, consistency, and determinism [10, 11].

However, even if a requirements specification is readable, understandable, and can be shown to be complete and consistent, designing and developing production quality code from such a black-box high-level specification can be a time consuming and error prone process.

To simplify and automate the design and implementation process, we have investigated the possibility of automatically generating code from RSML specifications. The semantics of RSML is relatively simple and is defined as a composition of simple mathematical functions defined by the state transitions in the model [11]. We have developed a tool that translates an RSML specification to executable code. The translation closely follows the formal semantics of RSML

and, thus, makes verification of the correctness of the generated code simple. For the translations where this straightforward approach generates obviously inefficient code, we have defined easily verifiable automated correctness preserving optimizations.

Our goal is to generate production quality code directly from RSML specifications with as little human intervention as possible. To determine if this is a realistic goal and to get some early feedback on our approach, we have applied our translation technique to several small sample RSML specifications, including a part of the TCAS II requirements specification. Initial results show that the generated code is approximately 5-10 times slower and about twice as large as highly optimized hand generated code. These results show that automatic code generation is feasible, at least in the domain for which RSML was developed, and that our easily verifiable translation is a promising start.

A program transformation approach provides a means of keeping a specification and its implementation consistent. When a change is needed, the change is made in the high-level specification rather than in the source code. The code is then simply regenerated from the specification. Furthermore, provided the transformations preserve correctness, the generated code is guaranteed to correctly implement the specification and satisfy all properties that have been proven about the specification.

Generating executable code from some higher level notation has been an active research area since the development of the first compiler. Although a transformational approach does not solve the fundamental problems of complex system development, it allows us to work in a notation suitable for the problem domain, in our case RSML and safety-critical embedded control systems, and in that way eliminate many accidental problems associated with obscure notation and manual coding.

Many systems that transform specifications into executable code have been developed over the years. The following overview is by no means an exhaustive description of such systems, but rather a brief summary of some of the main transformational approaches.

Early transformational systems, such as the TI (Transformational Implementation) system [1], CIP

210

(Computer-aided, Intuition-guided Programming) [2], and KIDS (Kestrel Interactive Development System) [16], were designed for use in general purpose programming. These types of systems are usually based upon wide spectrum languages that are iteratively transformed into lower level constructs. One of the major drawbacks to this approach is the need for the user to guide the transformational process and provide application-specific domain knowledge to the system. These systems can usually only perform a few simple transformations automatically, requiring the user to make decisions relating to complex data types and algorithms. The level of automation of such general-purpose systems varies, but it seems unlikely that a fully automatic transformational system can be built using this approach.

A different approach to code generation is taken by the Statemate system [6, 7, 12]. Statemate is a system for developing software for reactive systems that supports automatic generation of C, Ada, or VHDL code. A Statemate specification consists of three parts (1) a structural view defined by *module charts*, (2) a functional view defined by *activity charts*, and (3) a dynamic view defined by *Statecharts*.

The module charts are used to decompose a system into its main components. The functional decomposition of the components is captured using the activity charts. The modeling in Statemate is based on the stepwise refinement paradigm. The activity charts can be iteratively refined until atomic activities (or basic activities) can be described as simple transformations using a high-level programming language, for example, C or Ada. The activity charts fully capture the functionality of a system.

Statecharts are finite state machines augmented with hierarchy, parallelism, and modularity. In Statemate, the Statecharts are used to define the dynamic behavior of a system, that is, they determine when activities defined by the activity charts should be executed.

The transformation of a Statemate specification into code involves translating the activities to executable code. Activities with fully defined functionality will be directly translated to code, while activities with under-specified functionality will be translated into stubs that can be completed manually. The Statecharts are translated to a control program that determines when the code for representing the different activities will be invoked. The code generation is fully automated and results in what is considered to be prototype code.

The Statemate approach is similar to the RSML approach in some aspects. First, both approaches target embedded reactive systems and both are based on a state-based modeling approach. In fact, RSML is inspired by the Statecharts formalism and includes support for parallelism and hierarchical states.

Despite the similarities between Statecharts and RSML, the two modeling approaches are quite different. The main difference is that in RSML the full behavior of the system is captured using only the hier-

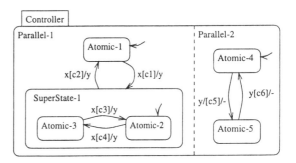

Figure 1: An example of a state machine.

archical state machine: RSML does not support activity charts and module charts. In the Statemate approach, the state machines are mainly used to control the scheduling of activities. Thus, the states reflect processing activity, that is, the states indicate which activities are currently executing. In RSML, on the other hand, the state machines are used to model the behavior of the physical components in a system: the states are used to visualize the system state.

During a large case study (TCAS) [3, 4, 15], we found that using the state machines to highlight the state of the system made it easy for the engineers (domain experts) such as avionics engineers, pilots, air frame manufacturers, and FAA representatives to understand and validate the requirements specification. The reason for this, we believe, is that an RSML specification is conceptually close to the application domain and becomes easier to validate. For a thorough discussion on this topic, the reader is referred to [15].

This difference in modeling approach leads to a different approach to code generation. While Statemate, for example, bases its code generation on stepwise refinement, we are generating all code directly from the state machines with fully automated transformations that can be proven correct.

The next section gives a brief description of the syntax and semantics of RSML. Sections 3 describes a provably correct mapping from RSML to executable code and Section 4 presents the results from our case study. Section 5 concludes.

2 The Semantics of RSML

RSML was developed as a requirements specification language for embedded systems. The language is based on hierarchical finite state machines and is in many ways similar to Statecharts by David Harel. For example, RSML supports parallelism, hierarchies, and guarded transitions borrowed from statecharts (Figure 1) [5, 8].

One of the main design goals of RSML was readability and understandability by non computer professionals such as, in our case, pilots, air frame manufacturers, and FAA representatives. During the TCAS project, we discovered that the guarding conditions re-

Transition(s): $\boxed{\text{ESL-4}} \longrightarrow \boxed{\text{ESL-2}}$

Location: Own-Aircraft ▷ Effective-SL$_{s\text{-}30}$

Trigger Event: Auto-SL-Evaluated-Event$_{e\text{-}279}$
Condition:

				OR	
Auto-SL$_{s\text{-}30}$ **in state** ASL-2		T	T	.
Auto-SL$_{s\text{-}30}$ **in one of** {ASL-2,ASL-3,ASL-4,ASL-5,ASL-6,ASL-7}		.	.	T
Lowest-Ground$_{f\text{-}241}$ = 2		.	.	T
Mode-Selector = **one of** {TA/RA,TA-Only,3,4,5,6,7}		T	.	T
Mode-Selector$_{v\text{-}34}$ = TA-Only		.	T	.

(*A N D* labels the rows on the left)

Output Action: Effective-SL-Evaluated-Event$_{e\text{-}279}$

Figure 2: A transition definition from TCAS II with the guarding condition expressed as an AND/OR table.

quired to accurately capture the requirements were often complex. The propositional logic notation traditionally used to define these conditions did not scale well to complex expressions and, thus, quickly became unreadable. To overcome this problem, we introduced a tabular notation for defining the guarding conditions (Figure 2). We call these tables AND/OR tables. The tables are read column-wise and were found to be very readable. To further increase the readability, we introduced many other syntactic conventions in RSML. For example, we allow expressions used in the predicates to be defined as mathematical functions (Other-Tracked-Relative-Alt-Rate$_{f\text{-}246}$), and familiar and frequently used conditions to be defined as macros (100-Ft-Crossing$_{m\text{-}195}$)[1]. A macro is simply a named AND/OR table defined elsewhere in the document. A detailed description of the full notation can be found in [15].

2.1 A Functional Framework

The behavior of a finite-state machine can be formally defined using a next-state relation. In RSML, this relation is modeled by transitions and the sequencing of events. Thus, one can view a graphical RSML specification as the definition of the mathematical next-state relation F.

Section 2.2 defines the static structure of the state hierarchies. Although we are using a different notation, the basic ideas for the formalization of the hierarchical structure of the state machines is borrowed from Harel *et al.* [9]. Section 2.3 describes how the dynamic behavior defined by the transitions and events in RSML

---

[1]The subscript is used to indicate the type of an identifier (f for functions, m for macros, and v for variables) and gives the page in the TCAS II requirements document where the identifier is defined.

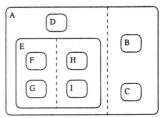

Figure 3: A sample state hierarchy

can be viewed as compositions of functions.

2.2 Hierarchical State Machines

An RSML state machine M can be described by a six-tuple: $M = (S, \leq, \sim, V, c_0, F)$ where:

S is a finite set of *states*.

$\leq$ is a tree-like partial ordering with a topmost point (called the *root*). This relation defines the hierarchy relation (or parent/child relation) on the states in S ($x \leq y$ meaning that x is a descendant of y, or x and y are equal).

In the graphical notation, this relation is visualized as containment (states are contained within superstates). In Figure 3, for example, B $\leq$ A, G $\leq$ A, I $\leq$ E, etc.

If the state x is a descendant of y ($x \leq y \wedge x \neq y$), denoted by $x < y$), and there is no z such that $x < z < y$, we say that the state x is a *child* of y (x *child* y).

Furthermore, we define $\sigma(y)$ as the set of all children of the state y, that is, $\sigma(y) = \{x \mid x \text{ child } y\}$

$\sim$ is an equivalence relation on the states in $S - \{root\}$ that satisfies one additional property: whenever

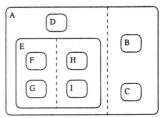

212

$x \sim y$, then x and y have the same parent.
$$x \sim y \Rightarrow \exists z \quad x, y \in \sigma(z)$$

The equivalence relation $\sim$ is used to partition the children of a state into disjoint sets called *parallel components*. In the graphical notation, this partition is indicated using dashed lines. In Figure 3, for example, the children of A are partitioned into two parallel components (equivalence classes) {B, C} and {D, E}.

V is a set containing the input and output histories of the model (the complete variable traces).

c_0 is the initial *global state* of the machine, $c_0 \in (Config \times V)$. A global state is an ordered pair consisting of a set of states, called the *configuration* of the machine, and a trace from V. The initial global state in Figure 3 is defined by the pair $(\{A, B, D\}, \emptyset)$.

F is a relation defining the global state changes in the machine M. F is a mapping $C \mapsto C$, where $C \subseteq (Config \times V)$. The relation F is also referred to as the *behavior* of M.

2.3 Next-State Mapping

The hierarchies and parallelism (defined by the relations $\leq$ and $\sim$), enforce a rigorous structure on the possible global states (the set C). The dynamic behavior (the possible global state changes) is defined by the next-state relation F ($C \mapsto C$). In a model of a system with nontrivial functionality this mapping will be complex. However, the mapping can be viewed as a composition of smaller, less complex mappings. Specifically, F can be viewed as composed of simple *functions*.

In the graphical notation, these simple functions are defined by transitions. The *domain* of a function is defined by the source, that is, the state that the tail of the transition is leaving, and the guarding condition on the transition. The *image* of a function is defined by the destination of a transition, that is, the state the transition enters, and possible changes to variables. The functions represented by the transitions are then composed depending on the structure of the particular state machine being considered and the events defined on the transition. The semantics of RSML are defined using three basic compositions:

Union: The union composition of two functions $(g \cup h)$ merges the domains of the functions.
Serial: Serial composition $g(h(c))$ (or $g \circ h$) corresponds to normal functional composition.
Parallel: Parallel application is denoted $\langle h, g \rangle(x)$. Parallelism is modeled as interleaving, that is, an arbitrary ordering of serial compositions. The notation $\langle X \rangle$, where X is a set of functions, will be used to denote the parallel composition of the functions in X.

In RSML, union composition occurs between non-parallel transitions triggered by the same event. For example, the functions representing the transitions t_1,

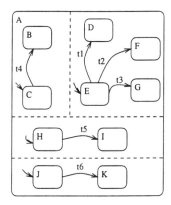

Figure 4: A sample state machine.

t_2, and t_3 in Figure 4 (assuming all are triggered by the same event) are composed in union.

Transitions triggered by the same event, but in parallel state components, are composed in parallel. In Figure 4, transitions t_3 and t_4 are composed in parallel (assuming they are triggered by the same event).

Finally, serial application is caused by the event propagation mechanism. Assume the transition t_5 is triggered by some external event and generates event e as an action. This event is picked up by transition t_6—that is, t_6 is triggered by e. Thus, transition t_5 is taken first and transition t_6 second. This sequencing is modeled as applying the functions representing t_5 and t_6 in series: $f_{t_6} \circ f_{t_5}(c)$.

A function f representing a transition can be textually defined by

$$f(c) = ((c.Conf - Q_s) \cup Q_d, \ v) \quad \text{if } (x \in c.Conf) \wedge p(c)$$

where Q_s and Q_d are sets of states, v is an updated variable trace, x is a state, and p is an arbitrary predicate over the global state c. The notation $c.Conf$ is used to refer to the *Configuration* part of the global state c. In the graphical representation, this function represents a transition starting in the state x and with the guarding condition p. If the transition is taken, the structure of the state machine may cause more states than x to be exited—for example, if x is a superstate. The set of states that is exited when the transition is taken is denoted by Q_s and the set that is entered by Q_d. Q_s and Q_d are the *source* ($Source(f)$) and *destination* ($Dest(f)$) of f respectively.

The functional definition of a complete RSML specification is recursively built from (composed of) the functional definitions of its components. To define this recursion we need to introduce some auxiliary concepts.

Let E be the set of all events in a model M. A transition is defined by a tuple $((C \mapsto C) \times E \times 2^E)$. The components of the tuple are denoted by *map*, *trigger*, and *actions* respectively. The *map* function is defined as outlined earlier in this section. Let T be the set of all transitions. Furthermore, let $Stages = S \cup \Pi$, where Π is the set of equivalence classes of $\sim$. Also, for

any $s \in S$, let $\pi(s)$ be the set of equivalence classes of children of s:
$$\pi(s) = \{x \in \Pi \mid x \subseteq \sigma(s)\}$$
For each $t \subseteq T$ we will define a function
$$g[t] : Stages \mapsto (C \mapsto C)$$
that defines the behavior of states and parallel components given a set of trigger events. The function g is defined by induction on $Stages$ over the relation $\ll$ defined as follows:
For all $s \in S, p \in \Pi$, and $st \in Stages$

$$s \ll st \quad \text{iff} \quad st \in \Pi \text{ and } s \in st \qquad (1)$$
$$p \ll st \quad \text{iff} \quad st \in S \text{ and } p \in \pi(st) \qquad (2)$$

Induction over $\ll$ is valid because it is well-founded: Whenever $s_1 \ll p \ll s_2$, it follows that $s_1 < s_2$. Therefore, $\ll$ does not contain an infinite descending chain.

The behavior of a composed state (a superstate) is defined as the parallel composition of its parallel state components. In Figure 4, for example, the behavior of the state A is defined as the parallel composition of its four parallel state components.

Definition 1 For any $p \in \Pi$ and $t \subseteq T$, the behavior (g) of a state $s \in S$:
$$g[t]_s = \langle \{g[t]_p \mid p \ll s\} \rangle$$

Informally, one can view the components of a composed state as processes, and the behavior of the composed state as the parallel execution of these processes.

The behavior of a set of states grouped in a parallel state component is defined as the union of (1) the behaviors of the states included in the component and (2) the behaviors introduced by the transitions between states at this level of abstraction. Let the notation $tr \sqsupset p$ (tr *belongs* to p) signify that a transition $tr \in T$ goes between two states in the parallel state component p. In Figure 4, the transition labeled with t_4 *belongs* to the parallel component $\{B, C\}$.

Definition 2 A transition tr belongs to the parallel component p of a state s, that is, $p \in \pi(s)$, (denoted by $tr \sqsupset p$) iff:
$$\exists x \in Source(tr.map) : x \in p$$

Definition 3 For any $s \in S$ and $t \subseteq T$, the behavior of a parallel state component $p \in \Pi$:

$$g[t]_p = \left(\bigcup_{m \ll p} g[t]_m \right) \cup \left(\bigcup_{tr \in t \wedge tr \sqsupset p} tr.map \right)$$

Informally, in a parallel state component, we will look for an applicable transition from the set of transitions between the states contained in this parallel component and among the transitions contained inside any of the states within the component.

Finally, the behavior of a model M under a specific event e can be defined. Let T_e be the set of all transition with the trigger $e \in E$, that is,
$$T_e = \{tr \in T \mid tr.trigger = e\}$$

Definition 4 The behavior of M under event $e \in E$ is defined as
$$F^e = g[T_e]_{root}$$

The rules defined above govern the behavior of M under one specific event, that is, all transitions in the model triggered by this one event are composed according to these rules. The behavior for all individual events in the model can now be modeled the same way. If an event e is generated, the function defined by the behavior under e (F^e), that is, the behavior generated by composing all transitions triggered by e, is applied, and a new system state is calculated. After a function has been applied and a new system state calculated, a new function is applied based on the output actions on the transitions used to construct the first function. We call the set of events generated as a result of output actions the *yield* of a next state calculation.

A sequence of function applications will follow. The next function is always determined by the yield of the previous function. Should the yield contain more than one event, the appropriate functions are composed in parallel. This sequence ultimately will stop when the next state calculation provides no yield.

3 Code Generation

The relative simplicity of the RSML semantics allows it to be translated into high-level programming constructs in a straightforward manner. In an attempt to make the generation of executable code easy to automate, easy to trace, and straightforward to verify, we have chosen to make a one-to-one mapping from the formal definition of the semantics to executable code. This translation provides a provably correct implementation of an RSML specification. However, such a direct approach introduces some inefficiencies in the generated code. Fortunately, these inefficiencies can be largely addressed through automated correctness preserving optimizations.

This section describes our translation approach and some optimizations are discussed in Section 4.

As mentioned in the previous section, the behavior of a single transition is defined by the function

$$f(c) = ((c.Conf - Q_s) \cup Q_d, v) \text{ if } (x \in c.Conf) \wedge p(c).$$

The behavior of this single transition can be trivially implemented by the following program:

```
P :   if
          ((x ∈ c.Conf) ∧ p(c)) → I
          ¬((x ∈ c.Conf) ∧ p(c)) → skip
      fi

I :   c.Conf = c.Conf − Q_s;
      c.Conf = c.Conf ∪ Q_d;
      v = modify(v);
```

```
bool ST()
{ if ((x ∈ c) and p){
    c = c - Q_s;
    c = c ∪ Q_d;
    v = modify(v);
    return TRUE; }
  return FALSE;
}
```

Figure 5: Pseudo-Code implementing a single transition.

```
void ge₁s()
{ ge₁p1();
  ge₁p2();
  ...
  ge₁pn();
}
```

Figure 6: Pseudo-Code implementing the behavior of a composed state under the event e_1.

The correctness of the above program is easily verifiable by a weakest precondition proof.

The program in Figure 5 is the equivalent (pseudo)code for the program P, which executes a single transition upon occurrence of the trigger event. The return value will be used later during code generation to determine whether or not the transition was actually taken.

In the following discussion of composed and parallel states, the functions ge_1s and ge_1p correspond to the functions $g[t]_s$ and $g[t]_p$ defined in the previous section. Here, these functions are viewed as responding to event e_1, rather than executing transitions from set T which have trigger e_1. Since the set of transitions can be determined at translation time there is no need to pass them as a parameter. The applicable transitions can simply be included directly into the code.

The behavior of a composed state is defined as the parallel execution of its parallel components, where parallelism is modeled as arbitrary interleaving. Thus, if (p1, p2, ..., pn) are the parallel components of state s that can respond to event e, then the program in Figure 6 is one possible implementation of the behavior of composed state s (compare with Definition 1).

Parallelism is deterministic only if the behavior of the system is equivalent for all possible serial orderings. If this is not the case, the behavior is nondeterministic, and this implementation simply executes one of the possible behaviors.

The rules governing the behavior of a parallel component (union composition) can be modeled as a sequential attempt to find a transition that can be taken,

```
void ge₁p()
{ if (ST1())
    return ;
  if (ST2())
    return ;
  ...
  if (STn())
    return ;
  ge₁s1();
  ge₁s2();
  ...
  ge₁sn();
}
```

Figure 7: Pseudo-Code implementing the behavior of a parallel component under the event e_1.

```
void execute(Event e)
{ switch (e) {
    case e₁ :  ge₁s_root(); break;
    case e₂ :  ge₂s_root(); break;
    ...
    case eₙ :  geₙs_root(); break; }
}
```

Figure 8: Pseudo-Code implementing the selection of function based on trigger event

that is, attempt to find a function that has the current state in its domain. By definition, a parallel component can successfully execute only one transition in response to a single event (Definition 3), and the program in Figure 7 enforces this behavior by immediately returning upon the successful completion of a single transition. (ST1, ST2, ... STn) are the corresponding functions for all transitions triggered by the event e_1 belonging to parallel component p, and $(ge_1s1, ge_1s2, ... ge_1sn)$ are the functions defining the behavior (under the event e_1) of the the states that are children of p. Here the return value of the single transition function (ST in Figure 5) is used to exit the function upon successful execution of a transition. It should be noted that should there be nondeterminism, this implementation gives precedence to transitions at the highest level in the state hierarchy, which is not a requirement of the language.

As discussed earlier, the behavior of an entire machine under a specific event e_1 is simply ge_1s_{root}. Thus, there is one ge_1s_{root} function for each possible event e_1, and a function that responds to an arbitrary event e need only select the proper function ges_{root} to call. The program in Figure 8 implements this behavior, where $(e_1, e_2, ..., e_n)$ are all events in the machine.

```
set⟨Event⟩ toProcess;
set⟨Event⟩ yield;
void input(Event e)
{ toProcess.insert(e);
  bool moreEvents = TRUE;
  while (moreEvents) {
    for (each event e in toProcess) {
      execute(e); }

    if (yield.empty())
      moreEvents = FALSE;
    else{
      toProcess = yield;
      yield = emptySet; } }
}
```

Figure 9: Pseudo-Code implementing the sequence of transitions caused by the event-action semantics.

```
bool ST()
{ if(x ∈ c)
    if (p){
```
$$c = c - Q_s;$$
$$c = c \cup Q_d;$$
$$v = modify(v);$$
$$yield.insert(actions);$$
```
      return TRUE; }
  return FALSE;
}
```

Figure 10: Modified code for a single transition

The event propagation (the sequence of function applications) can be modeled using two sets. One set (*toProcess*) to hold the events currently being evaluated and another set (*yield*) to collect the yield from the transitions taken. When all events in *toProcess* have been processed, the events in *yield* are copied to *toProcess* and the a new evaluation can start. This cycle is repeated until we have an empty yield (compare the discussion of yield in Section 2). A function implementing this behavior can be seen in Figure 9.

Note also that the function implementing the behavior of the individual transitions must include code to put their actions into the yield set. Thus, the revised function for a single transition can be seen in Figure 10.

The code described in this section corresponds one-to-one with the formal definition of RSML. This simple implementation makes proofs of consistency between specification and implementation trivial.

To get a fully executable system, the only missing pieces are the input and output interfaces to the environment. The input and output mechanisms are often highly system dependent and usually not modeled in detail in a high-level specification. We are currently investigating how RSML can be augmented with a communications model that will allow us to generate code for hardware independent input and output modules. To make a fully executable system, the hardware dependent code must be added separately. For now, however, we will just assume that modules exist for receiving messages from and sending messages to the environment. The input interfaces need simply receive a message from the environment, assign appropriate input variables, and pass the corresponding event to the input function to begin machine execution. Output interfaces are triggered by events generated while executing the machine. Thus, when an event is generated that is a trigger to an output interface, a call to the appropriate interface module can be made.

4 Case Studies

Two of the main impediments to widespread use of automatic code generation are code size and code efficiency: generated code is often unacceptably large and has poor run-time performance. The goal of our research is to develop a translation approach that will generate code of both acceptable size and acceptable performance. This is an ambitious goal. In many applications even using a compiler to translate from a high-level language such as C++ or Ada leads to unacceptably large and inefficient executables. Highly optimized hand coded assembly is still used in many time and space critical applications, for example, in the automotive industry [17]. Nevertheless, our goal is to be able to generate code that is efficient enough to be used in all but the most time and space critical applications.

To evaluate our approach and to get an indication of how automatically generated code compares to manually written code, we performed some small case studies.

We applied our technique to a subset of the TCAS II requirements specification [13, 15]. We translated a portion of the specification that determines the setting of a collection of aircraft parameters, for example, the effective sensitivity level (Effective-SL) that is a concept that determines how early to issue a collision warning to the pilot and Descend-Inhibit that determines if an aircraft is prohibited from descending (the state machine representing these parts of TCAS can be seen in Figure 11).

The pseudo-code outlined in the previous section can trivially be translated into executable code. In our case we chose C++, but any high-level language would be appropriate.

Consider the state machine for Descend-Inhibit in Figure 12 and the definition of the transition from the state Inhibited to the state Not-Inhibited (Figure 13). The C++ code corresponding to the state machine can be seen in Figure 14, and the code for the transition (or function) in Figure 13 is shown in Figure 15. The C++

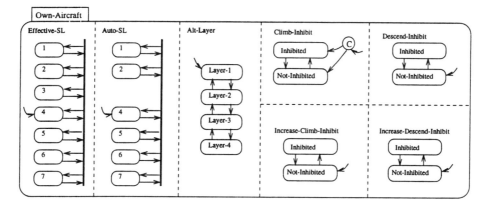

Figure 11: The state machine (Own-Aircraft) used in our case study.

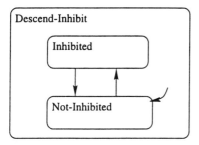

Figure 12: The state machine for Descend-Inhibit

code has a one-to-one correspondence to the pseudo-code in the previous section (compare the code with Figures 6 and 10).

We compared the efficiency and size of the automatically generated code with code derived from an English language and pseudo-code specification of the same system. The programs were compared on a set of test suites ranging from a few inputs to more than a million inputs.

While initial observations were somewhat disappointing (the generated code was approximately two times larger and approximately 30 times slower than the manually generated code), an analysis of the generated code uncovered several possibilities for automated optimizations which, when applied, significantly improved the efficiency of the generated code.

Optimizations: One of the first areas where the generated code demonstrates inefficiency is in the large number of function calls that must be made to find a transition that can be taken. The transitions triggered by an event are simply searched until an applicable transition is taken. A transition can only be taken, of course, if its source state is in the current configuration, and this test is done in the transition function (Figure 10). The first step in optimizing the code is thus to move this test for the source state's presence in the configuration into the calling function, the parallel component function (Figure 7). The resulting transi-

Transition(s): Inhibited → Not-Inhibited

Location: Own-Aircraft ▷ Descend-Inhibit$_{s\text{-}30}$

Trigger Event: Surveillance-Complete-Event$_{e\text{-}279}$
Condition:

(Own-Tracked-Alt − Ground-Level) < 12,00ft   T

Output Action: Descend-Inhibit-Evaluated-Event$_{e\text{-}279}$

Figure 13: The transitions out of Inhibited

```
void Descend_Inhibit()
{ if (Inhib_to_NotInhib()) return;
  if (NotInhib_to_Inhib()) return;
}
```

Figure 14: The C++ code for the state machine Descend-Inhibit

```
bool Inhib_to_NotInhib()
{ if (isInConfig(DI_INHIBITED))
    if (OTrackAlt_Minus_GL_GT_1200()){
      removeFromConfig(DI_INHIBITED);
      addToConfig(DI_NOT_INHIBITED);
      Yield.insert(DESC_IN_EVAL_EVENT);
      return true;}
  return false;
}
```

Figure 15: The C++ code for the transition from Inhibited to Not-Inhibited

```
bool ST()
{ if (p){
    c = c − Q_s;
    c = c ∪ Q_d;
    v = modify(v);
    yield.insert(actions);
    return TRUE; }
  return FALSE;
}
```

Figure 16: Code for single transition with check for source state removed.

```
void gep()
{ if ((x1 ∈ c)&& ST1())
    return ;
  if ((x2 ∈ c)&& ST2())
    return ;
  . . .
  if ((xn ∈ c)&& STn())
    return ;
  ges1();
  ges2();
  . . .
  gesn();
}
```

Figure 17: Code for parallel component including check for source state.

tion function and parallel component function can be seen in Figures 16 and 17.

This optimization allows the presence of a state in the configuration to be tested earlier and avoids a function call if the state is not in the configuration. The performance gain realized by this optimization is rather small, but it enables another optimization to be performed that will have a greater impact on the overall efficiency of the generated code.

The second optimization involves the repeated queries of the configuration to test for a state's presence. Looking at the modified code for a parallel component function (Figure 17), note that a query for the presence of a state in the configuration is made for each transition that is tested. If all queries could be reduced to one, a significant performance gain may be realized. This optimization was implemented by defining a function childActive(p) that returns the currently active child state of parallel component p. A parallel component can have only one active child state, and thus a single query of the configuration can determine which of the transitions may be applicable, that is, have their source states in the current configuration. The modified parallel component function can be seen in

```
void gep()
{ switch (childActive(p)) {
  case x1 :
    if (ST1()) return;
    ges1();
    break;
  case x2 :
    if (ST2()) return;
    ges2();
    break;
  . . .
  case xn :
    if (STn()) return;
    gesn();
    break; }
}
```

Figure 18: Further optimized code for parallel component.

Figure 18.

The optimized generated code provided significantly better results than the unoptimized code. The optimized code was still approximately twice as large as the hand-generated code, but execution time was now only five times slower than the highly optimized hand-generated code.

The approach has also been applied to two additional small RSML specifications. In these cases the code was approximately 8 and 13 times slower than highly optimized hand-generated code. The size of the automatically generated code was about 10% larger than the hand-generated code in each case.

To summarize, the basis of our approach–the one-to-one mapping from specification to code–generated easily verifiable but inefficient code. However, by applying some obvious, fully automated, correctness preserving optimizing transformations the efficiency of the code can be significantly increased. By taking advantage of additional optimizations, for example, using the lazy evaluation of Boolean expressions in C++ when evaluating large guarding conditions, we believe the efficiency of the generated code can be comparable to efficient hand written code.

5   Summary and Conclusion

Experience has shown RSML to be an excellent tool for the modeling of safety-critical embedded control systems. However, without a means of formally verifying consistency between specification and implementation, there is no guarantee that the final implementation matches the specification. Automatic code generation offers the guarantee of correctness, as well as a significantly simplified and shortened development cycle, allowing the developers to concentrate on correctly specifying the system and validate its behavior rather than trying to overcome obscure implementation details.

Several other systems that generate code from specifications require the user to extensively guide the translation process, thus reducing the benefits of code generation. Because of the relative simplicity of the formal semantics of RSML and RSML's focus on a particular problem domain, fully automatic, provably correct code generation is possible.

In this paper, we outlined our approach for code generation from an RSML requirements specification to an implementation in C++. Our approach is based on a one-to-one mapping between the formal semantics of RSML and an implementation in a high-level language, in our case C++. The simplicity of the translation makes it trivial to verify its correctness. However, such a straightforward translation introduces inefficiencies in the generated code. To overcome this problem we apply fully automated, correctness preserving optimizing transformations that eliminate obvious inefficiencies.

To evaluate the feasibility of the approach, we compared the efficiency and size of the generated code with highly optimized hand written code. In this small case study the generated code was approximately one order of magnitude slower and twice as large as hand generated code. We are currently evaluating additional optimizations and we are convinced that it is possible for the generated code to achieve comparable performance to hand written code.

In addition to further applications of optimizations to the generated code, other future investigations will include the formulation of strategies to generate code for the remaining parts of RSML including input and output interfaces and timing constructs. Furthermore, more rigorous case studies of the performance of generated code will be conducted, including its performance on machines of various sizes and structures.

Acknowledgment: We would like to thank Martin Feather of JPL for his valuable comments on an earlier version of this paper.

References

[1] R. Balzer. A fifteen year perspective on automatic programming. *IEEE Transactions on Software Engineering*, 11(11):1257–1267, November 1985.

[2] F. L. Bauer, B. Moller, M. Partsch, and P. Pepper. Formal program construction by transformations–computer-aided, intuition-guided programming. *IEEE Transactions on Software Engineering*, 15(2):165–180, February 1989.

[3] S. Gerhart, D. Craigen, and T. Ralston. Experience with formal methods in critical systems. *IEEE Software*, vol-11(1):21–39, January 1994.

[4] S. Gerhart, D. Craigen, and T. Ralston. Formal methods reality check: Industrial usage. *IEEE Transactions on Software Engineering*, 21(2):90–98, February 1995.

[5] D. Harel. Statecharts: A visual formalism for complex systems. *Science of Computer Programming*, 8:231–274, 1987.

[6] D. Harel, H. Lachover, A. Naamad, A. Pnueli, M. Politi, R. Sherman, A. Shtull-Trauring, and M. Trakhtenbrot. Statemate: A working environment for the development of complex reactive systems. *IEEE Transactions on Software Engineering*, 16(4), April 1990.

[7] D. Harel and A. Naamad. The STATEMATE semantics of Statecharts. Technical Report CS95-31, The Weizmann Institute of Science, October 1995.

[8] D. Harel and A. Pnueli. On the development of reactive systems. In K.R. Apt, editor, *Logics and Models of Concurrent Systems*, pages 477–498. Springer-Verlag, 1985.

[9] D. Harel, A. Pnueli, J.P. Schmidt, and R. Sherman. On the formal semantics of statecharts (extended abstract). In *2nd Symposium on Logic in Computer Science*, pages 54–64, Ithaca, NY, 1987.

[10] M. P.E. Heimdahl and N.G. Leveson. Completeness and Consistency Analysis of State-Based Requirements. Technical Report CPS-94-52, Michigan State University, October 1994. Accepted for publication in IEEE Transactions on Software Engineriring.

[11] M. P.E. Heimdahl and N.G. Leveson. Completeness and Consistency Analysis of State-Based Requirements. *IEEE Transactions on Software Engineering*, TSE-22(6):363–377, June 1996.

[12] i Logix. The languages of Statemate, March 1987.

[13] N. G. Leveson, M. Heimdahl, H. Hildreth, and J. Reese. TCAS II Requirements Specification.

[14] N. G. Leveson, M. P.E. Heimdahl, H. Hildreth, J. Reese, and R. Ortega. Experiences using Statecharts for a system requirements specification. In *Proceedings of the Sixth International Workshop on Software Specification and Design*, pages 31–41, 1991.

[15] N. G. Leveson, M. P.E. Heimdahl, H. Hildreth, and J. D. Reese. Requirements specification for process-control systems. *IEEE Transactions on Software Engineering*, 20(9), September 1994.

[16] D. R. Smith. Kids: A semiautomatic program development system. *IEEE Transactions on Software Engineering*, 16(9):1024–1043, September 1990.

[17] Interview with D. Bogden. Reliability in the real world of embedded automotive software.

Session 10B

Workshop

Organizer

Stephen Fickas
University of Oregon, USA

"Software on Demand: Issues for RE"

Workshop
Software on Demand: Issues for RE

Stephen Fickas
University of Oregon
fickas@cs.uoregon.edu

Introduction. Software on demand is a means of delivering software over the internet on an as-needed basis. The model is one where a user can download full apps or small plug-ins to complete the current task at hand. There are interesting features of this model:

- Throw away software. After it's used, software may be thrown away. The latest, most bug-free version can be retrieved again when needed. This opens up new payment schemes—software can be rented or even follow a pay-for-use approach.

- Niche marketing. It may be possible to develop software more tailored to individual needs.

- Software push/pull. The user may request new software (pull) or software vendors may attempt to sell the user on new versions (push).

A Strawman System. A prototype software on demand system, built by a team from the University of Oregon, will be used as a strawman for discussion. The prototype system was built with COTS in mind—an effort was made to use publically available software. The final system brought together the following pieces:

1. A base application. The example chosen was an information system for the University of Oregon campus. The base system, which could be given away free or at low cost, provided a map of campus and window panes for new components to work within. The expectation is that the base application would run on a user's home computer. The user would ask (demand) new components on an as-needed basis, e.g., if the user wanted information on computer labs on campus, she would ask for a specialized component to be added to the base app. The base application was written in Tcl/Tk for WIN95.

2. A component warehouse. Components were small pieces of functionality that could be added, on demand, to the base application. A general goal was that component addition would be without user intervention, e.g., no recompilation nor restart of the app would be necessary. The component warehouse was implemented as a database server on the Internet. Components were small Tcl/Tk programs.

3. A matchmaker service. This is the heart of the system. It is the piece that matches up a user's needs with the component that will meet those needs. Our approach was to use a modification of the reuse facets suggested by Prieto-Diaz. A user would select keywords from facet lists, and the matchmaker would index into the component warehouse for matches. The matchmaker was implemented as a web form and associated cgi scripts.

4. Online payment system. To close the loop, a means of paying component writers for their components was needed. The general goal was to transfer funds from user to component writer on download of a component. We chose to use First Virtual as the payment scheme.

We are able to pull all of these pieces together into a working system. We concluded that (1) the technology is largely in place to handle the "transport layer" (components, money) of a software on demand system, but (2) the question of matching user needs with available components is a wide open issue. Item (2) led to the organization of this workshop.

Some Workshop Questions. We list some questions that have been raised in our explorations of the software on demand idea.

1. How do end-consumers state their software requirements demands so that software can be found and brought to them? Such requirements will likely include a healthy dose of nonfunctional requirements having to do with price, guarantee and deliverability.

2. Does the software reuse community having anything to say in the software on demand area? We chose to use a reuse methodology for the prototype system. Should we continue to look to the reuse field for new results?

3. Can we eliminate humans from the RE process? Can software running on the user's machine detect the need for a new app or plug-in and write the requirements for it automatically?

4. How can software be organized on the net so that it can be found and downloaded on demand? How can vendors advertise the features

(or even existence) of their software? Is there a role for negotiation between user and vendor, maybe through intermediate brokers?

We expect other questions will surface at the workshop itself. In summary, this is a new area with little prior work to base judgements on. We expect a lively and open discussion.

Session 11A

Life-Cycle

Towards Modelling and Reasoning Support for Early-Phase Requirements Engineering
Eric S.K. Yu

"Early-phase" requirements analysis doesn't refer to the writing of the first words on the blank first page of a system specification, nor does it focus on the initial gathering of descriptions system functionality or behavior. Instead, the early phase explores the organizational environment in which the software system will operate, focusing on the actors in that environment and the meaningful dependencies that these actors share. These dependencies form the primary motivations for the functionality that will eventually be selected for the system, and also help to establish some of the important nonfunctional characteristics of the system. By examining the needs, goals, beliefs, and committments of the various actors in the system's environment, the motivations for system functionality and behavior can be better understood; that understanding should lead to a system that will fit well in its organizational environment and that will evolve effectively as the environment changes. This paper summarizes the i* (pronounced "eye star") framework, and introduces a new example that demonstrates how i* supports early-phase requirements gathering and analysis; it concludes by exploring some of the representation and reasoning support necessary to enable the capture and analysis of early phase requirements information.
— *Mark Feblowitz*

A Decision Making Methodology in Support of the Business Rules Lifecycle
Daniela Rosca, Sol Greenspan, Mark Feblowitz, and Chris Wild

The correct alignment of an operational system to the enterprise is an important consideration for managing change. This paper advocates an approach to RE centred on the explicit representation of business rules and introduces a methodology for the acquisition, deployment and evolution of such rules. Business rules, in the context of this paper, are considered as the link between the enterpise objectives and the way these objectives are realised in operational systems. The methodology presented makes use of a metamodel that integrates three views: enterprise modelling, rationale modelling and business rules modelling.
— *Periklis Loucopoulos*

A Logical Framework for Modeling and Reasoning about the Evolution of Requirements
Didar Zowghi and Ray Offen

The paper proposes a promising formal framework for modeling and reasoning about the evolution of requirements. The framework is based on default reasoning techniques developed in AI. Apart from presenting and discussing the framework, the paper describes a supporting tool and suggests how it can be used in order to manage changing requirements.
— *Eric Dubois*

Towards Modelling and Reasoning Support for Early-Phase Requirements Engineering

Eric S. K. Yu

Faculty of Information Studies, University of Toronto
Toronto, Ontario, Canada M5S 3G6

eric.yu@utoronto.ca

Abstract

Requirements are usually understood as stating what a system is supposed to do, as opposed to how it should do it. However, understanding the organizational context and rationales (the "Whys") that lead up to systems requirements can be just as important for the ongoing success of the system. Requirements modelling techniques can be used to help deal with the knowledge and reasoning needed in this earlier phase of requirements engineering. However, most existing requirements techniques are intended more for the later phase of requirements engineering, which focuses on completeness, consistency, and automated verification of requirements. In contrast, the early phase aims to model and analyze stakeholder interests and how they might be addressed, or compromised, by various system-and-environment alternatives. This paper argues, therefore, that a different kind of modelling and reasoning support is needed for the early phase. An outline of the i^* framework is given as an example of a step in this direction. Meeting scheduling is used as a domain example.

1   Introduction

Requirements engineering (RE) is receiving increasing attention as it is generally acknowledged that the early stages of the system development life cycle are crucial to the successful development and subsequent deployment and ongoing evolution of the system. As computer systems play increasingly important roles in organizations, it seems there is a need to pay more attention to the early stages of requirements engineering itself (e.g., [6]).

Much of requirements engineering research has taken as starting point the initial requirements statements, which express customer's wishes about what the system should do. Initial requirements are often ambiguous, incomplete, inconsistent, and usually expressed informally. Many requirements languages and frameworks have been proposed for helping make requirements precise, complete, and consistent (e.g., [4] [19] [13] [15]). Modelling techniques (from boxes-and-arrows diagrams to logical formalisms) with varying degrees of analytical support are offered to assist requirements engineers in these tasks. The objective, in these "late-phase" requirements engineering tasks, is to produce a requirements document to pass on ("downstream") to the developers, so that the resulting system would be adequately specified and constrained, often in a contractual setting.

Considerably less attention has been given to supporting the activities that *precede* the formulation of the initial requirements. These "early-phase" requirements engineering activities include those that consider how the intended system would meet organizational goals, why the system is needed, what alternatives might exist, what the implications of the alternatives are for various stakeholders, and how the stakeholders' interests and concerns might be addressed. The emphasis here is on understanding the "whys" that underlie system requirements [37], rather than on the precise and detailed specification of "what" the system should do.

This earlier phase of the requirements process can be just as important as that of refining initial requirements to a requirements specification, at least for the following reasons:

- System development involves a great many assumptions about the embedding environment and task domain. As discovered in empirical studies (e.g., [11]), poor understanding of the domain is a primary cause of project failure. To have a deep understanding about a domain, one needs to understand the interests and priorities and abilities of various actors and players, in addition to having a good grasp of the domain concepts and facts.

- Users need help in coming up with initial requirements in the first place. As technical systems increase in diversity and complexity, the number of technical alternatives and organizational configurations made possible by them constitute a vast space of options. A systematic framework is needed to help developers understand what users want and to help users understand what technical systems can do. Many systems that are technically sound have failed to address real needs (e.g., [21]).

- Systems personnel are increasingly expected to contribute to business process redesign. Instead of automating well-established business processes, systems are now viewed as "enablers" for innovative business solutions (e.g., [22]). More than ever before, requirements engineers need to relate systems to business and organizational objectives.

226

- Dealing with change is one of the major problems facing software engineering today. Having a representation of organizational issues and rationales in requirements models would allow software changes to be traced all the way to the originating source – the organizational changes that leads to requirements changes [18].

- Having well-organized bodies of organizational and strategic knowledge would allow such knowledge to be shared across domains at this high level, deepening understanding about relationships among domains. This would also facilitate the sharing and reuse of software (and other types of knowledge) across these domains.

- As more and more systems in organizations interconnect and interoperate, it is increasingly important to understand how systems cooperate (with each other and with human agents) to contribute to organizational goals. Early phase requirements models that deal with organizational goals and stakeholder interests cut across multiple systems and can provide a view of the cooperation among systems within an organizational context.

Support for early-phase RE. Early-phase RE activities have traditionally been done informally, and without much tool support. As the complexity of the problem domain increases, it is evident that tool support will be needed to leverage the efforts of the requirements engineer. A considerable body of knowledge would be built up during early-phase RE. This knowledge would be used to supporting reasoning about organizational objectives, system-and-environment alternatives, implications for stakeholders, etc. It is important to retain and maintain this body of knowledge in order to guide system development, and to deal with change throughout the system's life time.

A number of frameworks have been proposed to represent knowledge and to support reasoning in requirements engineering (e.g., [19] [13] [27] [17] [12] [5] [35]). However, these frameworks have not distinguished early-phase from late-phase RE. The question then is: Are there modelling and reasoning support needs that are especially relevant to early-phase RE? If there are specific needs, can these be met by adapting existing frameworks?

Most existing requirements techniques and frameworks are intended more for the later phase of requirements engineering, which focuses on completeness, consistency, and automated verification of requirements. In contrast, the early phase aims to model and analyze stakeholder interests and how they might be addressed, or compromised, by various system-and-environment alternatives. In this paper it is argued that, because early-phase RE activities have objectives and presuppositions that are different from those of the late phase, it would be appropriate to provide different modelling and reasoning support for the two phases. Nevertheless, a number of recently developed RE techniques, such as agent- and goal-oriented techniques (e.g., [16] [14] [17] [12] [7]) are relevant, and may be adapted for early-phase RE.

The recently proposed i^* framework [39] is used in this paper as an example to illustrate the kinds of modelling features and reasoning capabilities that might be appropriate for early-phase requirements engineering. It introduces an ontology and reasoning support features that are substantially different from those intended for late-phase RE (e.g., as developed in [15]).

Section 2 reviews the i^* framework and outlines some of its features, using meeting scheduling as a domain example. Section 3 discusses, in light of the experience of developing i^*, the modelling and support requirements for early-phase requirements engineering. Section 4 reviews related work. Section 5 draws some conclusions from the discussions and identifies future work.

2 The i^* modelling framework for early-phase requirements engineering

The i^* framework[1] was developed for modelling and reasoning about organizational environments and their information systems [39]. It consists of two main modelling components. The Strategic Dependency (SD) model is used to describe the dependency relationships among various actors in an organizational context. The Strategic Rationale (SR) model is used to describe stakeholder interests and concerns, and how they might be addressed by various configurations of systems and environments. The framework builds on a knowledge representation approach to information system development [27]. This section offers an overview of some of the features of i^*, using primarily a graphical representation. A more formal presentation of the framework appears in [39]. The i^* framework has also been applied to business process modelling and redesign [41] and to software process modelling [38].

The central concept in i^* is that of the intentional actor [36]. Organizational actors are viewed as having intentional properties such as goals, beliefs, abilities, and commitments. Actors depend on each other for goals to be achieved, tasks to be performed, and resources to be furnished. By depending on others, an actor may be able to achieve goals that are difficult or impossible to achieve on its own. On the other hand, an actor becomes vulnerable if the depended-on actors do not deliver. Actors are strategic in the sense that they are concerned about opportunities and vulnerabilities, and seek rearrangements of their environments that would better serve their interests. [2]

2.1 Modelling the embedding of systems in organizational environments – the Strategic Dependency model

Consider a computer-based meeting scheduler for supporting the setting up of meetings[3]. The requirements might state that for each meeting request, the meeting

---

[1]The name i^* refers to the notion of distributed intentionality which underlies the framework.

[2]An early version of the framework was presented in [36].

[3]The example used in this paper is a simplified version of the one provided in [34].

scheduler should try to determine a meeting date and location so that most of the intended participants will participate effectively. The system would find dates and locations that are as convenient as possible. The meeting initiator would ask all potential participants for information about their availability to meet during a date range, based on their personal agendas. This includes an exclusion set – dates on which a participant cannot attend the meeting, and a preference set – dates preferred by the participant for the meeting. The meeting scheduler comes up with a proposed date. The date must not be one of the exclusion dates, and should ideally belong to as many preference sets as possible. Participants would agree to a meeting date once an acceptable date has been found.

Many requirements engineering frameworks and techniques have been developed to help refine this kind of requirements statements to achieve better precision, completeness, and consistency. However, to develop systems that will truly meet the real needs of an organization, one often needs to have a deeper understanding of how the system is embedded in the organizational environment.

For example, the requirements engineer, before proceeding to refine the initial requirements, might do well to inquire:

- Why is it necessary to schedule meetings ahead of time?

- Why does the meeting initiator need to ask participants for exclusion dates and preferred dates?

- Why is a computer-based meeting scheduler desired? And whose interests does it serve?

- Is confirmation via the computer-based scheduler sufficient? If not, why not?

- Are important participants treated differently? If so, why?

Most requirements models are ill-equipped to help answer such questions. They tend to focus on the "what" rather than the "why". Having answers to these "why" questions are important not only to help develop successful systems in the first instance, but also to facilitate the development of cooperation with other systems (e.g., project management systems and other team coordination "groupware" for which meeting information may be relevant), as well as the ongoing evolution of these systems.

To provide a deeper level of understanding about how the proposed meeting scheduler might be embedded in the organizational environment, the Strategic Dependency model focuses on the *intentional* relationships among organizational actors. By noting the dependencies that actors have on one another, one can obtain a better understanding of the "whys".

Consider first the organizational configuration before the proposed system is introduced (Figure 1). The meeting initiator *depends* on meeting participants p to attend meeting m. If some participant does not attend the meeting, the meeting initiator may fail to achieve some goal (not made

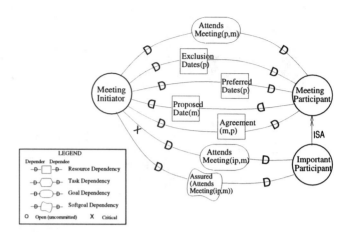

Figure 1: Strategic Dependency model for meeting scheduling, without computer-based scheduler

explicit in the SD model), or at least not succeed to the degree desired. This is the reason for wanting to schedule the meeting in advance. To schedule meetings, the initiator depends on participants to provide information about their availability – in terms of a set of exclusion dates and preferred dates. (For simplicity, we do not separately consider time of day or location.) To arrive at an agreeable date, participants depend on the initiator for date proposals. Once proposed, the initiator depends on participants to indicate whether they agree with the date. For important participants, the meeting initiator depends critically (marked with an "X" in the graphical notation) on their attendance, and thus also on their assurance that they will attend.

Dependency types are used to differentiate among the kinds of relationships between depender and dependee, involving different types of freedom and constraint. The meeting initiator's dependency on participant's attendance at the meeting (AttendsMeeting(p,m)) is a *goal dependency*. It is up to the participant how to attain that goal. An agreement on a proposed date Agreement(m,p) is modelled as a *resource dependency*. This means that the participant is expected only to give an agreement. If there is no agreement, it is the initiator who has to find other dates (do problem solving). For an important participant, the initiator critically depends on that participant's presence. The initiator wants the latter's attendance to be assured (Assured[AttendsMeeting(p,m)]). This is modelled as a *softgoal dependency*. It is up to the depender to decide what measures are enough for him to be assured, e.g., a telephone confirmation. These types of relationships cannot be expressed or distinguished in non-intentional models that are used in most existing requirements modelling frameworks.

Figure 2 shows an SD model of the meeting scheduling setting with a computer-based meeting scheduler. The meeting initiator delegates much of the work of meeting scheduling to the meeting scheduler. The initiator no longer needs to be bothered with collecting availability information from participants, or to obtain agreements about proposed dates from them. The meeting scheduler also determines what are the acceptable dates, given the avail-

228

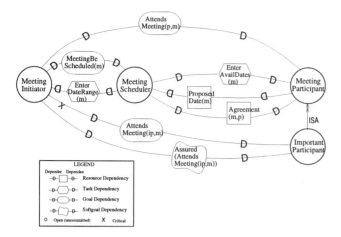

Figure 2: Strategic Dependency model for meeting scheduling with computer-based scheduler

ability information. The meeting initiator does not care how the scheduler does this, as longer as the acceptable dates are found. This is reflected in the *goal dependency* of MeetingBeScheduled from the initiator to the scheduler. The scheduler expects the meeting initiator to enter the date range by following a specific procedure. This is modelled via a *task dependency*.

Note that it is still the meeting initiator who *depends* on participants to attend the meeting. It is the meeting initiator (not the meeting scheduler) who has a stake in having participants attend the meeting. Assurance from important participants that they will attend the meeting is therefore not delegated to the scheduler, but retained as a dependency from meeting initiator to important participant.

The SD model models the meeting scheduling process in terms of intentional relationships among agents, instead of the flow of entities among activities. This allows analysis of opportunity and vulnerability. For example, the ability of a computer-based meeting scheduler to achieve the goal of MeetingBeScheduled represents an opportunity for the meeting initiator not to have to achieve this goal himself. On the other hand, the meeting initiator would become vulnerable to the failure of the meeting scheduler in achieving this goal.

2.2 Modelling stakeholder interests and rationales – the Strategic Rationale model

The Strategic Dependency model provides one level of abstraction for describing organizational environments and their embedded information systems. It shows external (but nevertheless intentional) relationships among actors, while hiding the intentional constructs within each actor. As illustrated in the preceding section, the SD model can be useful in helping understand organizational and systems configurations as they exist, or as proposed new configurations.

During early-phase RE, however, one would also like to have more explicit representation and reasoning about actors' interests, and how these interests might be addressed

or impacted by different system-and-environment configurations – existing or proposed.

In the i^* framework, the Strategic Rationale model provides a more detailed level of modelling by looking "inside" actors to model internal intentional relationships. Intentional elements (goals, tasks, resources, and softgoals) appear in the SR model not only as external dependencies, but also as internal elements linked by means-ends relationships and task-decompositions (Figure 3). The SR model in Figure 3 thus elaborates on the relationships between the meeting initiator and meeting participant as depicted in the SD model of Figure 1.

For example, for the meeting initiator, an internal goal is that of MeetingBeScheduled. This goal can be met (represented via a means-ends link) by scheduling meetings in a certain way, consisting of (represented via task-decomposition links): obtaining availability dates from participants, finding a suitable date (and time) slot, proposing a meeting date, and obtaining agreement from the participants.

These elements of the ScheduleMeeting task are represented as subgoals, subtasks, or resources depending on the type of freedom of choice as to how to accomplish them (analogous to the SD model). Thus FindSuitableSlot, being a subgoal, indicates that it can be achieved in different ways. On the other hand, ObtainAvailDates and ObtainAgreement refer to specific ways of accomplishing these tasks. Similarly, MeetingBeScheduled, being represented as a goal, indicates that the meeting initiator believes that there can be more than one way to achieve it (to be discussed in section 2.4, Figure 4).

MeetingBeScheduled is itself an element of the higher-level task of organizing a meeting. Other subgoals under that task might include equipment be ordered, or that reminders be sent (not shown). This task has two additional elements which specify that the organizing of meetings should be done quickly and not involve inordinate amounts of effort. These qualitative criteria are modelled as softgoals. These would be used to evaluate (and also to help identify) alternative means for achieving ends. In this example, we note that the existing way of scheduling meetings is viewed as contributing negatively towards the Quick and LowEffort softgoals.

On the side of the meeting participants, they are expected to do their part in arranging the meeting, and then to attend the meeting. For the participant, arranging the meeting consists primarily of arriving at an agreeable date. This requires them to supply availability information to the meeting initiator, and then to agree to the proposed dates. Participants want selected meeting times to be convenient, and want meeting arranging activities not to present too many interruptions.

The SR model thus provides a way of modelling stakeholder interests, and how they might be met, and the stakeholders evaluation of various alternatives with respect to their interests. *Task-decomposition links* provide a hierarchical description of intentional elements that make up a *routine*. The *means-ends links* in the SR provides understanding about *why* an actor would engage in some tasks,

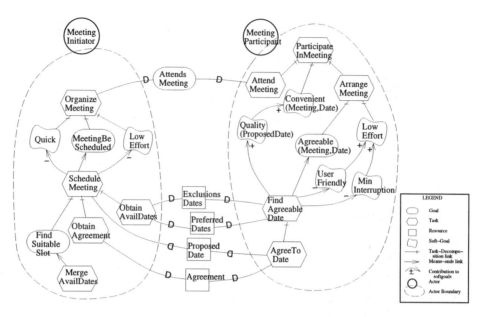

Figure 3: A Strategic Rationale model for meeting scheduling, before considering computer-based meeting scheduler

pursue a goal, need a resource, or want a softgoal. From the softgoals, one can tell *why* one alternative may be chosen over others. For example, availability information in the form of exclusion sets and preferred sets are collected so as to minimize the number of rounds and thus to minimize interruption to participants.

2.3 Supporting analysis during early-phase RE

While requirements analysis traditionally aims to identify and eliminate incompleteness, inconsistencies, and ambiguities in requirements specifications, the emphasis in early-phase RE is instead on helping stakeholders gain better understanding of the various possibilities for using information systems in their organization, and of the implications of different alternatives. The i^* models offer a number of levels of analysis, in terms of *ability, workability, viability* and *believability*. These are detailed in [39] and briefly outlined here.

When a meeting initiator has a routine to organize a meeting, we say that he is *able* to organize a meeting. An actor who is able to organize one type of meeting (say, a project group meeting) is not necessarily able to organize another type of meeting (e.g., the annual general meeting for the corporation). One needs to know what subtask, subgoals, resources are required, and what softgoals are pertinent.

Given a routine, one can analyze it for *workability* and *viability*. Organizing meeting is workable if there is a workable routine for doing so. To determine workability, one needs to look at the workability of each element – for example, that the meeting initiator can obtaining availability information from participants, can find agreeable dates, and can obtain agreements from participants. If the work-

ability of an element cannot be judged primitively by the actor, then it needs to be further reduced. If the subgoal FindSuitableSlot is not primitively workable, it will need to be elaborated in terms of a particular way for achieving it. For example, one possible means for achieving it is to do an intersection of the availability information from all participants. If this task is judged to be workable, then the FindSuitableSlot goal node would be workable. A task can be workable by way of external dependencies on other actors. The workability of ObtainAvailDates and ObtainAgreement are evaluated in terms of the workability of the *commitment* of meeting participants to provides availability information and agreement. A more detailed characterization of these concepts are given in [39].

A routine that is workable is not necessarily viable. Although computing intersection of time slots by hand is possible, it is slow and error-prone. Potentially good slots may be missed. When softgoals are not satisfied, the routine is not viable. Note that a routine which is not viable from one actor's perspective may be viable from another actor's perspective. For example, the existing way of arranging for meetings may be viable for participants, if the resulting meeting dates are convenient, and the meeting arrangement efforts do not involve too much interruption of work.

The assessment of workability and viability is based on many beliefs and assumptions. These can be provided as justifications for the assessment. The *believability* of the rationale network can be analyzed by checking the network of justifications for the beliefs. For example, the argument that "finding agreeable dates by merging available dates" is workable may be justified with the assertion that the meeting initiator has been doing it this way for years, and it works. The belief that meeting participants will supply availability information and agree to meeting dates may be

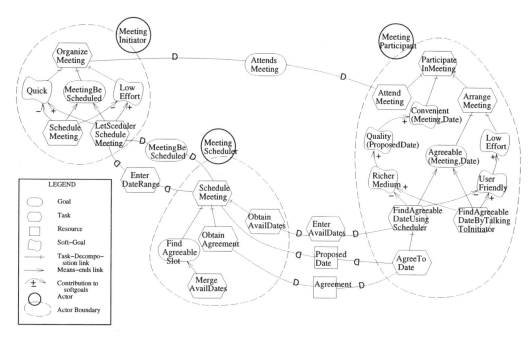

Figure 4: Strategic Rationale model for a computer-supported meeting scheduling configuration

justified by the belief that it is in their own interests to do so (e.g., programmers who want their code to pass a review). The evaluation of these goal graphs (or justification networks) is supported by graph propagation algorithms following a qualitative reasoning framework [8] [42].

2.4 Supporting design during early-phase RE

During early-phase RE, the requirements engineer assists stakeholders in identifying system-and-environment configurations that meet their needs. This is a process of design on a higher level than the design of the technical system per se. In analysis, alternatives are evaluated with respect to goals. In design, goals can be used to help generate potential solutions systematically.

In i^*, the SR model allows us to *raise* ability, workability, and viability as *issues* that need to be addressed. Using means-ends reasoning, these issues can be *addressed* systematically, resulting in new configurations that are then to be evaluated and compared. Means-ends rules that encode knowhow in the domain can be used to suggest possible alternatives. Issues and stakeholders that are *cross-impacted* may be discovered during this process, and can be raised so that trade-offs can be made. Issues are *settled* when they are deemed to adequately addressed by stakeholders. Once settled, one can then proceed from the descriptive model of the i^* framework to a prescriptive model that would serve as the requirements specification for systems development.[4] Believability can also be raised as an issue, so that assumptions would be justified.

In analyzing the SR model of Figure 3, it is found that the meeting initiator is dissatisfied with the

---

[4]One approach to this is described in [40].

amount of effort needed to schedule a meeting, and how quickly a meeting can be scheduled. These are raised as the issues Quick[MeetingScheduling] and LowEffort[MeetingScheduling].

Since the meeting initiator's existing routine for scheduling meetings is deemed unviable, one would need to look for new routines. This is done by raising the meeting initiator's ability to schedule meetings as an issue. To address this issue, one could try to come up with solutions without special assistance, or one could look up *rules* (in a knowledge base) that may be applicable. Suppose a rule is found whose *purpose* is MeetingBeScheduled and whose *how* attribute is LetSchedulerScheduleMeeting.

```
Class  CanLetSchedulerScheduleMeeting IN Rule WITH
  purpose
    ms:   MeetingBeScheduled
  how
    ssm: LetSchedulerScheduleMeeting
  applicabilityCond
    platform: HasAppropriatePlatform(team,
                          platform,scheduler)
END
```

This represents knowledge that the initiator has about software scheduler systems, their abilities, and their platform requirements. The rule helps discover that the meeting initiator can delegate the subgoal of meeting scheduling to the (computer-based) meeting scheduler. This constitutes a routine for the meeting initiator.

Using a meeting scheduler, however, requires participants to enter availability information in a particular format. This is modelled as a *task dependency* on participants (an SD link). A routine that provides for this is sought in the participant. Again, rules may be used to assist in this search.

When new configurations are proposed, they may bring in additional issues. The new alternatives may have associated softgoals. The discovery of these softgoals can also be assisted with means-ends rules. For example, using computer-based meeting scheduling may be discovered to be negative in terms of medium richness and user-friendliness. These in turn have implications for the effort involved for the participant, and the quality of the proposed dates. These newly raised issues also need to be addressed. Once new routines have been identified, they are analyzed for workability and viability. Further routines are searched for until workable and viable ones are found.

3 The modelling and reasoning support needs of early-phase RE

In the preceding section, the i^* framework was outlined in order to illustrate the kind of modelling and reasoning support that would be useful during the early phase of requirements engineering. This section summarizes and discusses these modelling and support needs in more general terms, drawing from the experience of the i^* framework.

Knowledge representation and reasoning. Although the example in the preceding section relies primarily on informal graphical notations, it is clear that a realistically-sized application domain would involve large numbers of concepts and relationships. A more formal knowledge representation scheme would be needed to support modelling, analysis, and design activities. Maintaining a knowledge base of the knowledge collected and used during early-phase RE is also crucial in order to reap benefits for supporting ongoing evolution (e.g., [8]), and for reuse across related domains.

Many of the knowledge-based techniques developed for other phases of software engineering are also applicable here. For example, knowledge structuring mechanisms such as classification, generalization, aggregation, and time [20] are equally relevant in early-phase as in late-phase RE. On the other hand, early-phase RE has certain needs that are quite distinct from late-phase RE.

Degree of formality. While representing knowledge formally has the advantage of amenability to computer-based tool support, the nature of the early-phase suggests that formality should be used judiciously. The early-phase RE process is likely to be a highly interactive one, with the stakeholders as the source of information as well as the decision maker. The requirements engineer acts primarily in a supporting role. The degree of formality for a support framework therefore needs to reflect this relationship. Use of knowledge representation can facilitate knowledge management and reasoning. However, one should not try to over formalize, as one may compromise the style of reasoning needed.

One approach is to introduce weaker constructs, such as softgoals, which requires judgemental inputs from time to time in the reasoning process, but which can be structured and managed nonetheless within the overall knowledge base [7] [39]. The notion of softgoal draws on the concept of satisficing [33], which refers to finding solutions that are "good enough".

Incorporating intentionality. One of the key needs in dealing with the subject matter in the early phase seems to be the incorporation of the concept of the intentional actor into the ontology. Without intentional concepts such as goals, one cannot easily deal with the "why" dimension in requirements.

A number of requirements engineering frameworks have introduced goal-oriented and agent-oriented techniques (e.g., [16] [14] [17] [12] [7]). In adapting these techniques for early-phase RE, one needs to recognize that the focus during the early phase is on *modelling* (i.e., describing) the intentionality of the stakeholders and players in the organizational environment. When new alternatives are being sought (the "design" component in early phase RE), it is the intentionality of the stakeholders that are being exercised. The requirements engineer is helping stakeholders find solutions to *their* problems. The decisions rests with the stakeholders.

In most goal-oriented frameworks in RE, the intentionality is assumed to be under the control of the requirements engineer. The requirements engineer manipulates the goals, and makes decisions on appropriate solutions to these goals. This may be appropriate for late-phase RE, but not for the early phase.

By the end of the early-phase, the stakeholders would have made the major decisions that affect their strategic interests. Requirements engineers and developers can then be given the responsibility to fill in the details and to realize the system.

One consequence of the early/late phase distinction is that intentionality is harder to extract and incorporate into a model in the early phase than in the late phase. Stakeholder interests and concerns are typically not readily accessible. The approach adopted in i^* is to introduce the notion of intentional dependencies to provide a level of abstraction that hides the internal intentional contents of an actor. The Strategic Dependency model provides a useful characterization of the relationships among actors that is at an intentional level (as opposed to non-intentional activities and flows), without requiring the modeller to know much about the actors' internal intentional dispositions. Only when one needs to reason about alternative configurations would one need to make explicit the goals and criteria for such deliberations (in the Strategic Rationale model). Even here, the model of internal intentionality is not assumed to be complete. The model typically contains only those concerns that are voiced by the stakeholders in order for them to achieve the changes they desire.

Multi-lateral intentional relationships. In modelling the embedding of a system in organizational environments, it is necessary to describe dependencies that the system has on its environment (human agents and possibly other systems), as well as the latter's dependencies on the system. When the system does not live up to the expectations of agents in its environment, the latter may fail to achieve certain goals. The reverse can also happen. During early-phase RE, one needs to reason about opportunities and vulnerabilities from both perspectives. Both the system and its environment are usually open to redesign, within limits. When opportunities or vulnerabilities are discovered, further changes can be introduced on either side to take advantage of them or to mitigate against them. A modelling framework for the early-phase thus needs to be able to express multi-lateral intentional relationships and to support reasoning about their consequences.

In most requirements frameworks, the requirements models are interpreted prescriptively. They state what a system is supposed to do. This is appropriate for late-phase RE. Requirements documents are often used in contractual settings – developers are obliged to design the systems in order to meet the specifications. Once the early-phase decisions have settled, a conversion from the multi-lateral dependency model to a unilateral prescriptive model for the late-phase can be made.

Distributed intentionality. Another distinctive feature of the early-phase subject matter is that the multiple actors in the domain all have their own intentionality. Actors exercise intentionality (e.g., they pursue goals) in the course of their daily routines. Actors have multiple, sometimes conflicting, sometimes complementary goals. The introduction of a computer system may make certain goals easier to achieve and others harder to achieve, thus perturbing the network of strategic dependencies. Different system-and-environment configurations can therefore be seen as different ways of re-distributing the pattern of intentionality[5]. The boundaries may shift (the responsibility for achieving certain goals may be delegated from some agents to other agents, some of which may be computer systems), but the actors remain intentional. The process of system-and-environment redesign does not solve all the problems (i.e., does not (completely) reduce intentional elements, such as goals, to non-intentional elements, such as actions). It merely rearranges the terrain in which problems appear and need to be addressed.

In contrast, in late-phase RE and in the rest of system development, one does attempt to fully reduce goals to implementable actions.

Means-ends reasoning. In order to model and support reasoning about "why", and to help come up with alternative solutions, some form of means-ends reasoning would appear necessary. However, a relatively weaker form of reasoning than customarily used in goal-oriented frameworks is needed. This is because of the higher degree of incompleteness in early phase RE. The emphasis is on modelling stakeholders' rationales. Alternative solutions may be put forth as suggestions, but it is the stakeholders who decide. The modelling may proceed both "upwards" and "downwards" (from means to ends or vice versa). There is no definitive "top" (since there may always be some higher goal) nor "bottom" (since there is no attempt to purge in-

tentionality entirely). It is the stakeholders' decision as to when the issues have been adequately explored and a sufficiently satisfactory solution found.

The type of reasoning support desired is therefore closer to those developed in issue-based information systems, argumentation frameworks, and design rationales (e.g., [10] [30] [26] [25]). The i^* approach is an adaptation of a framework developed for dealing with non-functional requirements [7], which draws on these earlier frameworks.

Organizational actors. In modelling organizational environments, a richer notion of actor is needed. i^* differentiates actors into agents, roles, and positions [39]. In late-phase requirements engineering, where the focus is on specifying behaviours rather than intentional relationships, such distinctions may not be as significant. Viewpoints has been recognized as an important topic in requirements engineering (e.g., [29]). In the early phase, the need to treat multiple viewpoints involving complex relationships among various types of actors is even more important.

4 Related work

In the requirements modelling area, the need to model the environment is well recognized (e.g., [4] [19] [23]). Organization and enterprise models have been developed in the areas of organizational computing (e.g., [1]) and enterprise integration (e.g., [9]). However, few of these models have considered the intentional, strategic aspects of actors. Their focus has primarily been on activities and entities rather than on goals and rationales (the "what" rather than the "why").

A number of requirements engineering frameworks have introduced concepts of agents or actors, and employ goal-oriented techniques. The framework of [5] uses multiple models to model actors, objectives, subject concepts and requirements separately, and is close in spirit to the i^* framework in many ways. The WinWin framework of [2] identifies stakeholder interests and links them to quality requirements. The notion of inquiry cycle in [31] is closely related to the early-phase RE notion, but takes a scenarios approach. The KAOS framework [12] [35] for requirements acquisition employs the notions of goals and agents, and provides a methodology for obtaining requirements specifications from global organizational goals.

However, these frameworks do not distinguish between the needs of early-phase vs. late-phase RE. For example, most of them assume a global perspective on goals, which are reduced, by requirements engineers, in a primarily top-down fashion, fully to actions. These may be contrasted with the notion of distributed intentionality in i^*, where agents are assumed to be strategic, whose intentionality are only partially revealed, who are concerned about opportunities and vulnerabilities, and who seek to advance or protect their strategic interests by restructuring intentional relationships.

---

[5] Hence the name i^*.

5 Conclusions

Understanding "why" has been considered an important part of requirements engineering since its early days [32]. Frameworks and techniques to explicitly support the modelling of and reasoning about agents' goals and rationales have recently been developed in RE. In this paper, it was argued that making a distinction between early-phase and late-phase RE could help clarify the ways in which these concepts and techniques could be applied to different RE activities.

The i^* framework was given as an example in which agent- and goal-oriented concepts and techniques were adapted to address some of the special needs of early-phase RE.

The proposal to use a modelling framework tailored specifically to early-phase RE and a separate framework for late-phase RE implies that a linkage between the two kinds of framework is needed [40]. As with other phases in the software development life cycle, the relationship between early and late phase RE is not strictly sequential or even temporal. Each phase generates and draw on a certain kind of knowledge, which needs to be maintained throughout the life cycle for maximum benefit [24] [20] [28]. The application of knowledge-based techniques to early-phase RE could potentially bring about a more systematic approach to this often *ad hoc*, under-supported phase of system development.

Preliminary assessments of the usefulness of i^* modelling in a real setting have been positive [3]. Supporting tools and usage methodologies are being developed in an on-going project [42].

Acknowledgments

The author gratefully acknowledges the many helpful suggestions from anonymous referees, Eric Dubois, Brian Nixon, and Lawrence Chung, as well as on-going guidance from John Mylopoulos, and financial support from the Information Technology Research Centre of Ontario, and the Natural Sciences and Engineering Research Council of Canada.

References

[1] A. J. C. Blythe, J. Chudge, J.E. Dobson and M.R. Strens, ORDIT: a new methodology to assist in the process of eliciting and modelling organizational requirements. *Proc. Conference on Organizational Computing Systems*, Milpitas CA, 1993. pp. 216–227.

[2] B. Boehm and H. In, Aids for Identifying Conflicts Among Quality Requirements, *IEEE Software*, March 1996.

[3] L. Briand, W. Melo, C. Seaman and V. Basili, Characterizing and Assessing a Large-Scale Software Maintenance Organization, *Proc. 17th Int. Conf. Software Engineering.* Seattle, WA. 1995.

[4] J. A. Bubenko, Information Modeling in the Context of System Development, *Proc. IFIP*, 1980, pp. 395-411.

[5] J. A. Bubenko, Extending the Scope of Information Modeling, *Proc. 4th Int. Workshop on the Deductive Approach to Information Systems and Databases,* Lloret-Costa Brava, Catalonia, Sept. 20-22, 1993, pp. 73-98.

[6] J. A. Bubenko, Challenges in Requirements Engineering, *Proc. 2nd IEEE Int. Symposium on Requirements Engineering,* York, England, March 1995, pp. 160-162.

[7] K. L. Chung, *Representing and Using Non-Functional Requirements for Information System Development: A Process-Oriented Approach*, Ph.D. Thesis, also Tech. Rpt. DKBS-TR-93-1, Dept. of Comp. Sci., Univ. of Toronto, June 1993.

[8] L. Chung, B. Nixon and E. Yu, Using Non-Functional Requirements to Systematically Support Change, *2nd IEEE Int. Symp. on Requirements Engineering (RE'95)*, York, England, March 1995.

[9] *CIMOSA – Open Systems Architecture for CIM*, ESPRIT Consortium AMICE, Springer-Verlag 1993.

[10] J. Conklin and M. L. Begeman, gIBIS: A Hypertext Tool for Explanatory Policy Discussions, *ACM Transactions on Office Information Systems*, 6(4), 1988, pp. 303-331.

[11] B. Curtis, H. Krasner and N. Iscoe, A Field Study of the Software Design Process for Large Systems, *Communications of the ACM*, 31(11), 1988, pp. 1268-1287.

[12] A. Dardenne, A. van Lamsweerde and S. Fickas, Goal-Directed Requirements Acquisition, *Science of Computer Programming*, 20, 1993, pp. 3-50.

[13] E. Dubois, J. Hagelstein, E. Lahou, F. Ponsaert and A. Rifaut, A Knowledge Representation Language for Requirements Engineering, *Proc. IEEE*, 74 (10), Oct. 1986, pp. 1431 –1444.

[14] E. Dubois, A Logic of Action for Supporting Goal-Oriented Elaborations of Requirements, *Proc. 5th International Workshop on Software Specification and Design*, Pittsburgh, PA, 1989, pp. 160-168.

[15] Ph. Du Bois, *The Albert II Language – On the Design and the Use of a Formal Specification Language for Requirements Analysis*, Ph.D. Thesis, Department of Computer Science, University of Namur, 1995.

[16] M. S. Feather, Language Support for the Specification and Development of Composite Systems, *ACM Trans. Prog. Lang. and Sys.* 9, 2, April 1987, pp. 198-234.

[17] S. Fickas and R. Helm, Knowledge Representation and Reasoning in the Design of Composite Systems, *IEEE Trans. Soft. Eng.*, 18, 6, June 1992, pp. 470-482.

[18] O.C.Z. Gotel and A.C.W. Finkelstein, An Analysis of the Requirements Traceability Problem, *Proc. IEEE Int. Conf. on Requirements Engineering*, Colorado Springs, April 1994, pp. 94-101.

[19] S. J. Greenspan, J. Mylopoulos, and A. Borgida, Capturing More World Knowledge in the Requirements Specification, *Proc. Int. Conf. on Software Eng.*, Tokyo, 1982.

[20] S. J. Greenspan, J. Mylopoulos and A. Borgida, On Formal Requirements Modeling Languages: RML Revisited, (invited plenary talk), *Proc. 16th Int. Conf. Software Engineering*, May 16-21 1994, Sorrento, Italy, pp. 135-147.

[21] J. Grudin, Why CSCW Applications Fail: Problems in the Design and Evaln of Organizational Interfaces, *Proc. Conference on Computer-Supported Cooperative Work* 1988, pp. 85-93.

[22] M. Hammer and J. Champy, *Reengineering the Corporation: A Manifesto for Business Revolution*, HarperBusiness, 1993.

[23] M. Jackson, *System Development*, Prentice-Hall, 1983.

[24] M. Jarke, J. Mylopoulos, J. W. Schmidt and Y. Vassiliou, DAIDA: An Environment for Evolving Information Systems, *ACM Trans. Information Systems*, vol. 10, no. 1, Jan 1992, pp. 1-50.

[25] J. Lee, *A Decision Rationale Management System: Capturing, Reusing, and Managing the Reasons for Decisions*, Ph.D. thesis, MIT, 1992.

[26] A. MacLean, R. Young, V. Bellotti and T. Moran, Questions, Options, and Criteria: Elements of Design Space Analysis, *Human-Computer Interaction*, vol. 6, 1991, pp. 201-250.

[27] J. Mylopoulos, A. Borgida, M. Jarke and M. Koubarakis, Telos: Representing Knowledge about Information Systems, *ACM Trans. Info. Sys.*, 8 (4), 1991.

[28] J. Mylopoulos, A. Borgida and E. Yu, Representing Software Engineering Knowledge, *Automated Software Engineering*, to appear.

[29] B. Nuseibeh, J. Kramer and A. Finkelstein, Expressing the Relationships Between Multiples Views in Requirements Specification, *Proc. 15th Int. Conf. on Software Engineering*, Baltimore, 1993, pp. 187-196.

[30] C. Potts and G. Bruns, Recording the Reasons for Design Decisions, *Proc. 10th Int. Conf. on Software Engineering*, 1988, pp. 418-427.

[31] C. Potts, K. Takahashi and A. Anton, Inquiry-Based Requirements Analysis, *IEEE Software*, March 1994, pp. 21-32.

[32] D. T. Ross and K. E. Shoman, Structured Analysis for Requirements Definition, *IEEE Trans. Soft. Eng.*, Vol. SE-3, No. 1, Jan. 1977.

[33] H. A. Simon, *The Sciences of the Artificial*, 2nd ed., Cambridge, MA: The MIT Press, 1981.

[34] A. Van Lamsweerde, R. Darimont and Ph. Massonet, The Meeting Scheduler Problem: Preliminary Definition. Copies may be obtained from Prof. Van Lamsweerde, Universite Catholique de Louvain, Unite d'Informatique, Place Sainte-Barbe, 2, B-1348 Louvain-la-Neuve, Belgium. (avl@info.ucl.ac.be)

[35] A. Van Lamsweerde, R. Darimont and Ph. Massonet, Goal-Directed Elaboration of Requirements for a Meeting Scheduler: Problems and Lessons Learnt, *Proceedings of 2nd IEEE Int. Symposium on Requirements Engineering*, York, England, March 1995, pp. 194-203.

[36] E. Yu, Modelling Organizations for Information Systems Requirements Engineering, *Proceedings of First IEEE Symposium on Requirements Engineering*, San Diego, Calif., January 1993, pp. 34-41.

[37] E. Yu and J. Mylopoulos, Understanding Why in Requirements Engineering – with an Example, *Workshop on System Requirements: Analysis, Management, and Exploitation*, Schloß Dagstuhl, Saarland, Germany, October 4–7, 1994.

[38] E. Yu and J. Mylopoulos, Understanding 'Why' in Software Process Modelling, Analysis, and Design, *Proc. 16th Int. Conf. on Software Engineering*, Sorrento, Italy, May 1994, pp. 159-168.

[39] E. Yu, *Modelling Strategic Relationships for Process Reengineering*, Ph.D. thesis, also Tech. Report DKBS-TR-94-6, Dept. of Computer Science, University of Toronto, 1995.

[40] E. Yu, P. Du Bois, E. Dubois and J. Mylopoulos, From Organization Models to System Requirements – A 'Cooperating Agents' Approach, *Proc. 3rd Int. Conf. on Cooperative Information Systems (CoopIS-95)*, Vienna, Austria, May 1995, pp. 194-204.

[41] E. Yu and J. Mylopoulos, From E-R to 'A-R' – Modelling Strategic Actor Relationships for Business Process Reengineering, *Int. Journal of Intelligent and Cooperative Information Systems*, vol. 4, no. 2 & 3, 1995, pp. 125-144.

[42] E. Yu, J. Mylopoulos and Y. Lesperance, AI Models for Business Process Reengineering, *IEEE Expert*, August 1996, pp. 16-23.

A Decision Making Methodology in Support of the Business Rules Lifecycle

Daniela Rosca*   Sol Greenspan†   Mark Feblowitz†   Chris Wild*

Abstract

The business rules that underlie an enterprise emerge as a new category of system requirements that represent decisions about how to run the business, and which are characterized by their business-orientation and their propensity for change. In this paper, we introduce a decision making methodology which addresses several aspects of the business rules lifecycle: acquisition, deployment and evolution. We describe a meta-model for representing business rules in terms of an enterprise model, and also a decision support submodel for reasoning about and deriving the rules. A technique for automatically extracting business rules from the decision structure is described and illustrated using business rules examples inspired by the London Ambulance Service case study. A system based on the metamodel has been implemented, including the extraction algorithm.

1 Introduction

Many requirements fall into a category known as *business rules*, which express computational requirements that determine or affect how a business is run. Business rules address how customers are to be treated, how resources are to be managed, or how special situations are to be handled in carrying out business processes. They represent decisions about how a business has decided to carry out its work to provide services to its customers. As such, they comprise an important set of requirements on any system being developed or procured for the enterprise. We prefer to view the combination of (legacy, COTS, and new) systems being employed to run the enterprise as a single operational system that is governed by business rules.

Business rules can be viewed as expressing functional and nonfunctional requirements which are, in principle, no different than other kinds of requirements. However, business rules are characterized by their strategic importance to the business and consequently deserve special consideration. They emphasize certain characteristics that imply how they should be dealt with.

For one thing, they are intended to be expressed and managed by business-level people and therefore need to be expressed in terms of an enterprise-level model [9, 1], not in terms of programs and databases. For another, the appropriate notion of "meeting the

requirements" is, for business rules, a rich and complex one, involving satisficing, tradeoffs, exceptions, or other coping strategies. The "implementation" (we prefer the term "deployment") of a business rule is not so straightforward, certainly not just the usual meaning of a system meeting its specifications.

The most demanding characteristic of business rules is that the business is likely to want to change them. Business rules are formulated to have planned consequences that contribute to the success of the business, and therefore it is expected that the rules will be continually re-evaluated and subject to change to improve the performance of the business. Business rules can also be generated by or in response to external sources, such as regulatory bodies, the law, market forces, or physical realities. Since the business does not have control over these external forces, the enterprise must be prepared to change the business rules that govern its operation.

When business rules change, the operational system needs to change, which can require major investment for maintenance and evolution. This is a source of the infamous "legacy" problem, which presents high risk not only due to costs but also due to difficulty of meeting time deadlines. Of course, this problem of adapting systems to changing requirements is a classic and pervasive one, but for business rules the problem is hopefully more focussed and more tractable than the general case.

The paper presents aspects of a methodology that was developed to cope with these issues by prescribing an approach for managing business rules over their lifecycle (i.e., over the lifecycle of the enterprise and its operational system).

The methodology relies on a conceptual modeling framework or metamodel, presented in Section 3, which prescribes representations for the enterprise model, the business rules, and a decision space. The needed support for requirements analysis is then defined in terms of relationships between the submodels.

The methodology consists of three overall activities, presented in Section 4, which are motivated by the issues raised above. These are:

• **Business rules acquisition** Facilitating expression and management of business rules at an enterprise level.

• **Business rules deployment** Getting the system to follow the rules.

• **Business rules evolution** Facilitating change without major re-development effort.

Acquisition includes not only acquiring the business

*Department of Computer Science, Old Dominion University, Norfolk, VA 23529, {rosca,wild}@cs.odu.edu

†GTE Laboratories Incorporated, 40 Sylvan Road, Waltham, MA 02254, {greenspan,feblowitz}@gte.com

rules and the enterprise model in terms of which they are stated, but also capturing the deliberation process that arrived at the rules. Deployment includes the resolution of issues such as determinism, conflicts, and ambiguity due to ungrounded terms, and also an evaluation of how well the business rules are expected to support enterprise goals.

Acquisition includes, in addition to populating the business rules environment, the automatic extraction of deployable business rules from the decision space. The *business rules extraction* method, in Section 6, is based on an induction algorithm that uses knowledge from the decision support system and statistical data about domain assumptions. Section 5 introduces the example (the London Ambulance Service) used to illustrate the method.

A system has been implemented for acquisition of models according to the framework described in this paper. While the particulars of this system are not the subject of this paper, it should be noted that this system has been populated with the models and data presented in this paper, and the extraction algorithm has been implemented. The results presented below were produced by running the system on the data.

More background on business rules is given in Section 2. There the reader will find more explanation of business rules, additional motivation for the approach taken by the methodology, and insights gained from studying a sampling of a few hundred business rules associated with three years of business process re-engineering projects.

2 Business Rules Perspective

2.1 Requirements Analysis for Business Rules

Business rules are requirements that arise from the business objectives of an enterprise. A bank requires a supervisor's signature before cashing a check over $5000 to minimize loss in case of fraud. A customer service organization, trying to resolve customer complaints by telephone, tries to avoid transferring the customer's phone call more than once and hopefully not at all (this is called the "two-touch" rule, and is intended to minimize customer inconvenience). A travel agent has been contracted to provide a client corporation with "the lowest possible airfares" which helps the client corporation achieve financial goals. A company requires all departments to remain within 10% of budget, provided the total company budget is met within 1%, also to meet financial goals. In general, business rules are statements about the enterprise's way of doing business. They reflect policies, procedures or other constraints on ways to satisfy customers, make good use of resources, conform to laws or business conventions, and the like.

Business rules are presumed to be of strategic/tactical importance to the success of the business. This means, among other things, that they are to be formulated and analyzed by business-oriented people, not by information technologists or software engineers. Decisions about cost, quality, responsibility, and good service are not supposed to be delegated to

system designers or programmers, but are to be implemented/deployed after they are decided upon by the appropriate parties. In reality, it is frequently the case that the only way to determine an organization's business rules is to look at the operational system; how does the Customer Service Rep schedule a repair visit? How does the software calculate the insurance reimbursement? You know there is a problem if you have to "look at the code," i.e., when there is no other explicit statement of what the business rules are. This has motivated discussion of "mining" business rules [11] through a kind of reverse engineering activity on the software. However, this can be from difficult to impossible, since the manifestation of the rules might be scattered all over the code, and since the belief that there were conscious business rules in the first place might be flawed.

Figure 1: Setting Business Rules in context

As shown in Figure 1, *business rules* are positioned between *enterprise objectives* and the *operational system* of the enterprise. The operational system is the combination of the hardware, software and humans that carry out the work of the enterprise. Enterprise objectives are the general aspirations of the enterprise: "make (more) money," "make customers happy," and "keep shareholders satisfied with their investment," etc. They are the ultimate source of all requirements on the operational system. The enterprise objectives can be decomposed and refined, eventually to the point where they can be translated into *business rules* that can achieve the goals. These business rules are intended to be carried out by the operational system, with the intent that the "deployed" or "operationalized" rules will realize the enterprise objectives.

Because of the ambiguities and conflicts inherent in the enterprise objectives, business rules are not simple refinements of the objectives; rather, business rules represent decisions that are made about how to achieve the enterprise objectives. Business rules become requirements that govern the operational system of the enterprise and determine how the business is actually run. To answer the central question, "does the operational system satisfy the business objectives (and to what degree)?", one must simultaneously address two subquestions, which are subjects for requirements analysis involving business rules:

- Decision-making: Do the business rules embody the right (or best) decisions about how to achieve the objectives of the business?
- Operationalization: Does the operational system satisfy the requirements represented in the business rules?

Decision-making requires one to choose among alternative business rules for achieving objectives, considering the expected influences or consequences - both in support of and in denial of the objectives, based on various assumptions. As described earlier in [14], we propose a decision support system/structure (DSS) for capturing the reasoning associated with business rules. (The DSS of [14] has since then been refined and implemented.) In addition to supporting the reasoning for choosing business rules, the DSS supports *business rules extraction*, presented below, which produces a business rule from the information in the DSS.

Operationalization of business rules requires a strategy (built into the enterprise infrastructure) and tactics (methods such as transformations) to deploy the business rules in the operational system. For each type of business rule in the taxonomy, an approach is needed for deployment, preferably with minimal time and expense. Some classes of business rule are directly deployable, while others require varying degrees of transformation or implementation; still others have no direct path into operation, and must be approximated in other ways. Deployment of business rules is an important part of the methodology, but further explanation is beyond the scope of this paper.

If the analysis is correct, i.e., if the business rules decisions and operationalizations are both done without error and with total insight, then the operational system should satisfy the enterprise objectives. This seems to be the presumption usually made in requirements engineering (RE), i.e. that the requirements are valid, and that if they are valid and the implementation is correct, then the overall objectives have been met. However, this picture is too simplistic for the real world, and we need to step outside of this picture to see what else is needed. In the first place, it cannot be assumed that enterprise objectives are devoid of conflicts; a set of rules that supports conflicting goals cannot be implemented without a hitch. The implementation will have to consider tradeoffs, not all of which have been identified by the time the system is in operation. The usual RE presumption is that all of the goal conflicts have been worked out as a prerequisite to system design. We posit that this will never be the case, at least for service-providing enterprises with which we are concerned.

Furthermore, the reasoning upon which the business rules are decided is always based on assumptions about what will influence what, and what will be the consequences. These assumptions could be wrong. Even if all of the assumptions were correct and all of the relevant business rules were followed (i.e., correctly deployed), there might be further reasons why the operational system does not meet some of the business objectives; there could be missing business rules or something missing from the reasoning process. And all of this reasoning and decision making rarely occurs in the context of a fully stable environment; objectives change, as do business conditions, and strategic and tactical decisions.

It should be recognized that, just as the system can never really completely meet its requirements, the state of the reasoning about the system will also never be complete, totally correct or consistent. There should be a constant attempt to revise the business rules and their operationalizations, based on observations of the operational system and its achievement of the objectives. To this end, the methodology prescribes instrumenting the system with monitoring to check achievement of enterprise objectives, evaluating the validity of the assumptions underlying the business rules.

The saying "Rules were made to be broken" applies here. Rules will indeed be broken, either because the rules conflict with other rules, or the implementation is not complete, or the data needed is not available in some situation.

2.2 A Study of Business Rules and Other Related Work

The motivation for our work on business rules came from three years of requirements efforts associated with business process re-engineering projects. On these projects, information referred to as business rules was considered important to gather and document, but there was no particular method for doing so. There were requirements modeling methods and tools being used, but business rules were not part of the modeling framework. They had to be added as annotations to business process models or just captured separately in notebooks.

We studied a sampling, from one project, of some 250 statements referred to as business rules to try to understand better what kinds of information were being captured. An attempt to classify them produced the taxonomy shown in Figure 2. This taxonomy is not claimed to have any scientific validity but only indicates what kinds of business rules were present in this particular application.

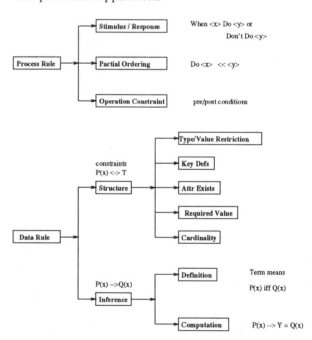

Figure 2: A Taxonomy of Business Rules

The taxonomy is divided into rules for processes and rules for data. A more detailed explanation and analysis of the taxonomy is beyond the scope of this paper, and we assume that the rule categories shown are sufficiently self-explanatory to the reader. The classification is quite close to the classification made in [10] and to that used in our implementation. In this sense, there seems to be some agreement on what types of business rules are to be encountered in everyday life.

However, a high percentage of the rules were extremely simple (attribute value constraints and cardinality constraints). These rules seemed too low-level to have an important impact on the business. We suspect that these types of rules dominated because of a reliance on an Entity-Relationship style of thinking. Process-oriented rules were less popular, but among them the stimulus response rules were the most numerous. These are also fairly simple rules, with a clear operational semantics. We suspect that the other kinds of rules are just as important but are more difficult for business-oriented people to specify and reason about. There are also other kinds of rules, not present in the taxonomy, that would be useful to have.

In our survey of the literature and trade press, we have found descriptions of business rules that were similarly narrow, tied to what can be achieved with a specific model construct or implementation mechanism. For example, just as in our study, cardinality constraints in Entity-Relationship models are commonly referred to as business rules, despite their extremely limited expressiveness. Other discussions of business rules talk about integrity constraints and trigger rules, which are expressible and enforceable with current database systems. These are good examples of common rule types with direct deployment strategies; the same needs to be done for the other rule types.

Business rules have received a lot of attention in the trade press and other literature as holding the answers to many information technology problems. Some of the literature simply advocates the idea that it is important to spend the effort to give conscious consideration to business rules (as opposed to not doing it at all). Even just gather a list of English language statements [16] is considered a good first step. Other proposals are more structured, relying on the syntax of E-R diagrams, or in the case of [13], a much more elaborate diagrammatic syntax. Still others propose a conceptual model approach for describing an enterprise, with a notation for specifying rules that further specify the requirements on the enterprise [9, 1]. This is closest to our approach, although our current enterprise model is limited to process and data/object rules, as in [10].

Some of the elements of our methodology have been discussed to some extent in the Requirements Engineering literature. For example, in [2] nonfunctional requirements are treated as goals that have to be met through a decision-making process in which change is expected. In [8], a goal-directed requirements elaboration methodology attempts to cope with the "deidealization of unachievable goals" and also models assumptions attached to goals. In [4], there is a discussion of requirements monitoring to instrument the running system to determine whether, and to what degree, requirements are being met by the system. These are examples of the work we are trying to bring to bear on business rules.

3   A Metamodel for the Business Rules Environment

The methodology supporting the business rules lifecycle consists of
- a modeling framework (metamodel),
- a prescription of activities for populating the models, and
- techniques for using the information for requirements analysis, including continuous lifecycle support.

This section describes the metamodel, in preparation for describing the methodology activities in the next section. The metamodel described here has been implemented in our experimental requirements modeling/analysis environment for supporting the methodology. The environment consists of three submodels: the *Enterprise Model*, the *Business Rules*, and the *Decision Space*. The *Enterprise Model* represents the world to which the business rules apply. It defines the domain concepts about which the rules are expressed. The *Business Rules* submodel represents the business rules themselves. The *Decision Space* submodel offers information about the enterprise objectives that comprise the origin of business rules and captures the reasoning leading to the selection and ultimate generation of the business rules. (see Figure 3).

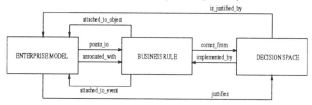

Figure 3: The Business Rules Environment

3.1   Enterprise Model

For the representation of the *Enterprise Model* we have chosen the paradigm of the LiveModel modeling environment [7]. In LiveModel, an enterprise is represented in terms of "objects" and "processes" (see [10]). Objects are represented by a set of Object Diagrams that are essentially Entity-Relationship diagrams (the Object diagram for a fragment of the LAS example can be seen in Figure 4). The business processes are represented by a set of Event Diagrams which define the sequence of operations for process execution (the corresponding Event Diagram for a fragment of the LAS example can be seen in Figure 5.) These Event Diagrams model a hierarchy of business processes, decomposing each operation in a diagram, if necessary, into a more detailed diagram. The Event Diagrams are executable specifications of a process as soon as: 1) input and output variables to operations are specified; 2) trigger rules are created to define branching and control conditions; 3) procedures to define operations are written. LiveModel allows the attachment of rules to event diagrams, with a particular operational semantics based on those of the Object and Process diagrams.

3.2   Decision Space Submodel

The Decision Space submodel is shown in figure 6. It represents the primitives of an issue-based decision

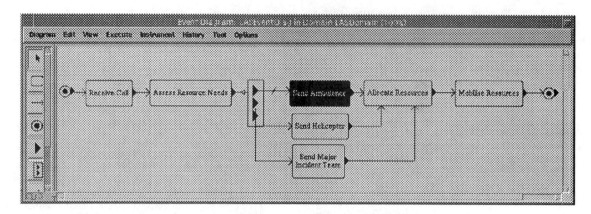

Figure 5: An Event Diagram for a fragment of the LAS example

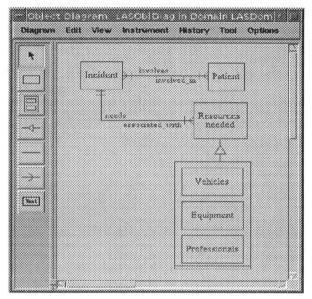

Figure 4: An Object Diagram for a fragment of the LAS example

support model. We can interpret our decision support model as follows. Both functional and non-functional requirements generate *issues* that need to be solved. These issues are refined during the deliberation process. In order to solve an issue different alternative solutions are considered for evaluation. The *alternatives* are evaluated against a set of *criteria* in order to decide which gives the best solution. A decision involves assessing the degree to which each alternative meets the entire set of criteria and choosing that alternate which best satisfies this set. *Arguments* and counterarguments based on various *assumptions* are recorded to document the evaluation of the alternatives or the creation of new issues that may follow after making a decision. The best alternative solution is reflected in the resulting artifact, which in our case is represented by *DSS Business Rules*, a set of business rules in decision support system (DSS) format. All of the information content of the above primitives can be retrieved from the *decision matrix* associated with a specific is-

sue. A more detailed description of this model and related work on decision support structures is given in [14]. Here we show an augmented model with the links to other submodels of the business rules environment: the *Enterprise Model* and the *Business Rules*.

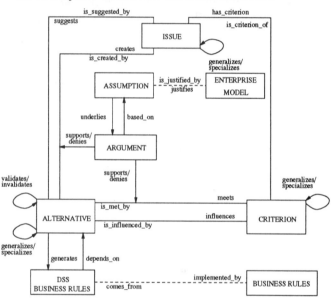

Figure 6: Decision Space Submodel

3.3 Business Rules Submodel

Business rules take the form of event-condition-action (ECA) rules, which we adopted from [6]. The ontology of the Enterprise Model, when examined at a more detailed level, contains events, conditions and actions. Whether a process/object enterprise model is used, as in this paper, or the extended SOS model described in [5] is used, ECA rules provide a convenient "assembly language" into which most kinds of rules can be translated. Since the ECA rules have a well-defined operational semantics, it has been straightforward to build an interpreter for them.

For the simplest form of an ECA rule

    WHEN *event*
    IF *condition*
    DO *action*

when the event occurs, if at that time the condition is found to hold, then the action is initiated.

The events, conditions, and actions are formulated as expressions on the objects in the Enterprise Model.

ECA rules are more generally applicable than they might first appear. As discussed in [1], where similar rules are used, judicious interpretations of special cases (such as default meanings for omitting one of the components of the rule) allow ECA rules to express several of the business rules types in the taxonomy. Additionally, an ECA rule can be used to express non-operational semantics, such as the situated enterprise objectives expressed at the *criteria level* in section 6.2.

3.4 Intermodel Relationships

Requirements analysis can be done by analyzing interrelationships between the submodels. Based on the links between these submodels the following types of analysis can be performed:

Business Rules ⟶ Enterprise Model:

Which process component(s) does a business rule define/constrain/govern? Which (event/action) operations operationalize a business rule? What object types are referred by a business rule? This information can be used for an impact/sensitivity analysis when a rule changes.

Enterprise Model ⟶ Business Rules:

In which business rules does a specific object type participate? This information can be used for impact/sensitivity analysis when the status of an object changes.

What business rules define/constrain/govern a specific process component? This information can be used for business processes improvement.

Business Rules ⟶ Decision Space:

Where does the rule come from? This links a business rule to the issue that has generated it. Thus one can have a comprehensive picture of the business rule rationale by looking at the alternatives, criteria, arguments and assumptions that have been stated during the deliberation of that business rule.

Decision Space ⟶ Business Rules:

What business rules address a specific issue? This information allows an impact/sensitivity analysis when factors like Government regulations, company policies, etc. change. It is also a useful source of information for a reuse process.

Decision Space ⟶ Enterprise Model:

What object types/attributes are addressed by a decision/issue? This can be useful for an impact/sensitivity analysis when a decision is changed.

Enterprise Model ⟶ Decision Space:

What decisions/issues involve this object or attribute? What decisions are affected when an object changes?

4 Methodology

The methodology we propose spans all phases of the business rules lifecycle: acquisition, deployment, change in response to changes in internal or external influences and change based on evaluation of the degree of requirements satisfaction.

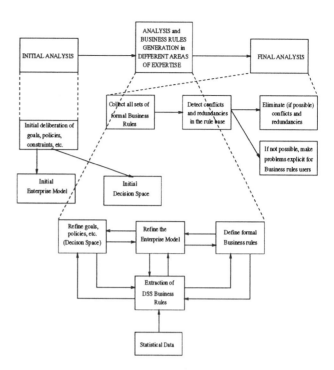

Figure 7: Business Rules Acquisition

4.1 Business Rules Acquisition

We see three major steps in the acquisition of business rules: the initial analysis, the analysis and generation of business rules in different areas of expertise, and final analysis (see Figure 7). During the initial analysis, brainstorming sessions take place for deliberating which are the goals, policies and constraints of the business that need to be modeled. As a result of these deliberations, initial versions of the enterprise model and decision space are sketched and also a first set of business rules that specify how the business should be run is defined. Because they define the goals of the enterprise, these are *strategic business rules* (in the next section, we will find these rules at the *criteria level* rules) that express very high level decisions. These rules need to be refined in order to become operational.

The first step in the refinement of business rules is the analysis and rule generation in different areas of expertise. In this phase business analysis is carried out by separate groups of people, with different areas of expertise, for refining the understanding of business entities, processes and business rules. These activities imply more detailed discussions on the ways of achieving the goals, policies and constraints of the business. They can be complemented with interviews with domain experts and/or reading existing documentation and information related to the subject of analysis. As the understanding of the enterprise objectives becomes clearer the Enterprise Model and the Decision Space are updated.

Based on the entities and processes stated in the Enterprise Model, on the decision structures captured in the Decision Space, and on statistical data from the enterprise's way of doing business, business rules (decision support system level rules, or DSS rules) can be automatically extracted following an algorithm described in detail in the next section. These rules (called *arguments*, respective *assumptions level rules*) are more concrete than the strategic rules. They underly the structure of operational rules in ECA format that are expressed in a formal rule language, like the one used in the Livemodel tool. At the end of this step, a formal business rule will be defined for each alternative solution in the Decision Space and will be available for deployment. For process simulations, these rules can be attached to operation triggers in process diagrams, like the ones defined in Livemodel. See the slash mark on the Figure 5 for an example of rules attachment to an event diagram and Figure 8 for an example of stimulus/response (trigger) rule implemented in Livemodel (the example has been oversimplified for presentation purposes).

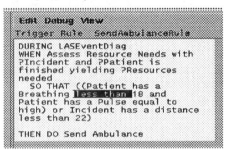

Figure 8: Example of stimulus/response (trigger) rule in the LAS case study

There can be multiple iterations on each operation of this step until a stable set of business rules, as well as a clear and comprehensive Enterprise Model for each specific area of expertise, are obtained.

During the final analysis of the business rules acquisition all of the existing sets of business rules, Decision Spaces, and parts of the Enterprise Model are put together, leading to the detection of redundancies and conflicts. The detection is facilitated by the types of analyses discussed in section 3. These redundancies and conflicts are either eliminated, or if not possible, made explicit to the designers, developers and users of the information system that will underlie the business and that will incorporate these business rules, or to the users of people oriented business rules.

4.2 Business Rules Deployment

After the business rules are defined and integrated into the enterprise process model they become operational. Therefore, whenever a new case is run through a process model inside the enterprise, there are a couple of situations that can arise in the application of business rules (see Figure 9):

1. The situation is deterministic, e.g. characterized by a single business rule and the data referenced by the rule are known with certainty. Therefore that rule can be automatically applied by the underlying information system, either by people or by machine.

2. The situation is characterized by multiple, conflicting business rules. In these cases we can show the decision matrix associated with those rules and let the

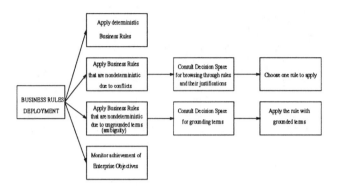

Figure 9: Business Rules Deployment

user browse through it, analyze the information contained in the decision structures and assess the merit of each alternative associated with each rule. The user can choose one of the proposed rules or apply his own judgment and select another rule. This way the decision of which rule is the best has been shifted from the analysis time to operation time, when concrete data about a case is available.

3. If the applicable rule(s) is(are) ambiguous, e.g. they contain ungrounded terms whose grounding couldn't be done with certitude at analysis time we can show the interpretation of these terms using the links among the business rules, the Decision Space and the Enterprise Model. This way the definitional business rules attached to various attributes of the entities in the business can be made available for consulting. The user can choose one of the legal values of an ungrounded term based on the definitional business rules or can disagree with those rules and choose a value according to his own judgment. This approach permits development to move forward even when requirements are not fully understood.

4. For evaluating how well the enterprise objectives are achieved we propose instrumenting the system with monitoring to check whether the assumptions underlying the business rules are valid. This information is fed back to the system for updating the business rules, enterprise model and decision space.

4.3 Business Rules Evolution

There are several possibilities for improving the business rules based on the information captured in the methodology framework. Data obtained through monitoring of the operational system can be used to study the validity of assumptions recorded in the Decision Space, leading to changed rules. New sources of information, both inside and outside of the enterprise, may arise. New solutions may be chosen by users for resolving conflicting or ambiguous situations. For example, by studying the Decision Space one can detect that some criteria, alternatives or arguments could be added/eliminated, or that their current weights were wrong. Or, by tracing back the rules applied, we can detect that some attributes are missing or should be added for more accurate business rules.

Therefore the Enterprise Model and the Decision Space is continuously updated for keeping the pace with the constant changes that occur both inside the enterprise and in the outside world.

Depending on the nature of change observed, business rules are changed by either choosing other existing rules, modifying existing rules or creating new rules if none of the existing ones meet the new context coordinates.

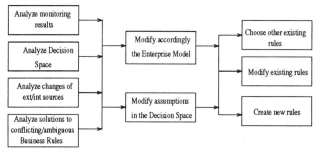

Figure 10: Business Rules Evolution

5 Example

Here we introduce the example used in the next section for the extraction of business rules algorithm. Our example is inspired from the London Ambulance Service (LAS) case study proposed as a common example at the 1996 International Workshop on Software Specification and Design [3].

Issue: Assess Resource Needs

Criteria \ Alternatives	Quick Response Time	Importance	Effective Resources usage	Importance	Alternative Merit
Send Ambulance	0.351	0.6	0.376	0.4	0.18
Send Helicopter					
Send Major Incident Team					

	Arg. Merit	Arg. Import.	Arg. names	Arguments Description
	18.8%	-1.0	Slow Response	This alternative is too slow in lifethreatening cases.
	81.2%	0.3	Acceptable Response	This alternative is acceptable in lifethreatening cases.
	99.9%	1.0	Quick Response	This alternative is quick in non-lifethreatening cases.

Assumptions

Systolic Blood Pressure	Diastolic Blood Pressure	Pulse	Breathing	Temp.	Distance	Class
very high	very high	high	30	high	15	False
high	very high	normal	21	normal	65	False
very low	very low	low	8	normal	16	False
high	high	normal	21	high	65	True
high	very high	normal	22	high	5	True
...	...	...	...	...	...	...
high	normal	high	19	normal	18	True

Figure 11: Decision matrix and data base of assumptions for the Assess Resource Needs issue. The assumptions table contains examples referring to the *Send Ambulance is Quick in non-life-threatening cases* argument.

LAS is in charge of dispatching resources (ambulances, helicopters, etc.) to incident scenes. Some of the key objectives of the system are: to get the most appropriate resource(s) to the scene of an incident (e.g., a heart attack, a traffic accident, a terrorist attack, etc.) and to get it there as quickly as possible (i.e., according to standards that describe acceptable response times). In the format of the manual LAS system that existed before introducing the Computer Aided Dispatch system, a call taker writes down the details on a form and identifies the location of the caller when an incident is reported. All the incident forms go to a central collection point where a staff member reviews the details on each of them and decides what type of resources are needed for each incident and which resource allocator should deal with it. The resource allocator decides which resource should be mobilized and gives this information to a dispatcher who will pass the mobilization information to the appropriate resources (vehicles, crews, hospitals, etc.).

In this process we are interested in analyzing the situations where human judgment is involved, in particular, what business rules are applied and what is the motivation behind them. One such situation is the assessment of the type of resources needed to solve a particular incident. The information necessary to make this decision is represented in the decision matrix and the set of data illustrated in Figure 11. The content of the matrix was populated with information gathered from the decision makers. The numbers showed in the matrix are computed by the rule extraction algorithm (see the next section). In this step of the LAS process a staff member has to decide whether to send an ambulance, a helicopter or a major incident team based on the details mentioned in the incident form. In order to make this decision the staff member takes into account criteria such as quick response time since the call was received and an effective use of the resources available. The figure shows the pro (0.3, 1.0) and counter (-1.0) arguments taken into account for assessing the degree of satisfaction (0.954) of the *Quick Response Time* criterion by the *Send Ambulance* alternative. Also, we show a fragment of statistical data recording how the various assumption values support the argument "This alternative is quick in non-life-threatening cases." These assumptions are attributes that characterize the non-life-threatening cases (we have considered just a few of them for this example) and the distance of the incident location from the nearest hospital.

6 Extraction of Business Rules from Decision Structures and Examples

This section discusses how we support the automatic extraction of business rules. Our objective is to generate a set of business rules that preserve the information in the decision matrix and to reflect knowledge from statistical data about domain assumptions.

6.1 Decision Structure Knowledge

Next we will formalize the knowledge contained in the decision structures represented by a decision matrix. This knowledge is used in the rules extraction algorithm, and corresponds to the primitives described in the decision support submodel in Section 3. These primitives are naturally expressed in terms of variables and constraints on their values. Thus we distinguish the following variables:

- The *issue* (*Iss*) that needs to be solved: *Assess Resource Needs* in our example.
- A number of *alternatives* (Alt_i) that are proposed as solutions to the issue *Iss*. For example, $Alt_1 = Send\ Ambulance$.
- A set of *criteria* ($Crit_j$) against which all the alternative solutions are evaluated in order to decide upon the best alternative. For example, $Crit_1 = Quick\ Response\ Time$.
- A number of pro and counter *arguments* that correspond to each pair (alternative, criterion) ($Arg_k(Alt_i, Crit_j)$). For example, $Arg_1(Alt_1, Crit_1) = Slow\ Response$, related to the more intuitive description "This alternative is too slow in life-threatening cases".
- A set of *assumptions* that support each argument ($Attr_l(Arg_k)$). Assumptions represent groundable attributes of one or more objects in the Enterprise Model that are relevant to the issue under consideration. For example, $Attr_1(Arg_3(Alt_1, Crit_1)) = Systolic\ Blood\ Pressure$. Some attributes may be common for various arguments.

6.2 Format of the Business Rules Extracted from Decision Structures and Statistical Data

We are proposing three types of business rules that correspond to different levels of detail. They are the rules obtained at the criteria level, at the arguments level and at the assumptions level of a decision matrix. These different business rule types correspond to different levels of decision making in the hierarchy of an enterprise.

1. Criteria level:

The rules obtained at this level are the most general type of rules. They correspond to high level decision making and express enterprise objectives in very general terms. Therefore these objectives will need to be refined to the point where they can be translated into operational business rules. These objectives will serve as criteria for evaluating the alternatives of various solutions proposed for solving various problems.

For example, in the rule

   WHEN *Assess Resource Needs*
   IF *Quick Response Time* [0.6] ∨ *Effective Resource Usage* [0.4]
   THEN *Send Ambulance*

that expresses what solution (alternative) to choose (*Send Ambulance*) for solving the problem *Assess Resource Needs* for an incident, the criteria established by the enterprise for solving this problem are *Quick Response Time* of the resources sent to an incident site and *Effective Resource Usage*. As we can see, there is no precise definition yet of what the non-functional requirements *Quick Response Time* and *Effective Resource Usage* really mean. The only thing we know at this stage is the importance of each criterion (0.6 and 0.4) in evaluating the alternatives.

The general format of this type of rule is:

   WHEN *Iss*
   IF $Crit_1[w_1] \lor Crit_2[w_2] \lor \cdots \lor Crit_n[w_n]$
   THEN Alt_i

where w_i represent the weights or *Importance* (see Figure 11) of criterion $Crit_j$ in the process of alternative Alt_i evaluation. The w_i values are given by the decision makers.

Whenever this type of rule is applied we compute the merit of the alternative given in the action part of the rule (Alt_i), based on the weight of each criterion (w_{Crit_j}) and the merit $Merit_{Crit_j}$ of the alternative Alt_i in satisfying each criterion $Crit_j$:

$$Merit_{Alt_i} = \frac{\sum_{j=1}^{n_{crit}} w_{Crit_j} * Merit_{Crit_j}}{n_{crit}} \qquad (1)$$

2. The arguments level:

The rules at this level express the heuristics used in deciding how well an alternative satisfies a criterion when several arguments are presented for, or against a solution (alternative). They combine the evidence about the merit of each argument Arg_k correlated with a pair ($Alt_i, Crit_j$), in order to compute the $Merit_{Crit_j}$ of the alternative Alt_i against criterion $Crit_j$. These rules express the fact that the meaning of the enterprise objectives is not always obvious, and therefore requires negotiation among stakeholders.

For example, in the rule

   WHEN *Send Ambulance*
   IF *Slow Response* = True [-1.0] ∨ *Acceptable Response* = True [0.3] ∨ *Quick Response* = True [1.0]
   THEN *Quick Response Time*

we express the fact that in the process of refining the meaning of the *Quick Response Time* objective there have been brought up three arguments with different weights in the context of sending an ambulance to an incident. These arguments correspond to different situations perceived by various stakeholders as being plausible: 1) sending an ambulance may be too slow in life-threatening cases where the location of the incident is far away from a hospital (and therefore, this is a counterargument for sending an ambulance in these cases); 2) sending an ambulance may be acceptable in life-threatening cases if they are close to a hospital; 3) sending an ambulance is a quick solution in non-life-threatening situations, regardless of the distance from the hospital (and therefore gives a stronger support to the alternative than argument 2)). All these arguments (and others corresponding to other criteria in the level 1 rule) will be taken into consideration when making the decision about whether sending an ambulance.

The general format of this type of rule is:

   WHEN Alt_i
   IF $Arg_1[w_1] \lor Arg_2[w_2] \lor \cdots \lor Arg_n[w_n]$
   THEN $Crit_j$

w_i represent the weight of each argument in the evaluation process. It can take values on a scale [-1.0, 1.0] meaning: when $w_i = -1.0$, Arg_i is a counterargument, while when $w_i = 1.0$, Arg_i is a strong supporting argument.

The computation of $Merit_{Crit_j}$ is based on the weight of each argument (w_{Arg_i}) and the predicted accuracy of the truth value of that argument ($Merit_{Arg_i}$):

$$Merit_{Crit_j} = \frac{\sum_{i=1}^{n_{arg}} w_{Arg_i} * Merit_{Arg_i}}{n_{arg}} \quad (2)$$

3. The assumptions level:
This is the most detailed level of rules where the business objectives find their operational meanings. Even though they correspond to operational (low level) decision making, there might still be situations that require grounding of some terms inside the rules.

These rules express the operational conditions that need to be met in order for the alternative Alt_i to meet the criteria $Crit_j$ (an enterprise goal), e.g. they assess the truth value of the arguments Arg_i associated with the pair $(Alt_i, Crit_j)$ based on various domain assumptions. The subconditions in the antecedent part of the rules are obtained either automatically by induction from statistical data, or, when this data is not available, by asking the decision maker. Even though these rules might look like the rules of an expert system, they are in fact business rules that achieve a goal.

For example, the rule

IF (*Diastolic Blood Pressure* = High ∧ *Systolic Blood Pressure* = High) ∨ (*Distance from Hospital* < 28)
THEN *Quick Response* = True [99.9%]

expresses the operational conditions for the argument *Send ambulance is a quick solution in non-life-threatening situations* to be true. More than that, it shows the perceived accuracy (99.9%) of this assessment.

The condition part of this type of rules is a disjunctive normal form (DNF) formula that contains various assumptions of the argument under consideration. An assumption has the format $As \equiv (Attr < op > value)$, where $< op > = \{=, <=, >\}$. For instance, a rule whose condition has two disjunctive terms is

IF $(As_1 \wedge As_2 \wedge As_3) \vee (As_4 \wedge As_5)$
THEN $Arg_i [Merit_{Arg_i}]$

where $Attr_i$ represent the attributes whose values v_i need to be checked in order to assess the truth value of the argument Arg_i. The $Merit_{Arg_i}$ describes the certainty factor about the truth value of argument Arg_i.

6.3 Automatic Generation of Business Rules

The method for the automatic generation of business rules uses the knowledge structure provided by the decision support system through decision matrices. Also it uses statistical data recording how domain assumptions support various arguments (see Fig. 11).

We distinguish two types of automatically created business rules. The first type are the business rules that capture the heuristic knowledge from the decision matrix. These rules correspond to levels one and two from above. The second type are business rules that are extracted from decision trees induced from statistical data by applying inductive learning techniques. They correspond to level three rules from above.

Decision tree learning is a supervised machine learning technique that uses a collection of training examples and outputs a compact decision structure called a decision tree. The internal nodes in a decision tree correspond to tests on the values of particular attributes, while the leaves correspond to class values.

Decision trees logically correspond to a disjunction of conjunctions. They can be further generalized into business rules of level three using a technique of extraction of rules from decision trees (see [12, 15]).

The rules at levels two and one can be automatically generated from the decision structures represented in a decision matrix. The merit of each alternative in the decision matrix can be computed according to the formulae 1 and 2.

For obtaining the rules at level three we apply an algorithm for the induction of decision trees and rules from statistical data [12, 15]. The attributes used for classification are the domain assumptions that underlie an argument. The algorithm tolerates missing values for some of the attributes, therefore allowing for imprecise information about a case. The classes resulted after the application of the algorithm over a data base of assumptions values are the truth values for an argument. Each class has associated the certainty factor of that classification, which represents the likelihood that an argument value is true or false. From an induced decision tree we extract and optimize rules (both at the rule level and rule sets level) that will represent the level three type of rules.

The result of the antecedent phase is a set of rules (DSS rules) associated with every issue ($DSSR(Iss)$). Each such set of rules can be transformed into an operational business rule that follows the ECA format. An operational rule associated with an issue Iss is obtained from $DSSR(Iss)$ by applying the operation of consequent expansion at levels one and two. Consequent expansion means replacing a condition on a variable v in the antecedent of a rule, with the antecedent of the rule that has v as a consequent. If several such rules exist they are combined in an OR logical operation. The weights of the new conditions generated through consequent expansion depend on their initial weights, certainty factors of the rules expanded, and the number of rules expanded.

This way the enterprise objectives (stated in level one DSS rules) are refined to the point where they can be translated into operational business rules that achieve the enterprise goals.

7 Summary and Conclusions

A methodology has been presented that provides a fair degree of guidance for the activities of acquiring, deploying and evolving business rules of an enterprise. A requirements modeling framework has been introduced for representing the information needed to conduct the methodology. The meaning of requirements analysis in this context has been articulated. We have emphasized the importance of several aspects of the methodology and its associated framework.

First, we studied a project where business rules were recorded to understand the types of business rules and the factors that determined them. We were motivated by the desire to include business rules in the requirements modeling framework.

Second, we have emphasized the importance of an appropriate enterprise model, in terms of which the business rules can be expressed. The Process/Object enterprise model suffices for the purposes of this paper. However, ultimately, a richer enterprise model will be needed in order to facilitate the expression of additional types of business rules. The Service-Oriented Systems model [5] will be extended to include the types of reasoning proposed in [17], in terms of goals, tasks, and resource dependencies. Also, a model of services is needed to express business rules about how services are provided between the participants in a service (customers and providers), the roles of the processes, and the positions of the organization that bear capabilities and responsibilities.

Third, we have highlighted the special nature of business rules as *decisions* whose consequences are speculative and subject to change. The sense in which business rules are "satisfied" is also defined flexibly to correspond to how they must be treated and tolerated in practice. Although we have focussed on business rules in this paper, the ideas in this paper are applicable, in principle, to requirements in general.

Fourth, we described the use of a decision support framework for capturing deliberations that can be used to continually evaluate and regenerate the business rules as the enterprise and its operational systems evolve. It should be clear that the information used in deliberations of this sort will be needed through cycles in which the business rules are re-examined and revised. This is in contrast to the usual requirements engineering scenario where the requirements documentation may not contain the right kinds of information that are needed later in the lifecycle or are difficult to keep up to date.

Fifth, the possibility for lifecycle automated assistance has been demonstrated in terms of the automatic extraction of business rules using the decision support framework. The extraction algorithm applies a technique that is well-known in the field of Machine Learning to a new field, Requirements Engineering, by using it in the business rules context.

The contribution of this paper is in the synthesis of the elements of the methodology. The methodology has not been used in practice, so there is no experience to report on its overall effectiveness. The metamodel has been implemented and some example models have been acquired (enterprise models, decision spaces, business rules). The business rules extraction algorithm has been implemented and some of the results produced have been shown in this paper.

Acknowledgments

We acknowledge Howard Reubenstein for his work on the taxonomy of business rules.

References

[1] J. Bubenko Jr. and B. Wangler. Objectives driven capture of business rules and of information systems requirements. In *Proceedings of the International Conference on Systems, Man and Cybernetics*, pages 670–677, 1993.

[2] L. Chung, B. Nixon, and E. Yu. Using non-functional requirements to systematically support change. In *Second IEEE International Symposium on Requirements Engineering.* 1995.

[3] M. Feblowitz, S. Greenspan, H. Reubenstein, and R. Walford. ACME/PRIME: Requirements acquisition for process-driven systems. In *International Workshop on Software Specifications and Design*, pages 36–45. IEEE Computer Society Press, 1996.

[4] S. Fickas and M. Feather. Requirements monitoring in dynamic environments. In *Proceedings of IEEE International Symposium on Requirements Engineering*, 1995.

[5] S. Greenspan and M. Feblowitz. Requirements engineering using the SOS paradigm. In *Proceedings of IEEE International Symposium on Requirements Engineering, San Diego*, 1993.

[6] H. Herbst. A meta-model for specifying business rules in system analysis. In *Proceedings of CAiSE'95*, pages 186–199, 1995.

[7] Intellicorp. *Livemodel User's Guide, Betaversion*, 1995.

[8] A. v. Lamsveerde, R. Darimont, and P. Massonet. Goal-directed elaboration of requirements for a meeting scheduler:problems and lessons learnt. In *Proceedings of RE'95*. IEEE Computer Society Press, 1995.

[9] P. Loucopoulos and E. Katsouli. Modelling business rules in an office environment. *SIGOIS Bulletin*, 13(2):28–37, 1992.

[10] J. Martin and J. Odell. *Object-Oriented Methods: A Foundation*, chapter 20. Prentice Hall, 1995.

[11] L. Paul. Hidden assets. *PC Week*, February 20 1995.

[12] J. Quinlan. *C4.5: Programs for Machine Learning*. Morgan Kaufmann, 1993.

[13] R. Roland. *The Business Rules Handbook*. Database Research Group, Inc., Boston, MA, 1994.

[14] D. Rosca, S. Greenspan, C. Wild, H. Reubenstein, K. Maly, and M. Feblowitz. Application of a decision support mechanism to the business rules lifecycle. In *Proceedings of the KBSE95 Conference*, pages 114–122, 1995.

[15] J. Rosca and D. Rosca. Knowledge acquissition facilities within an expert system toolkit. In *Proceedings of the Seventh International Symposium on Computer Science*, Jassy,Romania, 1989.

[16] A. Sandifer and B. V. Halle. *Business Rules: Capturing the most elusive information asset*. Auerbach Publications, 1993.

[17] E. Yu. *Modelling Strategic Relationships for Process Reengineering*. PhD thesis, University of Toronto, December 1994.

A Logical Framework for Modeling and Reasoning about the Evolution of Requirements

Didar Zowghi and Ray Offen
CSIRO-Macquarie University Joint Research Centre for
Advanced Systems Engineering (JRCASE), Macquarie University,
NSW 2109, Australia. {didar,roffen}@mpce.mq.edu.au

Abstract

We present a logical framework for modeling and reasoning about the evolution of requirements. We demonstrate how a sufficiently rich meta-level logic can formally capture intuitive aspects of managing changes to requirements models, while maintaining completeness and consistency. We consider a theory as the deductive closure of a given set of axioms and conclude that software engineering is concerned, in essence, with, building and managing large theories. This theory construction commences with the development of the requirements model which we view as a theory of some nonmonotonic logic. Requirements evolution then involves the mapping of one such theory to another. Exploiting the deductive power of the theory of belief revision and nonmonotonic reasoning we develop a formal description of this mapping, as well as the requirements engineering process itself. This work thus offers a rigorous approach to reasoning about requirements evolution and a important focus for defining semantically well-founded methods and tools for the effective management of changing requirements.

1. Introduction

Software engineering is concerned with defining, constructing and maintaining computer-based systems. If we regard a theory as the deductive closure of a given set of axioms, then in essence software engineering is nothing more than the building and managing of large theories. Construction typically commences with a collection of activities referred to as Requirements Engineering (RE), which is concerned with eliciting real-world goals for, the function of, and the constraints on software systems [20]. The major objectives of requirements engineering are defining the purpose of a system and capturing its external behaviour.

Software development is characterised by continuous evolution. Requirements evolve because requirements engineers and users cannot possibly envision all the ways in which a system can be utilised and the fact remains that software will be used by different people with differing goals and differing needs. The environment where the software is situated frequently changes and so do the software boundaries and business rules governing the utilisation of that software. Correspondingly, designs change because the requirements change. Implementation has to be changed because designs evolve and defects have to be fixed [25]. Clearly, support for evolutionary processes is needed at all stages of software construction but especially in requirements engineering since it is the requirements modifications that typically initiate this cycle of change throughout the software development life cycle.

At the heart of requirements engineering research are basic issues of representation and reasoning about the knowledge and information captured during the requirements elicitation phase, a subject referred to as *requirements modeling*. [23]

The focus of this paper is on the question of how best to manage the evolution of requirements models. This involves handling continuous changes in requirements while maintaining the consistency and completeness of the requirements model. We argue that requirements management is most effective when it is based on a formal underlying framework. The major contributions of this paper are twofold. Firstly, we argue that the most suitable form of reasoning needed for RE processes is a combination of monotonic and nonmonotonic reasoning. We then propose a framework for modeling and reasoning about the evolution of requirements. This framework provides us with an opportunity to present a simple, yet complete model for the process of requirements engineering. At a meta-level, a requirements model is fundamentally viewed as a nonmonotonic theory, specifically a *default* theory as formalised in [1] [3]. Requirements evolution maps one such theory to another through a process of rational *belief revision*, which mediates our understanding of the changing system context. We apply belief revision techniques, as formalised in the so-called AGM theory [24], to define operations through which new requirements may be added or current requirements may be retracted from a default-based representation of a requirements model.

We illustrate how a sufficiently rich meta-level logic can formally and accurately capture intuitive ways of handling incompleteness and inconsistency in requirements and how operators that map between theories of this logic can provide a formal basis for requirements evolution. We argue that the AGM theory of belief change [24] should provide the formal basis for the theory change component. We demonstrate our ideas by using the THEORIST [1] system for nonmonotonic reasoning. The proposed framework presupposes a formal notation but does not commit to a particular representation language and therefore can be viewed as a language independent framework for modeling and reasoning about requirements engineering.

2. Reasoning about the evolution of requirements

Managing requirements for computer-based systems is an evolutionary and multi-participant communication process. It consists of a quasi-sequence of activities that can be divided into five major categories: *elicitation, modeling, specification, validation* and *management*. There is, however, little uniformity in terminology used by researchers and practitioners to describe these activities [13]. Requirements engineering is typically initiated by the identification of a problem and the expression of a need for a possibly computerised solution. Requirements engineers normally examine this brief problem statement and based on past experience, and possibly limited knowledge of the application domain, make a set of *assumptions* and use *defaults* to expand it into a more complete expression of the requirements. This expanded version of requirements which includes the assumptions made and defaults used by the requirements engineer are then conveyed to users for validation and correction.

During requirements analysis and modeling, alternative models for the system are elaborated and a conceptual model of the enterprise as seen by the system's eventual users is produced. This model is meant to capture as much of the semantics of the real world as possible and is used as the foundation for the detailed and precise description of the requirements that is to be derived and formally specified. The specifications are validated and analysed against *correctness* properties (such as

completeness and consistency), and *feasibility* properties (such as cost and resources needed). The final product of requirements engineering is normally a document written almost entirely in natural language, and referred to as the Requirements Specification (RS). Although this document is probably the most important document produced during software development the focus of our attention here is on the *requirements model*, from which the RS is generated.

During requirements elicitation and modeling, the descriptions of the problem, the *machine* and the *world* (we use Jackson's terminology here [26,21]), have to be elaborated so that the implicit facts about them become represented explicitly. Inference rules of First-Order Logic such as *modus ponens* and *generalisation* are able to do just that for us. This is the monotonic aspect of reasoning about requirements. As participants understanding of the problem increases, the statement of requirements begins to change. We would then like to modify the requirements (ie. the set of beliefs about the theory we are going to build), by adding new information (formulae), to the requirements model that states new (or revised) things about the problem, the machine and the world. Ordinary logic offers us no hint as to how to do this. We need nonmonotonic methods for reasoning with tentative statements. Requirements engineers are expected to draw reasonable conclusions based on possibly inconsistent and incomplete user requirements that are gradually captured. Brewka [8] states that logic is based on ontological assumptions which means the existence of a domain of identifiable nonchanging objects and nonchanging relations between these objects. In software development objects and their relationships can be very volatile. Clearly, classical logic on its own is not sufficient for reasoning about the RE process. Both monotonic and nonmonotonic reasoning must be utilised in requirements engineering

Our account of the requirement engineering process this far suggests that any formal framework for reasoning about requirements must satisfy the following three properties [2]:

- It must include an explicit notion of defaults, so the tentative assumptions and default knowledge about the domain can be brought to bear on an incompletely specified initial set of assumptions in order to obtain a more complete requirements model.
- It must permit the identification of consistent alternative models which resolve any contradictions arising in an initial set of potentially contradictory and incompletely specified requirements.
- It must provide an adequate account of how a requirements model evolves as a consequence of new requirements being added or existing requirements being retracted.

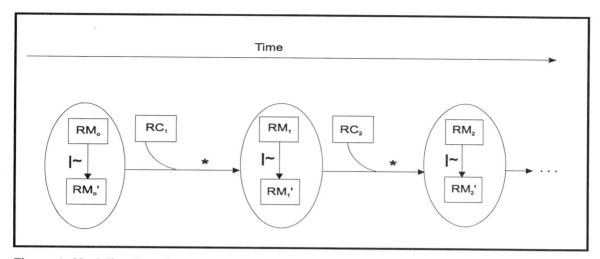

Figure 1: Modeling Requirements Evolution - RE starts with the expression of a set of incomplete goals, the incomplete Requirements Model (RM_0). Requirements engineers use *defaults* and assumptions to convert this incomplete set of sentences into a more complete requirements model (RM_0'). Each time the model is presented to problem owners, a new set of Requirements Changes (RC_i) is brought to bear. Through a series of *revisions* the model is refined and completed in each step.

The model we present, in schematic form, in figure 1 achieves precisely these three objectives. Fundamentally, a requirements model is viewed as a theory (or a belief set), in some nonmonotonic logic. Requirements evolution maps one such theory to another through a process of rational belief revision. The two operators |~ and * encode the two basic, orthogonal processes in our framework [2].

- The operator |~ encodes the process of nonmonotonic inference. We start with an initial incomplete set of requirements and apply relevant defaults or tentative assumptions about the problem domain to complete the requirements model. By applying the |~ operator, we obtain the set of nonmonotonic consequences sanctioned by the requirements model represented as a nonmonotonic theory. There can be possibly multiple, mutually contradictory, sets of nonmonotonic consequences (also called extensions). A choice is made from amongst these extensions through a process of iterative interaction between users and requirements engineers. For the purpose of exposition in this paper, we shall commit to a specific nonmonotonic formalism, namely the THEORIST system [1], which we shall describe in the next section.

- The operator * encodes the process of belief revision. The AGM theory of belief revision [24 , 4], which we describe in the next section, provides a semantic basis for *rational* belief revision. We maintain that any account of requirements evolution should be based on this semantically well-founded theory.

3. Formal Preliminaries

3.1 Nonmonotonic Reasoning

Forms of reasoning which allow additional information to invalidate pre-existing conclusions are called *nonmonotonic reasoning* (NMR). It involves adopting assumptions or defaults that may have to be abandoned in light of new information. In classical logics, the addition of new axioms to a set of axioms can never reduce the set of theorems. This property is known as *monotonicity*. In nonmonotonic logic, however, when new axioms are added, the set of theorems may lose members as well as gain some. The need for NMR arises whenever our knowledge about the world is incomplete and does not allow for the sound derivation of the conclusions necessary to base our decisions, plans and actions [8]. One such case of NMR that have been studied is *default reasoning*. Defaults are statements or rules according to which some statement is to be believed in the absence of any information to the contrary and until and unless otherwise demonstrated.

Consider the initial set of requirements expressed by the problem owners as a first order formalisation of what we know about the machine we are going to build in software development. Since it is impossible to know everything about the problem, world and the machine, this first-order theory is potentially incomplete, that is there are gaps in our knowledge base. In order to progress further in the process of engineering requirements, we need to draw some inferences despite this incompleteness. Defaults therefore act like meta-rules that enable us to fill in some of the gaps in the current requirements model. In other words, they allow us to further complete the underlying first-order theory. They are instructions about how to create an *extension* of this incomplete theory. Those formulae sanctioned by the defaults and which extend the theory can then be viewed as a set of beliefs about the problem, machine or the world. [9] The major issue addressed by nonmonotonic logic is that it provides some well-defined semantics for defaults which allow a single set of axioms and defaults to have several coherent interpretation.

The THEORIST framework for default reasoning and representation [1], considers defaults to be possible hypotheses which can be used in an explanation. A default is something that an agent is prepared to accept tentatively as part of an explanation of why something is expected to be true. When considering possible hypotheses to be defaults, the thing to be explained is intended to be something which we are predicting, rather than something we are actually observing [3]. The representation language in Theorist is the clausal form of first-order logic. The knowledge base consists of a set of closed formulas, called facts, that we know are true about the world we intend to model, and a set of possibly open formulas that are tentatively true, called hypotheses. Default reasoning in Theorist involves identifying *maximal scenarios* (extensions), where a scenario consists of the set of facts together with some subset of the set of ground instances of the hypotheses which is consistent with the set of facts. The framework can be augmented with *constraints*, which are closed formulas such that every Theorist scenario is required to be consistent with the set of constraints [2]. Following [3], we can present the following definition of a maximal scenario.

Definition 3.1 For a THEORIST specification (F,H,C) where F is the set of facts, H is the set of hypotheses and C is the set of constraints, such that $F \cup C$ is satisfiable, a *maximal scenario* is a set $F \cup h$ such that $h \subseteq H$ and $F \cup h \cup C$ is satisfiable and there exist no h' such that $h \subset h'$ and $F \cup h' \cup C$ is satisfiable.

In viewing requirements models as default theories, specifically THEORIST knowledge bases, requirements that are known to be true of the domain, together with domain knowledge are treated as *facts* (ie. elements of F) while defaults or tentative requirements are treated as elements of H. We shall not discuss the role of the set of constraints C now, but shall point out that it plays a crucial role when requirements are retracted [2].

A default theory includes explicit language constructs for asserting tentative or default knowledge. The accompanying machinery for generating extensions

resolve potential contradictions among defaults and provide consistent views of the world sanctioned by the default theory. The model we presented in Section 2 commits to a nonmonotonic logic as a meta-level representation language for requirements. Our choice of the Theorist nonmonotonic reasoning framework for modeling requirements is not entirely arbitrary. Although we could potentially use any formalism such as those in [9,28,29], with an explicit notion of defaults for this purpose, the Theorist framework provides the features necessary for our purpose and is, at the same time, simple enough to facilitate ease of exposition. Additionally, it provides several advantages over Reiter's default logic [9], such as avoiding situations where Reiter's default logic is too strong or too weak, guaranteeing the existence of extensions, semi-monotonicity and a constructive definition for extensions (see [17] for a detailed discussion of how a closely related default logic improves over Reiter's default logic).

3.2 Belief Revision

Recently, belief revision has received a great deal of attention from AI researchers. The core of research in this area is based on the work of Alchourron, Gärdenfors and Makinson [4]. They developed a framework (widely known as the AGM framework), for investigating the process of rational belief change. Belief revision is concerned with the issue of revising one's state of belief with new information (possibly contradicting the current belief state), while preserving consistency. The AGM framework represents belief states as theories of a logical language **L**, the new information is modelled by sentences of the language **L**. It is assumes that **L** is a propositional language over the standard sentential connectives ($\neg$, $\wedge$, $\vee$, $\rightarrow$, $\leftrightarrow$), individual propositions are denoted by x, y, and z, and sets of propositions by A, B, and C. We write $\vdash$ to express classical propositional derivability, and write Cn to mean the corresponding closure operator

$Cn(A) = \{ x \in \mathbf{L} \mid A \vdash x \}$

Instead of $Cn(\{x\})$, we will also write $Cn(x)$. A propositional theory K of **L** is any set of sentences of **L** closed under logical implications, ie. $K = Cn(K)$.

The process of belief revision is modelled as a function * over theories, called a *revision function*, satisfying eight postulates (known as the AGM *postulates*) that capture the essence of rational belief revision. The basic idea used in the formulation of the AGM postulates is that when we, for some reason, change our beliefs, we would like to retain as much as possible of our old beliefs in the new belief state. This is known as the *principle of minimal change* (also referred to as the criterion of *information economy*). Beliefs are generally considered to be valuable (useful in arguments, expensive to acquire or infer), so unnecessary loss of beliefs is irrational. When we receive new information which is consistent with our current belief state, this requirement does not cause any problems, since we can then preserve all the old beliefs and add the new one. If we are ideally rational, we must also accept all the logical consequences of the new belief. On the contrary, if the new information is inconsistent with the current state of belief, then some of the old beliefs must be retracted in order to preserve the consistency of the belief state.

The AGM framework [24] provides three types of operations on belief states. For each belief state K and proposition α we may have:

- **Expansion** - Expanding K with α, written as K_α^+, which means to add α to K and close under logical entailment, which may produce an inconsistent belief state.

- **Contraction** - Retracting α form K, to obtain K_α^-, such that the result is a consistent belief state.

- **Revision** - Revising K with respect to α, written as K_α^*, which means adding α to K such that the result is consistent belief state.

The set of all consistent theories of **L** is denoted by K_L. A revision function * is any function from K_L x **L** to K_L, mapping <K , α> to K_α^*, which satisfies the AGM revision postulates. For the sake of brevity, we do not present these postulates here. Essentially, what these postulates demand is the requirement that the result of a revision operation must be a logically closed consistent theory, that the revision operation must succeed, the outcome should be independent of the syntactic form of the input and the operation involve *minimal change*. The AGM framework also offers a parallel set of postulates for contraction and expansion, which we do not present here. The initial major result of the AGM theory is that the postulates for contraction are complementary to the revision postulates if the revision K_α^* is defined by using the *Levi Identity* [5]:

$$K_\alpha^* = (K_{\neg\alpha}^-)_\alpha^+$$

Which means revision by α is equivalent to contracting by ¬α in order to remove any inconsistent beliefs and then expanding with α.

In the model we presented in Figure 1. we introduced operator * to encode the process of revising requirements rationally whenever new requirements are to be added to and/or pre-existing ones retracted from the requirements model. The AGM theory of belief change provides an excellent theoretical underpinning for this process based on the principle of minimal change. In our framework we regard an instance of a requirements model (RM), as a belief state K and elements of the set of requirements

changes (RC), as *epistemic[†] input* α. When requirements evolve, in accordance with the theory of belief revision, if RM is a consistent requirements model, then for any new requirement α ∈ RC, one of three types of *epistemic attitudes* can be expressed:

(1) α ∈ K : α is accepted,
(2) ¬α ∈ K : α is rejected,
(3) α ∈ K and ¬α ∈ K : α is indetermined, ie. α is neither accepted nor rejected.

We maintain that as long as we restrict ourselves to the representation of requirements models as belief sets and the three kinds of epistemic attitudes that are expressible for such models, then *expansion, contraction* and *revision* are the only relevant forms of requirements change in requirements evolution.

3.3 Epistemic Entrenchment

The AGM axioms for *expansion* uniquely represent the addition of new information to a set of beliefs that is consistent with the contents of the set. The revision postulates, however, confine, but do not *uniquely* determine the result K_α^* of revising K by α. This is because the exact outcome of the revision process depends on a certain *extra-logical* factors that are domain-dependent. These factors can be represented with a complete preorder (a transitive and reflexive relation) over sentences of K, called *epistemic entrenchment*. This ordering, which incidentally may differ from belief state to belief state, models the relative epistemic importance of the sentences in K. The effect of this ordering over revision is the requirement that revision operation should preserve more entrenched beliefs in preference to less entrenched ones. If x and y are sentences of belief set K, we write x ≤ y to mean that y is at least as important as x. The strict part of this order is represented as x < y to mean that y is more entrenched than x. What this means informally is that for any two formulas x and y such that x < y, whenever we have a choice between giving up x or y, the former is removed in order to minimise the epistemic loss. This powerful procedure conveys all the required information to uniquely determine the result of K_α^*. The following postulates characterise the qualitative structure of this order [6]:

(≤1) If x ≤y and y ≤ z, then x ≤ z ; (transitivity)
(≤2) If x ⊢ y , then x ≤y ;   (dominance)
(≤3) Either x ≤ x ∧ y or y ≤ x ∧ y ; (conjunctiveness)
(≤4) If K is a consistent theory, then x ≤ y for all y iff x ∉ K ;   (minimality)
(≤5) If x ≤ y for all x, then ⊢ y.   (maximality)

[†] relating to knowledge or degree of acceptance

251

For brevity, we do not describe the exact meaning of the above axioms (see [24] for details), suffice it to say that the major result of the epistemic entrenchment theory is that this notion is fundamentally equivalent to the previously postulated notion of belief revision of theories. That is, rational contraction functions may be constructed from orderings of epistemic entrenchment, and that the ordering may be constructed from rational contraction functions.

When new requirements become available, some of the previously stated requirements that have been expressed as the sentences of the current belief state may have to be given up to accommodate the new requirement while maintaining the consistency and completeness of the requirements model. The outcome of this revision process depends on a domain-dependent epistemic ordering on the sentences of the existing requirements model as well as the ordering of the new set of requirements. These orderings determine that the epistemologically least entrenched requirements be retracted first in accordance with the principle of minimal change.

While the AGM framework provides a useful abstraction for the belief change process, it does not lend itself to implementation in a straightforward way. Several studies (such as [7]) have therefore focussed on *belief bases*, which are finite sets of sentences, instead of logically closed theories, as representations of belief states. Belief base approaches consider priority relations on the belief base, instead of entrenchment relations defined on the entire language, in determining the outcome of a belief change step. In the rest of this paper, we shall only consider AGM-rational operators (ie., operators which satisfy the relevant AGM postulates) for belief bases [2].

4. A Formal Model for the Evolution of Requirements

In this section, we shall provide a model for requirements evolution based on a revision scheme for default theories [2]. This model can be viewed as a specific instance of the general framework presented in Section 2 and we shall use it to demonstrate that it is possible to define operators for mapping between requirements models which are well-founded in the semantics of rational belief change.

Following the approach in the previous section, we shall view a requirements model abstractly as a THEORIST specification (F, H, C). Thus, the basis for our model will be a set of operators for mapping between default theories. In general, a default theory may have several extensions (or maximal scenarios). Thus, in general, a mapping between default theories translates to a mapping between sets of theories (each theory

corresponding to a potential default extension). The operators we present in this section are loosely based on similar operators defined for a belief revision scheme for default theories in [16], which in turn are closely related to operators defined in [11].

We consider only operations which revise a requirements model with a new requirement, or retract an existing requirement from a requirements model. In other words, we do not consider expansion operations, given that there are rarely any guarantees that the new requirement being added will not conflict with some existing requirement.

We establish some notational conventions first. Following the literature on belief change, we shall denote the revision operation with the symbol * and the contraction operation with the symbol -. Thus $(F, H, C)^*_\phi$ and $(F, H, C)^-_\phi$ denote the outcome of a revision with ϕ and a contraction of ϕ, respectively, starting with a THEORIST specification (F, H, C). Moreover, let E(Δ) denote the set of maximal scenarios of a THEORIST specification Δ.

Following the spirit of the AGM postulates for belief change, we motivate our definition of revision/contraction operators by noting the following requirements for these operations:

- The outcome should be a consistent default theory. In the context of a THEORIST specification (F, H, C), this implies that $F \cup C$ should be satisfiable.

- The operation should be successful. Thus, every maximal scenario of $(F, H, C)^-_\phi$ must not include ϕ, while every maximal scenario of $(F, H, C)^*_\phi$ must include ϕ.

- The outcome should be independent of the syntax of the input. If $\phi \equiv \phi'$, then
$$E((F, H, C)^-_{\phi'}) = E((F, H, C)^-_\phi) \text{ and}$$
$$E((F, H, C)^*_{\phi'}) = E((F, H, C)^*_\phi)$$

- The operation should involve minimal change to the default theory. For classical theories, minimal change is ensured by establishing conformance with the AGM postulates. For default theories, this is not very straight forward, as we shall see later.

We define the revision operator as follows:
$$(F, H, C)^*_\phi = (F', H', C') \quad \text{where:}$$
$$F' = F^{*'}_\phi , \ H' = H \cup (F - F') , \ C' = C^{-'}_{\neg F'}$$

Where *' and -' are AGM-rational revision and contraction operations respectively for classical theories.

Theorem 1. Let $(F,H,C)^*_\phi = (F',H',C')$. Then:

1. $F' \cup C'$ is satisfiable.
2. $\forall e: e \in E((F',H',C')) \rightarrow e \models \phi$.
3. If $\phi \equiv \phi'$, then $E((F,H,C)^*_\phi) = E((F,H,C)^*_{\phi'})$

In the result above, the first condition guarantees that the outcome is a consistent default theory. The second condition guarantees that the revision operation succeeds, ie., every extension of the resulting default theory contains the new input. The third condition ensures that the operation is independent of the syntax of the input. We define the contraction operator as follows:

$$(F,H,C)^-_\phi = (F',H',C') \text{ where:}$$

$$F' = F'_{\neg C'} \, , \; H' = H \cup (F - F') \, , \; C' = C^{*'}_{\neg \phi}$$

Where $*'$ and $-'$ are AGM-rational revision and contraction operations respectively for classical theories. We now establish a set of properties for the contraction operation similar to those for the revision operation.

Theorem 2. Let $(F,H,C)^-_\phi = (F',H',C')$. Then:

1. $F' \cup C'$ is satisfiable.
2. There is no $e \in E((F',H',C'))$ such that $e \models \phi$.
3. If $\phi \equiv \phi'$, then $E((F,H,C)^-_\phi) = E((F,H,C)^-_{\phi'})$

A crucial question is the extent to which these operators satisfy the requirement that belief revision involve minimal change. For classical theories, conformance with the AGM postulates ensures that the minimal change requirement is satisfied. However, the AGM postulates cease to be useful in the case of default theories since they do not apply to situations where the belief change operators map between sets of classical theories (default extensions). One obvious feature of the operators we define is that the belief change operations on the F and C components of a THEORIST specification are guaranteed to be rational with respect to the AGM postulates.

It has been shown in [16] that a closely related set of operators satisfy a reformulated version of the AGM postulates, under a set of reasonable conditions. Briefly, the reformulation involves replacing every statement of the form *belief set K is a subset of belief set K'* to a sentence of the form *for every extension e of a default theory Δ there exists an extension e' of the default theory Δ' such that e is a subset of e'*. This specific instance of the general framework presented in Section 2 is thus well-founded with respect to the semantics of rational belief change (as formulated in the AGM framework).

In addition to guaranteeing a consistent and rational mapping between requirements models, the revision and contraction operators defined above offer several additional benefits which assume special importance in the context of requirements engineering.

- The explicit representation of contractions in the set of THEORIST constraints C ensures that revisions and contractions are treated in a symmetric manner. In the AGM framework, as well in approaches inspired by it (including Nebel's operators for belief bases [7]), contractions are never represented explicitly, while revisions are. Consequently, the effects of a contraction operation are not guaranteed to persist beyond a single step. Requirements evolution operations such as the *Telos* UNTELL operation [27], must persist over iterated evolution steps, which is only possible given an explicit representation for contractions, such as in our approach.

- Our model ensures that requirements are never discarded. A requirement r1, once added via a revision operation, is contained in either F or H at all future times. Thus, if a new requirement r2 contradicts an existing requirement r1 contained in F, then r1 is demoted to the status of a default (ie., it becomes an element of H). Moreover, no maximal scenario of the THEORIST specification will contain r1. If, however, r2 is later retracted, r1 can reappear in a maximal scenario of the resulting THEORIST specification.

- One may view every element of F and C as representing a prior requirements evolution step. Every element of F represents a prior revision, while every element of C represents a prior contraction. Thus, priority relations on F and C, which are the only two prerequisites necessary for generating a consistent outcome of a requirements evolution step, can be obtained by merely requiring an ordering on the belief change steps. Results from [16] show that such an approach provides an elegant and easily implementable solution to a well-known problem with the AGM framework and related systems, namely, the absence of a definition of belief change operations beyond a single step (often referred to as the problem of iterated belief change).

4.1 The Notion of Epistemic Entrenchment in RE

Representing the degree of epistemic entrenchment of individual requirements in a requirements model is a non-trivial task. There are many factors in RE that may contribute to the construction of this ordering. Perhaps the most important ones are the cost and project schedule constraints. Often, when stakeholders are faced with the choice of giving up a requirement, they choose to surrender the one that costs more to build or may upset the project schedule drastically. In safety critical systems, for example, safety-related factors may be considered to be more epistemically entrenched than cost. In real-time systems, on the other hand, response time might be more important to the users than functionality. The issue of social order (discussed in section 6), may be considered as a significant factor in determining this ordering during requirements elicitation.

Clearly, this ordering is application domain dependent. One possible method of deriving such ordering is by using a matrix oriented approach such as Quality Function Deployment (QFD) [27]. The postulates provided by AGM theory of belief revision and epistemic entrenchment enable us to construct rational revision functions from orderings expressed via the QFD method. Furthermore, the ordering of epistemic entrenchment in requirements modeling has to be driven by the choice of how epistemic states are modelled. For example one of the common paradigms in philosophical literature is to use Bayesian models where a state of belief is represented by a *probability measure* defined over some object language or over some space of events [24]. Here we have assumed that the epistemic states are represented as a set of propositions expressed by sentences from some given object language. This kind of model is simpler than Bayesian but also less informative. More recently, Ryan's Ordered Theory Presentation [29], offers a possible formalism for this ordering on a finite set of propositional sentences that could also be exploited in the AGM theory.

5. Previous work

Previous research that has recognised the relevance of default theory, such as Ryan [28,29] and Schobbens [30], focuses mainly on the use and structuring of defaults in specifications. Ryan has developed a formalism called Ordered Theory Presentation for representation of defaults in specifications. He has argued that the same formalism can be used in belief revision. Schobbens has concentrated on Algebraic specifications and their structuring mechanisms. The composition strategy he proposed for defaults and exceptions handling is nonmonotonic.

In the research presented here, however, there is a deliberate move away from the previous approaches that are focused primarily on the specification of requirements. We are addressing the other important and often neglected non-technical issues such as support for effective management of requirements throughout the development life cycle and the user related aspects of the requirements change. We believe that our ideas about the application of nonmonotonic reasoning and belief revision to the problem of managing requirements change are more general and at a more abstract level in RE than those that are strictly associated with proposing new requirements specification languages. A large body of requirements

engineering research in the last two decades has been dedicated to solving some of its fundamental problems, but because of the lack of a coherent foundation for the requirements engineering process, these too frequently isolated pieces of research do not relate well to one another. The aim of this paper is to put requirements engineering in context and to provide means for some of the previous research effort to be exploited in that context. In this section, for the sake of brevity, we review only two of the related RE research contributions in order to identify their relationship to the current work.

5.1 Requirements Apprentice

The Requirements Apprentice (RA) [20], is an intelligent assistant for requirements acquisition and analysis. The focus of RA is on bridging the gap between informal and formal specifications. RA uses *semantic networks* to model epistemic states in requirements. Its support for the evolution of requirements is based on a reasoning system called CAKE which is a truth maintenance facility for retracting the obsolete statements and asserting new ones. According to Reubenstein [20], CAKE is not capable of making all possible deductions from a set of facts and hence RA is not capable of detecting every contradiction in the requirements model. This way RA is allowed to continue to make deductions even in presence of a contradiction. The reasoning mechanism used in CAKE is based on first-order monotonic Truth Maintenance System (TMS) [18]. The reasoning procedure in all layers of CAKE are required to record their dependencies in a single uniform network provided by TMS [25]. Because nodes can only be added to the so called dependency network, this evolution cycle is never complete. The decision on which requirement to retract in the presence of contradiction is left to the analyst user and therefore not based on a specific formalism.

In our framework, we propose to complement the monotonic reasoning machinery offered by classical logic with nonmonotonic reasoning provided by the theory of belief revision and the THEORIST default theory. Therefore tools that will be built using our framework will be able to perform the requirements revision and retraction minimally and automatically once the ordering on the requirements is formally defined. So all possible contradictions can and will be detected and resolved at each step of requirements evolution.

5.2 RML and TELOS

The RML requirements modeling language [10], and its successor *Telos* [19] provide a rich repertoire of language constructs for representing knowledge about an information systems. Features of Telos include an object-oriented framework which support aggregation, generalisation, and classification; a novel treatment of attributes; an explicit notion of time; and machinery for specification of integrity constraints and deductive rules. [23] The Telos language provides a set of three operations for updating a knowledge base: TELL, UNTELL, and RETELL. Superficially, TELL, UNTELL and RETELL operators appear to be implementations of the three basic belief change operators, expansion, contraction and revision, respectively. Like expansion, TELL adds new beliefs to a belief state with no provision for inconsistency handling. Like contraction, UNTELL retracts beliefs and RETELL, like revision, involves the addition of new beliefs with a committant removal of prior beliefs which are inconsistent with the new one. RETELL is defined in manner analogous to the Levi identity [5]. Despite these apparent similarities, we have demonstrated elsewhere [2] that there are crucial differences between these operators and the basic belief change operators of AGM.

Clearly, a Telos knowledge base is a default theory, and any account of the dynamics of such a knowledge base must be formulated in terms of the dynamics of a default theory. Since the literature on Telos provides no such account, we reconstructed the Telos requirements modeling system based on Poole's THEORIST approach [1] and the AGM account of theory revision [24] to reveal several shortcomings of Telos. [2] In this logical reconstruction of the dynamics of a Telos knowledge base, we viewed it as a THEORIST specification (F, H, C). The TELL, UNTELL and RETELL operations were expressed as operations on the specification (F, H, C). Furthermore, we demonstrated how these shortcomings can be effectively handled by a new set of operators (expressed in section 4 here), in our framework.

6. Social aspects of RE

Elicitation of requirements is the most communications intensive part of requirements engineering and as such most of the techniques and methods developed do not originate from computer science, but from areas such as group interaction research, organisational theory, social sciences, ethnomethodology, sociolinguistics and so on. The fact remains that social models cannot easily be integrated and broken down into more manageable pieces. Consequently, refining and analysing such models is very difficult. The other problem associated with these approaches is that they lack the extent of formality needed for the construction of large theories (software). Although they can address the so called "softer" issues of software development, for most part, they do not express these concerns in a formal way. By "formal" we mean one whose semantics is based on a mathematical formulation. Formal descriptions have the advantage of providing the high degree of precision and clarity which is required for software development.

Goguen and Linde [12] provide a noteworthy survey of techniques for requirements elicitation focusing on how these techniques can deal with the social aspects of this activity. They pose the important question of *social order* in requirements elicitation. They claim that in all of their surveyed methods, the requirements analyst's order is always imposed on the social world of requirements gathering sessions. The issue of whose social order is assumed can be very important where people from possibly very different communities work together to construct a model of a system that is satisfactory for all of them. They conclude that the requirements elicitation problem is fundamentally social and unsolvable if we use methods that are based entirely around an individual cognition.

Goguen and Linde[12] presuppose that the social world is ordered and that this order may not be readily describable by common sense. They state that social order cannot be assumed to have an *a priori* structure and that it can only be determined by concentrating in the actual unfolding of social phenomena. This view suggests that social order possesses both static and dynamic properties. Clearly, an accurate understanding of how this social order is constructed and maintained must complement any methodology for requirements elicitation. There is, however, no mention of how one might formally talk about the issue of social order in elicitation and no methodological guidelines is offered as to how to model this phenomena or how to construct and maintain a desirable social order in requirements gathering activities. As a result, this work like other social approaches to the problem lack a sound scientific foundation that is much needed in requirements engineering.

One of the premises of the framework we present is that it is possible to model a social order with the notion of epistemic entrenchment. That is, by modeling the requirements of an individual participant in requirements elicitation as a set of beliefs that she has about the universe of discourse, one can then construct a partial ordering that would satisfy the postulates for epistemic entrenchment. Therefore the problem of building and maintaining the social order is then reduced to updating this partial ordering that incorporates the social order of all participants. For example, when some technical aspects of the construction of machine in a RS are being captured in a requirements elicitation session, we demand that the requirements engineers beliefs are more epistemically entrenched than those stated by the users because they have better knowledge of what is technically feasible. Conversely, when aspects of application domain are being explored, the social order has to be biased towards the users beliefs since their understanding of the business

255

domain is more valid than that of the requirements engineers. In this way both the static and dynamic aspects of social order can be modelled by belief revision.

Jirotka and Goguen [14] refer to the term "construction" of requirements to suggest that requirements may not be readily "out there" and that they are an arbitrary result of the RE process. This idea is also quite in line with the fact that we propose to use Nonmonotonic Reasoning in the initial (and perhaps in intermediate), stages of the elicitation to construct those requirements that are not explicit or have not yet been captured. In our framework, default logic is used to create the deductive closure of the incomplete requirements. What this implies is that there is nothing arbitrary about this "construction", rather it can be achieved formally and systematically. Extending the user's initial state of belief by nonmonotonic reasoning allows us to build a complete and finite belief set that may be revised after further discussions with the users.

7. Conclusion

In this paper we have addressed two important issues in requirements modeling, namely, representation and reasoning. We have presented a logical framework for modeling and reasoning about the evolution of requirements for computer-based systems. We have argued that the management of requirements models is most effective when it is based on a combination of monotonic and nonmonotonic reasoning. We proposed a reasoning framework based on the theory of belief revision and default theory. The techniques from these theories were then applied to define logical operators through which new requirements may be added or current requirements may be retracted from a default based representation of a requirements model. In order to illuminate and test our theoretical research on formal RE frameworks, we are about to undertake the post-pilot-survey phases of a detailed empirical study of requirements volatility in industrial software development projects. Further, we believe that theoretical frameworks require effective methodologies and accessible tools if they are to be useful and acceptable to practitioners. Consequently, we are in the process of defining a prototype toolset, based on the graph-theoretic *hypernode* meta-data model [15] with its supporting logic language *Hyperlog*. Subsequently, the toolset, with its associated methodology, will be validated in realistic industrial contexts.

Acknowledgments

We would like to gratefully acknowledge the assistance provided by Dr. Pavlos Peppas and Dr. Aditya Ghose in formalising this framework and for sharing with us their knowledge of belief revision and nonmonotonic reasoning.

8. References

[1] D. Poole, R. Goebel, and R. Aleliunas, "Theorist: A Logical Reasoning System for Defaults and Diagnosis," in The Knowledge Frontier: Essays in the Representation of Knowledge, N. J. Cercone and G. McCalla, Eds. New York: Springer-Verlag, pp. 331-352, 1987.

[2] D. Zowghi, A. K. Ghose, and P. Peppas, "A Framework for Reasoning about Requirements Evolution," Proc. of Fourth Pacific Rim International Conference on Artificial Intelligence, PRICAI-96, Cairns, Australia, August 1996.

[3] D. Poole, "A Logical Framework for Default Reasoning," Artificial Intelligence, vol. 36, pp. 27-47, 1988.

[4] C. E. Alchourron, P. Gardenfors, and D. Makinson, "On the Logic of Theory Change: Partial Meet Contraction and Revision Functions," Journal of Symbolic Logic, vol. 50, pp. 510-530, 1985.

[5] I. Levi, "Subjunctives, Dispositions and Chances," Synthese, vol. 34, pp 423-455, 1977.

[6] J. Doyle, "Rational Belief Revision, (Preliminary Report)," Proc. of Second International Conference on Principles of Knowledge Representation and Reasoning, pp. 163-174, 1991.

[7] B. Nebel, "Belief Revision and Default Reasoning: Syntax-Based Approaches," Proc. of Second International Conference on Principles of Knowledge Representation and Reasoning , pp. 417-428, 1991.

[8] G. Brewka, Nonmonotonic Reasoning: Logical Foundations of Commonsense. Cambridge, Great Britain: Cambridge University Press, 1991.

[9] R. Reiter, "A Logic for Default Reasoning," Artificial Intelligence, vol. 13, pp. 81-132, 1980.

[10] S. J. Greenspan, J. Mylopoulos, and A. Borgida, "Capturing More World Knowledge in The Requirements Specification," Proc. of the sixth International Conference on Software Engineering (ICSE6), IEEE Computer Society Press, Japan, 1982.

[11] G. Brewka, "Belief Revision in a Framework for Default Reasoning," Proc. of the Konstanz Workshop on Belief Revision, pp. 206-222, 1989.

[12] J. A. Goguen and C. Linde, "Techniques for Requirements Elicitation," Proc. Of First IEEE International Symposium on Requirements Engineering (RE93), IEEE Computer Society Press, pp.152-164, 1993.

[13] A. M. Davis, Software Requirements Analysis & Specification , Second ed. USA: Prentice-Hall, 1993.

[14]   M. Jirotka and J. A. Goguen, "Introduction," in Requirements Engineering Social and Technical Issues , M. Jirotka and J. A. Goguen, Eds. Great Britain: Academic Press, 1994, pp. 1-13.

[15]   M. Levene and G. Loizou, "A Graph Based Data Model and its Ramifications", IEEE Trans. Knowledge and Data Engineering, pp. 809-823, October 1995.

[16]   A. K. Ghose, "Practical Belief Change", PhD thesis, Department of Computing Science, University of Alberta, 1995.

[17]   J. P. Delgrande, T. Schanb, W. K. Jackson, "Alternative approaches to Default Logic," Artificial Intelligence, vol. 70, pp. 167-237, 1994.

[18]   J. Doyle, "A Truth Maintenance System," Artificial Intelligence , vol. 12, pp. 231-272, 1979.

[19]   J. Mylopoulos, A. Borgida, M. Jarke, and M. Koubarakis, "Telos: Representing Knowledge About Information Systems," ACM Transactions on Information Systems , vol. 8 No 4, pp. 325-362, 1990.

[20]   H. B. Reubenstein and R. C. Waters, "The Requirements Apprentice: Automated Assistance for Requirements Acquisition," IEEE Trans Soft Eng, vol. 17 No. 3, pp. 226-240, 1991.

[21]   M. Jackson and P. Zave, "Deriving Specifications from Requirements: an Example," Proc. Of the seventeenth International Conference on Software Engineering (ICSE17), IEEE Computer Society Press, pp. 15-24, 1995.

[22]   A. K. Ghose, A. Sattar, R. Goebel. "Pragmatic Belief Change: Computational efficiency and approximability," In Proc. of the Workshop on 'Belief Revision: Bridging the Gap between Theory and Practice', Held in conjunction with the sixth Australian Joint Conference on Artificial Intelligence, 1993.

[23]   S. Greenspan, J. Mylopoulos, and a Borgida, "On Formal Requirements Modeling Languages: RML Revisited," Proc. Of International Conference on Software Engineering (ICSE16), pp. 135-147, 1994.

[24]   P. Gardenfors, "Knowledge in Flux: Modeling the Dynamics of Epistemic States," Cambridge, Massachusetts: MIT press, 1988.

[25]   C. Rich and Y. Feldman, "Seven Layers of Knowledge Representation and Reasoning in Support of Software Development," IEEE Trans Soft Eng , vol. 18 No 6, pp. 451-469, 1992.

[26]   M. Jackson, "The World and the Machine," Proc. Of the seventeenth International Conference on Software Engineering (ICSE17), IEEE Computer Society Press, pp. 283-292, 1995.

[27]   R. E. Zultner, "Quality Function Deployment (QFD) for Software", American Programmer, February 1992.

[28]   M. Ryan, "Representing Defaults as Sentences with Reduced Priority", Proc. Of Third International Conference on Knowledge Representation and Reasoning (KR'92), Morgan Kaufmann, 1992.

[29]   M. Ryan, "Defaults in Specifications", Proc. Of IEEE International Symposium on Requirements Engineering (RE'93), 142-149, 1993

[30]   P. Schobbens, "Exceptions for Algebraic Specifications: On the Meaning of 'but'", Science of Computer Programming 20(1-2), North Holland, pg73, 1993.

Appendix 1: A simple example

Consider a requirements model for developing a wordprocessor [2]. Let the requirement for a wordprocessor be denoted by the proposition *wordproc*. Let *colour* denote the requirement that the monitor screen be colour and *mono* denote the requirement that the monitor screen be monochrome. Let *adults* denote that the target market for the wordprocessor is adults while *children* denote that the target market for the wordprocessor is young children. Let the domain knowledge (DK), include the following propositional sentences:

{colour $\leftrightarrow$ *¬mono, children* $\rightarrow$ *colour}.* Consider revising the requirements model initially with *wordproc, adults* respectively. The requirements model (F_0, H_0, C_0) is given by:

$F_0 = DK \cup \{wordproc, adults\}$, $H_0 = \{\}$, $C_0 = \{\}$

Revision with *mono* generates a requirements model (F_1, H_1, C_1) where:

$F_1 = F_0 \cup \{mono\}$

$H_1 = \{\}$

$C_1 = \{\}$

Finally, revision with *children* will generate a requirements model (F_2, H_2, C_2) where

$F_2 = (F_1)_{children}^{*'}$

$H_2 = F_1 - F_2$ and $C_2 = \{\}$

At this point, because of the DK, *children* $\rightarrow$ *colour*, a requirement *colour* is generated, which contradicts the *mono* requirement on account of the other DK, *colour* $\leftrightarrow$ *¬mono*. The revision process resolves this contradiction and the precise outcome is determined by the priority relation used by the $*'$ operator. One possible outcome is given by:

$F_2 = DK \cup \{wordproc, adults, children\}$

$H_2 = \{mono\}$

$C_2 = \{\}.$

Session 11B

Panel

Chair

Steve Miller
Collins Commercial Aviation, Rockwell, USA

"How Can Requirements Engineering Research Become Requirements Engineering Practice?"

Panel:
How Can Requirements Engineering Research Become Requirements Engineering Practice?

Steve Miller

Collins Commercial Avionics

Rockwell

Cedar Rapids, IA 52498

spmiller@cca.rockwell.com

The path from conceptualization of a good idea to its widespread use in industry is usually long, complicated, and fraught with peril. Too often, research justified as satisfying the needs of industry begins with a wrong or simplified understanding of industry's problems. Even given a real solution to a real problem, successful transfer of that solution into practice depends on many other factors such as funding, the emergence of champions, availability of tools, education, integration with existing methods, and all too often, plain luck.

This panel will explore how methods for requirements engineering for real-time and embedded systems can be moved into practice. Representatives from industry will discuss their needs and problems using existing methods, members of the research community will discuss current research trends, and tool vendors will discuss the difficulties of moving a good solution to a real problem into practice.

Each speaker will be asked to briefly state, with respect to requirements engineering for real-time, embedded systems:

- What their role should be in transferring innovation to industry

- What their role actually is

- Why the two are different, i.e., what their needs are

- How the other members of the panel could help meet those needs

- How they could help meet the needs of the other panel members

Differences in roles, needs, and contributions will be discussed among by the panel members, followed by a question and answer session open to the audience.

Members of the audience should leave with a better understanding of industry needs for requirements engineering for real-time embedded systems, where current research trends are headed, and why many good ideas will never make it into practice.

Tutorials

Making Requirements Measurable
B. Nuseibeh and S. Robertson

The SCR Approach to Requirements Specification and Analysis
S. Faulk and C. Heitmeyer

Software Requirements Specification and System Safety
Mats P.E. Heimdahl and J.D. Reese

Requirements Traceability
A. Finkelstein

Advanced Object-Oriented Requirements Specification Method
R. Wiering

Making Requirements Measurable

Bashar Nuseibeh

Department of Computing
Imperial College
London SW7 2BZ, UK
ban@doc.ic.ac.uk

Suzanne Robertson

Atlantic Systems Guild Ltd.
11 St. Marys Terrace
London W2 1SU, UK
100065.2304@compuserve.com

Background

Eliciting and specifying customer requirements in a precise and unambiguous way is critical to the success of a project. However anyone who has done any requirements engineering also knows that it is a very difficult activity involving many diverse skills. An important reason for the degree of difficulty is that requirements engineering involves many different people. Each person has his own opinion of what is or is not a requirement. Customers often find it difficult to articulate their requirements and for large, complex systems these requirements are often conflicting.

Tutorial Content

This is a *full-day* tutorial that focuses on guiding participants through the *requirements definition process*. After presenting an overview of requirements engineering activities, the emphasis of this exercise is on making requirements measurable so that they can be negotiated, communicated and traced throughout the project. A requirement is measurable if there is an unambiguous way of determining whether a given solution fits that requirement.

Participants in this tutorial *examine requirements measurability by building a requirements specification* for a familiar (but nevertheless complex) system. A requirements template is used as a guide for the tutorial. The tutorial concludes with a discussion of how measurable requirements can be used to build a requirements quality filter. The tutorial combines a presentation of issues and techniques in an interactive, participative format that encourages "hands-on" learning of requirements definition and specification.

The tutorial has five main components:

(1) An introduction to the tutorial and to the field of requirements engineering. *[60 minutes]*

(2) An introduction to making requirements measurable and to the case study. *[60 minutes]*

(3) Interactive, participative exercise of requirements specification based on case study. *[120 minutes]*

(4) Review of exercise by both participants and instructors. *[60 minutes]*

(5) Discussion and conclusion of issues raised and lessons learned. *[60 minutes]*

Participants are provided with notes, copies of transparencies and a copy of the requirements template.

After the tutorial, the instructors will amalgamate all the requirements specified by the different participants into a composite requirements specification which is then made available to all attendees.

Instructors' Biographies

Bashar Nuseibeh is a Lecturer and Head of the Software Engineering Laboratory in the Department of Computing, Imperial College, London. He is the Chairman of the BCS Requirements Engineering Specialist Group and an Editor-in-Chief of the Automated Software Engineering Journal. His research interest are in Distributed Software Engineering, including requirements engineering, process modelling and technology, and technology transfer. His current work is on supporting multiple views and managing inconsistencies in software development. Bashar holds a B.Sc. in Computer Systems Engineering from the University of Sussex, and an M.Sc. and PhD in Software Engineering from Imperial College.

Suzanne Robertson is a teacher and consultant specialising in modelling techniques for system development. She has over 30 years of development and consulting experience and has co-authored courses on systems analysis and software design for both procedural and object-oriented systems, requirements engineering, quality assessment and problem solving. Suzanne is currently developing techniques for identifying and reusing requirements patterns. She is one of seven Principals of the Atlantic Systems Guild - a New York and London based think-tank, researching and communicating system development techniques. She studied Information Processing at the New South Wales Institute of Technology.

Acknowledgements

Bashar Nuseibeh would like to acknowledge the financial support of the British Council, EU and EPSRC. Parts of the tutorial are based on past seminars presented by the instructors and on material published in [1, 2].

References

1. Nuseibeh, B., From Requirements to Satisfied Customer: A road map, *Keynote Address, DTI workshop on "RE - Connecting with the Customer"*, Gloucester, UK, Sep 95.

2. Robertson, S. and Robertson, J. *Complete Systems Analysis: The Workbook, The Textbook, the Answers*, Dorset House, UK, 1994.

The SCR Approach to Requirements Specification and Analysis

Stuart Faulk
University of Oregon
faulk@cs.uoregon.edu

Connie Heitmeyer
Naval Research Laboratory
heitmeyer@itd.nrl.navy.mil

The Software Cost Reduction (SCR) requirements method is a practical, industrial-strength approach to requirements that leads to precise, unambiguous, and testable requirements specifications. The method scales to large applications, producing specifications that are both easy to understand and easy to change. Effectiveness of the SCR approach has been demonstrated in a variety of industrial, safety-critical applications. These include software for military aircraft, commercial aircraft, and the shutdown system of a nuclear power plant. Recent work has extended the method to include mechanical support for creation, validation, and verification of formal requirements specifications. This support is based on a formal requirements model.

Researchers have claimed that formal methods have the potential to address many of industry's problems with requirements, including ambiguity, incompleteness, and imprecision. Nonetheless, industry has been slow to adopt formal techniques because they are perceived as impractical for large, complex applications.

The SCR method was developed to provide the benefits of formal methods to industrial developers of large-scale real-time systems. It has been effective in meeting industry needs because the technical approach addresses constraints and concerns common to industrial software developers, including ease of use, scalability, and cost-effectiveness.

This tutorial gives an overview of the SCR method, its rationale, and empirical results on its effectiveness. It includes the following topics:

- The current industrial perspective on the requirements problem and why the available methods and tools are inadequate.

- Necessary characteristics of methods and tools appropriate for industrial development of requirements and why industry is skeptical of formal methods.

- How the SCR method addresses common industrial concerns, such as scalability, ease of use, fit within the software development life-cycle, and technology transfer.

- Overview of the SCR formal model and the constructs and notation useful for representing the requirements of industrial-strength systems.

- Description of prototype tool support for SCR. The tools support simulation, analysis of completeness and consistency, and analysis of application properties such as safety.

- Discussion of technology transfer efforts and the results of empirical studies

- Results and lessons learned from application of the SCR method to a commercial software development effort (Operational Flight Program for the C-130J aircraft).

The goal of the tutorial is for attendees to develop an understanding of the current problems with requirements in industry and how software engineering principles were applied to produce a practical, cost-effective approach to formal specification of requirements. The goal is not to teach the method but to convey an understanding of the underlying technical approach, how fundamental software engineering principles can be applied to address industrial problems, key features of the formal model, and the results of applying the model to the requirements specification of industrial systems and software. Principles and results will be illustrated with examples.

Instructors. Stuart Faulk is a member of the University of Oregon's Computer Science Department. Previously, he worked on the SCR project at the Naval Research Laboratory (NRL). He also headed the development of the Consortium Requirements Engineering (CoRE) method at the Software Productivity Consortium, where he successfully applied CoRE to the Lockheed C-130J. Connie Heitmeyer heads the Software Engineering Section of NRL's Center for High Assurance Computing. She leads NRL's research program in Software Requirements which is developing a formal requirements model and formal techniques and software tools for specifying and analyzing requirements based on the SCR method. Ms. Heitmeyer recently published a new book entitled "Formal Methods for Real-Time Computing."

References

[1] Stuart R. Faulk, "Software requirements: A tutorial," NRL report 7775, Naval Research Lab, Wash., DC, 1995.

[2] Constance L. Heitmeyer, Ralph D. Jeffords, and Bruce G. Labaw, "Automated consistency checking of requirements specifications," *ACM Trans. Software Eng. and Methodology 5*, 3, July, 1996, 231-261.

Software Requirements Specification and System Safety

Mats P.E. Heimdahl

University of Minnesota
Institute of Technology
Dept. of Computer Science
Minneapolis, MN 55455
heimdahl@cs.umn.edu

Jon Damon Reese

Safeware Engineering Corporation
7200 Lower Ridge Road, Unit B
Everett, WA 98203-4925

jdreese@cs.washington.edu

Computer software is playing an increasingly important role in safety-critical embedded computer systems, where incorrect operation of the software could lead to loss of life, substantial material or environmental damage, or large monetary losses. Such diverse technologies as avionics, automobile drive trains, power plants, and medical equipment are relying more and more on the computer to control system parameters. Although software is a powerful and flexible tool for industry, these very advantages have contributed to a corresponding increase in system complexity. Traditional approaches to system development have not successfully handled the problems of increased system complexity. The fatal accidents caused by software in the Therac-25 radiation therapy machine, as well as other incidents, have brought public attention to these problems. Ironically, it is becoming clear that the powerful control logic that software can bring to a system can also impair the ability of the systems analyst to study and understand, and hence safely control, the system's behavior.

Safety is a property of the physical system. It is the physical system that has the ability to uncontrollably release sufficient energy to inflict harm and injury. Thus, safety issues must be addressed during the system design phase. All safety-related functions that are allocated to software must be accurately captured in the software requirements specification and the application experts, for example, systems engineers, must be intimately involved in the requirements specification process; safety issues *can not* be deferred until software development commences. Most software-related accidents have been traced back to errors in the requirements specification rather than coding errors.

In this tutorial we will provide a short introduction to system safety, discuss how the use of software control affects the safety analysis of a system, and outline the root causes of safety-related problems. We will present a formal state-based modeling language called RSML (Requirements State Machine Language) suitable for requirements specification of safety-critical control systems. RSML has been used to capture the requirements for several safety-critical commercial systems, most notably TCAS II (Traffic alert and Collision avoidance System II). Furthermore, using a formal requirements specification language, for example, RSML, enables several types of automated or semi automated analysis techniques that can be used to detect and eliminate potential safety problems from the specification. In this tutorial, we will discuss techniques for automatically detecting incomplete, inconsistent, and nondeterministic requirements, show how fault tree analysis can be used in the RSML framework, and demonstrate how a new analysis technique called deviation analysis can be used to evaluate the effects on the system if the inputs to the system deviates from expected value.

Mats P.E. Heimdahl is an assistant professor of computer science at the University of Minnesota, Twin Cities, and he is Vice President of Safeware Engineering Corporation. Dr. Heimdahl received his Ph.D. in computer science from the University of California, Irvine working with Nancy Leveson.

Jon Damon Reese received his Ph.D. in Information and Computer Science from the University of California, Irvine, under the guidance of software safety pioneer Nancy Leveson. Dr. Reese is the president of Safeware Engineering Corporation and a researcher at the University of Washington.

Requirements Traceability

Anthony Finkelstein
City Universrity of London, UK

Richard Stevens
QSS, UK

Advanced Object-Oriented Requirements Specification Methods

Roel Wieringa

Faculty of Mathematics and Computer Science, Free University

De Boelelaan 1081a

1081HV Amsterdam, the Netherlands

roelw@cs.vu.nl

This tutorial presents the latest developments in the field of object-oriented requirements specification and places them into perspective by comparing them to recent developments in structured analysis. The following four techniques and methods are treated:

- the Unified Modeling Language (UML) by Rumbaugh, Booch and Jacobson,

- the 1996 version of Fusion extended with Use cases,

- the 1996 version of OOA (Shlaer-Mellor), and

- the 1993 version of the Yourdon Systems Method.

The treatment is based upon published material as well as information publicly accessible at WWW sites. The presentation thus excludes developments of these methods that are internal to the companies that market the technique or method.

The techniques and methods are analysed in terms of a framework for requirements specifications that is derived from systems engineering. The framework defines several views on software products. First, it distinguishes externally observable software product behavior from internal software product decomposition. External behavior is often described by listing required system functions. Second, the internal decomposition is further divided into a conceptual decomposition, in which the components have a meaning in terms of the environment in which the software product will operate, and a physical decomposition that has a meaning in terms of the implementation on which the software will run. The tutorial is restricted to what the techniques and methods have to say about ways to specify external behavior and the conceptual decomposition, as well as the relationship between the two.

The techniques used by the methods use to specify these different views of a software product are reviewed. Roughly, the 1996 version of the object-oriented methods treated in this tutorial specify external behavior by means of use cases and the conceptual decomposition as a collection of communicating objects. Communication may be synchronous or asynchronous, and each object may perform its behavior according to a life cycle. The 1993 version of the Yourdon Systems Method specifies external behavior by means of a list of events to which the software product must respond, the initiator of the events, the desired system response and the data entering and leaving the product during the event or its response. The conceptual components of the system are data processes, data stores, event stores and control processes. The tutorial goes into some detail to show exactly how external behavior and conceptual decomposition are specified in each of the methods. In particular, the elements in the notations of the methods are listed and compared to each other.

In general, the object-oriented techniques and methods tend to be strong in defining a coherent, modular conceptual architecture for the software product, where the structured methods tend to be strong in the definition of functional requirements on external software behavior. An obvious possibility for combining parts of the two approaches is to use heuristics and techniques from structured analysis for the specification of external behavior requirements and object-oriented techniques for the specification of a conceptual decomposition of the system. It turns out that the structured techniques for specifying external behavior can readily be combined with use case specification, but that the techniques of structured and object-oriented conceptual decomposition are incompatible.

If we compare the techniques for conceptual decomposition of object-oriented methods with those of structured analysis, two major differences stand out. First, structured methods separate data from control, whereas object-oriented methods encapsulate data and control into objects. Second, structured methods encapsulates control around functions whereas object-oriented methods encapsulate control around objects. The first difference may lead to conceptual difficulties when transitioning from structured to object-oriented decomposition. On the other hand, it is shown that the second difference is more apparent than real and should not lead to major conceptual difficulties in the transition.

R.J. Wieringa is associate professor in computer science at the Free University, Amsterdam. He recently wrote a book about Requirements Engineering Methods and Frameworks, published by Wiley, and is currently preparing a second book on Semantic, Real-Time and Object-Oriented Requirements Engineering Methods.

Author Index

Notes

Notes

Notes

Notes

IEEE Computer Society Press Publications

The world-renowned Computer Society Press publishes, promotes, and distributes a wide variety of authoritative computer science and engineering texts. These books are available in two formats: 100 percent original material by authors preeminent in their field who focus on relevant topics and cutting-edge research, and reprint collections consisting of carefully selected groups of previously published papers with accompanying original introductory and explanatory text.

Submission of proposals: For guidelines and information on CS Press books, send e-mail to cs.books@computer.org or write to the Acquisitions Editor, IEEE Computer Society Press, P.O. Box 3014, 10662 Los Vaqueros Circle, Los Alamitos, CA 90720-1314. Telephone +1 714-821-8380. FAX +1 714-761-1784.

IEEE Computer Society Press Proceedings

The Computer Society Press also produces and actively promotes the proceedings of more than 130 acclaimed international conferences each year in multimedia formats that include hard and softcover books, CD-ROMs, videos, and on-line publications.

For information on CS Press proceedings, send e-mail to cs.books@computer.org or write to Proceedings, IEEE Computer Society Press, P.O. Box 3014, 10662 Los Vaqueros Circle, Los Alamitos, CA 90720-1314. Telephone +1 714-821-8380. FAX +1 714-761-1784.

Additional information regarding the Computer Society, conferences and proceedings, CD-ROMs, videos, and books can also be accessed from our web site at www.computer.org.

9/20/96